Methods in Neurosciences

Volume 28

Quantitative Neuroendocrinology

Methods in Neurosciences

Editor-in-Chief

P. Michael Conn

Methods in Neurosciences

Volume 28
Quantitative Neuroendocrinology

Edited by
Michael L. Johnson
Departments of Pharmacology and Internal Medicine
University of Virginia Health Sciences Center
Charlottesville, Virginia

Johannes D. Veldhuis
Department of Internal Medicine
Division of Endocrinology and Metabolism
University of Virginia Health Sciences Center
Charlottesville, Virginia

ACADEMIC PRESS
San Diego   New York   Boston   London   Sydney   Tokyo   Toronto

Front cover illustration: Network of physiological regulatory interactions involved in controlling secretory activity of the growth hormone neuroendocrine axis. Stimulatory interactions are indicated in green and inhibitory ones in red. This interaction scheme is the basis for construction of a dynamical network model of coupled nonlinear ordinary differential equations discussed by Chen, Veldhuis, Johnson, and Straume in Chapter [11]. Courtesy of Hal Noakes, University of Virginia, Center for Biological Timing, Charlottesville, VA.

Academic Press, Inc.
A Division of Harcourt Brace & Company
525 B Street, Suite 1900, San Diego, California 92101-4495

United Kingdom Edition published by
Academic Press Limited
24-28 Oval Road, London NW1 7DX

International Standard Serial Number: 1043-9471

International Standard Book Number: 0-12-185298-9

Printed and bound in the United Kingdom
Transferred to Digital Printing, 2011

Table of Contents

Contributors to Volume 28

Article numbers are in parentheses following the names of contributors. Affiliations listed are current.

MICHAEL J. BERRIDGE (16), Babraham Institute Laboratory of Molecular Signalling and Department of Zoology, University of Cambridge, Cambridge CB2 3EJ, United Kingdom

B.G. BOVING (8), Reproductive Sciences Program, Bioengineering Program, and National Center for Infertility Research at Michigan, The University of Michigan, Ann Arbor, Michigan 48109

R. M. BRAND (8), Departments of Endocrinology and Metabolism, University of Nebraska, Omaha, Nebraska 68144

H. C. CANTOR (8), Reproductive Sciences Program, Bioengineering Program, and National Center for Infertility Research at Michigan, The University of Michigan, Ann Arbor, Michigan 48109

MOLLY CARNES (7), Department of Medicine, University of Wisconsin Hospital, and Geriatric Research Education and Clinical Center, William S. Middleton Memorial Veterans Hospital, Madison, Wisconsin 53706

TIMOTHY R. CHEEK (16), Babraham Institute Laboratory of Molecular Signalling, and Department of Zoology, University of Cambridge, Cambridge CB2 3EJ, United Kingdom

LUBIN CHEN (11), Interdisciplinary Biophysics Program, University of Virginia, Charlottesville, Virginia 22903

WILLIAM S. EVANS (5), Department of Internal Medicine, Division of Endocrinology and Metabolism, National Science Foundation Center for Biological Timing, University of Virginia Health Sciences Center, Charlottesville, Virginia 22908

P. A. FAVREAU (8), Reproductive Sciences Program, Bioengineering Program, and National Center for Infertility Research at Michigan, The University of Michigan, Ann Arbor, Michigan 48109

W. OTTO FRIESEN (16), Department of Biology, National Science Foundation Center for Biological Timing, University of Virginia, Charlottesville, Virginia 22903

M. N. GHAZZI (8), Parke-Davis Pharmaceuticals, Ann Arbor, Michigan 48106

BRIAN GOODMAN (7), Department of Medicine, and Geriatric Research Education and Clinical Center, University of Wisconsin Hospital, William S. Middleton Memorial Veterans Hospital, Madison, Wisconsin 53705

DANIEL K. HARTLINE (10), Bekesy Laboratory of Neurobiology, Pacific Biomedical Research Center, University of Hawaii, Honolulu, Hawaii 96822

MICHAEL L. JOHNSON (1–5, 9, 11, 15), Departments of Pharmacology and Internal Medicine, Division of Endocrinology and Metabolism, Interdisciplinary Biophysics Program, National Science Foundation Center for Biological Timing, University of Virginia Health Sciences Center, Charlottesville, Virginia 22908

DANIEL KEENAN (12), Department of Mathematics, University of Virginia, Charlottesville, Virginia 22903

MICHELLE LAMPL (15), Department of Anthropology, Emory University, Atlanta, Georgia 30322

ORLA M. MCGUINNESS (16), Babraham Institute Laboratory of Molecular Signalling and, Department of Zoology, University of Cambridge, Cambridge CB2 3EJ, United Kingdom

JAMES E. A. MCINTOSH (13), School of Biological Sciences, Victoria University of Wellington, Wellington 6000, New Zealand

A. R. MIDGLEY (8), Reproductive Sciences Program, Bioengineering Program, and National Center for Infertility Research at Michigan, The University of Michigan, Ann Arbor, Michigan 48109

ROGER B. MORETON (16), Babraham Institute Laboratory of Molecular Signalling and Department of Zoology, University of Cambridge, Cambridge, CB2 3EJ, United Kingdom

THOMAS MULLIGAN (3, 4), Division of Geriatric Medicine, Virginia Commonwealth University, Medical College of Virginia, and Hunter Holmes McGuire Veterans Affairs Medical Center, Richmond, Virginia 23249

ROSALIND P. MURRAY-MCINTOSH (13), Department of Medicine, Wellington School of Medicine, University of Otago, Wellington 6000, New Zealand

ROBERT WHITNEY NEWCOMB (10), Neurex, Inc., Menlo Park, California 94025

ROBERT WAYNE NEWCOMB (10), Microsystems Laboratory, Department of Electrical Engineering, University of Maryland, College Park, Maryland 20742

V. PADMANABHAN (8), Reproductive Sciences Program, Bioengineering Program, and National Center for Infertility Research at Michigan, The University of Michigan, Ann Arbor, Michigan 48109

STEVEN M. PINCUS (14), Guilford, Connecticut 06437

MARTIN STRAUME (9, 11), Department of Medicine, Division of Endocrinology and Metabolism, Interdisciplinary Biophysics Program, National Science Foundation Center for Biological Timing, University of Virginia Health Sciences Center, Charlottesville, Virginia 22908

JEPPE STURIS (6), Department of Medicine, The University of Chicago, Chicago, Illinois 60637

EVE VAN CAUTER (6), Department of Medicine, The University of Chicago, Chicago, Illinois 60637

JOHANNES D. VELDHUIS (1–5, 9, 11), Departments of Internal Medicine, Division of Endocrinology and Metabolism, Interdisciplinary Biophysics Program, National Science Foundation Center for Biological Timing, University of Virginia Health Sciences Center, Charlottesville, Virginia 22908

E. A. YOUNG (8), Mental Health Research Institute, The University of Michigan, Ann Arbor, Michigan 48109

V. P. VAN SARAN (9), Reproductive Sciences Program, Bioengineering Program, and National Center for Infertility Research at Michigan, The University of Michigan, Ann Arbor, Michigan 48109

STEVEN M. PINCUS (1A), Guilford, Connecticut 06437

MARTIN STRAUMAN (9, 11), Department of Medicine, Division of Endocrinology and Metabolism, Interdisciplinary Biophysics Program, National Science Foundation Center for Biological Timing, University of Virginia Health Sciences Center, Charlottesville, Virginia 22908

JEPPE STRUIS (6), Department of Medicine, The University of Chicago, Chicago, Illinois 60637

EVE VAN CAUTER (6), Department of Medicine, The University of Chicago, Chicago, Illinois 60637

JOHANNES D. VELDHUIS (1–5, 9, 11), Departments of Internal Medicine, Division of Endocrinology and Metabolism, Interdisciplinary Biophysics Program, National Science Foundation Center for Biological Timing, University of Virginia Health Sciences Center, Charlottesville, Virginia 22908

E. A. YOUNG (8), Mental Health Research Institute, The University of Michigan, Ann Arbor, Michigan 48109

Preface

Early efforts in quantitating neuroendocrine phenomena involved visual inspection of dynamic variations in serum concentrations of anterior and posterior pituitary hormones over time in experimental animals. For example, Gay and Midgely observed that release of luteinizing hormone in castrate rats is episodic rather than continuous or explicitly rhythmic. Some quantitative effort to detect discrete peaks of pituitary hormone release was provided by Santen and Bardin in 1973, who defined a neurohormone pulse as an abrupt 20% increase in concentrations of that agonist in plasma (nominally, a 3-fold change over the intra-assay coefficient of variation). The complexity of neuroendocrine control mechanisms, however, became more clearly evident following greater appreciation of underlying neuroendocrine physiology, stochastic components in the data, and feedback facilitation and suppression inferred by stimulation or interruption of neuroendocrine axes. As experimental strategies have become more sophisticated, high-speed computing has been required for the formulation and solution of more elaborate algebraic statements of neuroendocrine pulsatility, such that desktop microprocessors are now used to solve matrices defining 100–300 equations, each containing 10–30 variables.

In this volume, we address contemporary methods in the neurosciences designed to provide quantitative insights into, mathematical structure for, and predictive inferences about neuroendocrine control mechanisms. Multiple quantitative themes beyond that of neurohormone pulse detection, which was one of the earlier points of emphasis in quantitative neuroendocrinology, are included. For example, Van Cauter, Mulligan, and others discuss quantitative modeling of neuroendocrine axes with an effort to examine the validity, sensitivity, specificity, and test–retest reliability of quantitative neuroendocrine methods proposed to enumerate neurohormone pulsatility. Other work, e.g., by Hartline *et al.*, develops models of quantitative neuroendocrinology at the level of granule exocytosis and hormone release processes. Techniques of cell biology, such as cell perifusion models (see chapters by Midgley *et al.* and the McIntosh's), employ short-term agonist delivery in perifusion systems to evaluate the impact of a pulse signal on target tissue responses. Other recent approaches, including physiological networking, namely a connectionistic feedback model defined by available experimental data, are described by Straume *et al.*, on the one hand, for the growth hormone axis in the rat, and by Friesen *et al.*, on the other, for neural networks that control the locomotor function of freshwater flukes. More mathematical treatment of system regularity, namely serial pattern persistence, is developed by Pincus using among other concepts the approximate entropy statistic. Keenan discusses a stochastic differential equation approach to formulate a statistical model of episodic hormone release with some deterministic trends with longer periods of observation, e.g., over 24 hours.

Special problems remaining in the field of quantitative neuroendocrinology are also highlighted. For example, networking with feedback connectivity in a well-defined model should ultimately allow *in vivo* estimates of some physiological rate constants, e.g., of hormone diffusion, metabolic clearance, distribution volumes, and association or dissociation rate constants. This challenge is made substantial by the large number of parameters required to define complex physiological systems, which therefore imposes a risk of statistical overdetermining of the model. A second aspiration is to model the neurohormone secretory burst at a molecular level, accounting for the intracellular biosynthesis of organic molecules, their transport and chemical modification within the cell, vectorial movement to the plasma membrane, and then exocytosis or transmembrane diffusion to generate a macroscopic release episode as integrated over cells and an entire neuroendocrine unit. Third, models of the anticipated biochemical impact of intermittent agonist signals on target-cell responses will also require considerably more study. And last, among other issues that quantitative neuroendocrinology must address further is the extent to which the observer can estimate mixed modes of neurohormone secretion *in vivo,* e.g., combined basal and pulsatile neurohormone release. A beginning effort toward defining the scope of this problem is made in this volume in comments by the editors and other contributors.

The field of quantitative neuroendocrinology in the last two decades has emerged as a multipartite discipline with outstanding supporting contributions by the fields of mathematical probability, systems engineering, stochastic differential equations, and the experimental natural sciences such as cell biology, molecular biology, and other approaches to subcellular analyses. Consequently, we anticipate and predict that quantitative methods in neuroendocrinology will continue to burgeon as opportunity and need coexist for explicit quantitative appraisals of the time-specified behavior of neuroendocrine axes whether studied as an intact ensemble in the whole animal, as an individual neuroendocrine unit *in vivo* or *in vitro,* and/or as cellular and molecular processes under distinct regulation.

MICHAEL L. JOHNSON
JOHANNES D. VELDHUIS

Methods in Neurosciences

[1] Evolution of Deconvolution Analysis as a Hormone Pulse Detection Method

Michael L. Johnson and Johannes D. Veldhuis

Introduction

It is well known that the serum concentration of some hormones change by orders of magnitude multiple times within each day. Luteinizing hormone (LH), growth hormone (GH), prolactin (PRL), thyrotropin (TSH), and adrenocorticotropic hormone (ACTH) are examples of hormones that exhibit large, short duration fluctuations in concentration. The temporal variation in serum hormone concentration is believed to be a significant portion of the signaling pathway by which endocrine glands communicate with remote target organs (Desjardins, 1981; Urban *et al.*, 1988a; Evans *et al.*, 1992). It is thus important to be able to identify and quantify the pulsatile nature of time series of serum concentrations. Many algorithms have been developed for analysis of such data (Veldhuis and Johnson, 1986; Urban *et al.*, 1988a,b; Evans *et al.*, 1992).

In 1987 we proposed that the temporal shape of endocrine hormone pulses is a convolution integral of

$$C(t) = C_0 + \int_0^t S(z)E(t - z)\, dz, \tag{1}$$

where $C(t)$ is the concentration of serum hormone at any positive time, $t \geq 0$; C_0 is the concentration of hormone before any secretion at the first time point; $S(z)$ is the amount of hormone secreted at time z per unit time and unit distribution volume; and $E(t - z)$ is the amount of hormone elimination that has occurred in the time interval $t - z$ (Veldhuis *et al.*, 1987). This convolution process is shown in Fig. 1. The top curve in Fig. 1 depicts the rate of hormone secretion into the serum for a typical hormone pulse. The middle curve depicts a typical elimination pattern for an instantaneous bolus of hormone. This is the so-called instrument response function. The bottom curve in Fig. 1 depicts the convolution integral of the secretion and the elimination, that is, the hormone concentration as a function of time that results from the secretory event shown in the top curve of Fig. 1.

Our definition of the temporal shape of these hormone pulses has provided a significant and different prospective with which to identify and quantify

Methods in Neurosciences, Volume 28

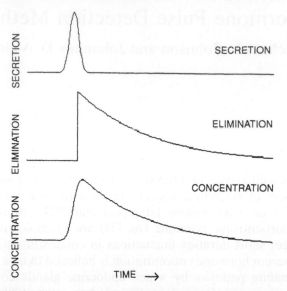

Fig. 1 Graphical depiction of the convolution process. The time course of the rate of hormone secretion into the serum for a typical secretory event is shown at top. The middle curve is a typical "instrument/impulse response function" (i.e., the elimination function). The instrument response function is simply the concentration as a function of time that would result from an instantaneous bolus injection of the hormone into the serum. The bottom curve is the resulting concentration as a function of time (i.e., the convolution integral of the top two curves).

endocrine pulses. With a deconvolution process it is now possible to evaluate the time course of the secretion of hormone into the serum. The temporal nature of the secretion can now be directly used to identify and characterize hormone pulses within the serum.

The deconvolution process is simply an algorithm to separate an observed concentration time series into its component parts, the secretion and the elimination as a function of time. However, the application of deconvolution algorithms to hormone time series data is not simple. Most deconvolution methods were developed for spectroscopic and/or engineering applications where data are abundant and the signal to noise ratio is high (i.e., low experimental uncertainties) (Jansson, 1984). The radioimmunoassays and other assays utilized to evaluate hormone concentrations in blood allow fewer observations and provide a substantially lower signal to noise ratio (i.e., significantly higher experimental uncertainties).

For the spectroscopic and/or engineering applications the convolution integral provides the observed signal as a combination of the actual signal and an instrument response function. Generally for the spectroscopic and/or engineering applications the instrument response function is easily characterized and defined, and thus the deconvolution process is comparatively simple. The analogous quantities for the hormone case are the observed concentrations, which are a convolution integral of the secretion (the actual signal) and the elimination (the bodily response function). The rate of hormone elimination is, of course, both hormone and patient specific and within a given patient can be a function of many variables such as diet and the presence of various drugs. Thus the deconvolution process for the hormone case is comparatively more difficult because both the secretion and elimination processes must be estimated simultaneously.

The purpose of this chapter is to outline and define the methodologies and algorithms that we have developed for the deconvolution analysis of hormone concentrations series. We also describe how the deconvolution process can be utilized as a pulse detection algorithm.

Deconvolution Methods

Since we introduced the concept of deconvolution analysis (Veldhuis et al., 1987) we have developed three different deconvolution algorithms for the evaluation of secretory rate as a function of time: DECONV, PULSE, and PULSE2. The primary difference among the algorithms is the assumed functional form of the secretory function $S(t)$. Each of these strategies is unique with advantages and disadvantages. This section develops each of these methods in detail and explains the advantages and disadvantages of each.

All three methods (DECONV, PULSE, and PULSE2) are based on fitting Eq. (1) to the actual experimental data. The details of the various weighted nonlinear least-squares algorithms (damped Gauss–Newton, steepest descent, and Gauss–Newton) are presented elsewhere (Johnson and Frasier, 1985; Straume et al., 1991; Johnson, 1992; Johnson and Faunt, 1992) and are not repeated here. The algorithms for the evaluation of confidence intervals of estimated parameters (Johnson, 1983; Johnson and Frasier, 1985; Straume et al., 1991; Johnson and Faunt, 1992; Straume and Johnson, 1992a) and goodness-of-fit criteria (Straume et al., 1991; Straume and Johnson, 1992b) are also presented elsewhere.

Technically Eq. (1) is not a convolution integral because the limits of integration are written as 0 to t instead of the minus infinity to plus infinity that is characteristic of convolution integrals. To be rigorously correct Eq. (1) should be written as a convolution integral:

$$C(t) = \int_{-\infty}^{\infty} S(z)H(t-z)E(t-z)\,dz = S(t) \otimes [H(t)E(t)]. \tag{2}$$

$H(t-z)$ is a Heaviside step function:

$$H(t-z) = \begin{pmatrix} 1, & \text{if } t-z \geq 0 \\ 0, & \text{if } t-z < 0 \end{pmatrix}. \tag{3}$$

The purpose of the Heaviside step function is to eliminate the possibility of hormone being eliminated before it is secreted. In Eq. (1) this is accomplished by stopping the integration at an upper limit of t instead of infinity. The C_0 in Eq. (1) corresponds to the integral in Eq. (2) integrated from minus infinity to zero:

$$C_0 = \int_{-\infty}^{0} S(z)[H(t-z)E(t-z)]\,dz. \tag{4}$$

Elimination Function: $H(t-z)E(t-z)$

The elimination function that we utilize for all the deconvolution methods is based on the classic one-compartment [Eq. (5)] or two-compartment [Eq. (6)] pharmacokinetic models for elimination:

$$H(t-z)E(t-z) = \begin{pmatrix} e^{-[\ln 2(t-z)]/HL}, & \text{if } t-z \geq 0 \\ 0, & \text{if } t-z < 0 \end{pmatrix}, \tag{5}$$

where HL is the one compartment elimination half-life and

$$
\begin{aligned}
&H(t-z)E(t-z) \\
&= \begin{pmatrix} (1-f_2)e^{-[\ln 2(t-z)]/HL_1} + f_2 e^{-[\ln 2(t-z)]/HL_2}, & \text{if } t-z \geq 0 \\ 0, & \text{if } t-z < 0 \end{pmatrix},
\end{aligned} \tag{6}
$$

where HL_1 and HL_2 are the two half-lives of the two compartment model and f_2 is the fractional amplitude corresponding to HL_2. For simplicity the Heaviside step function has been included in Eqs. (5) and (6).

DECONV

Our original deconvolution algorithm DECONV (Veldhuis *et al.*, 1987) modeled the secretory event as a sum of a series of Gaussian shaped secretory events:

$$S(z) = S_0 + \sum_{l=1}^{n} e^{\ln H_i - (1/2)[(z - PP_i)/SD]^2}, \qquad (7)$$

where $\ln H_i$ is the natural logarithm of the amplitude of the ith secretory event (in units of mass per unit time per unit distribution volume), PP_i is the position of the ith secretory peak, SD is the standard deviation of the secretory events, and S_0 corresponds to the basal secretion. The half-width at half-height of the secretory event is equal to 2.354 times the SD.

Once the mathematical forms of the secretion events, Eq. (7), and the elimination process, Eqs. (5) or (6), are defined then the convolution integral, Eq. (2), can be fit to a time series of experimental data. The actual fitting is done by a damped Gauss–Newton weighted nonlinear least-squares technique (Johnson, 1983, 1992; Johnson and Frasier, 1985; Straume *et al.*, 1991; Johnson and Faunt, 1992). The least-squares procedure provides numerical values of the various parameters of the model, namely, S_0, SD, $\ln H_i$, PP_i, and either HL or HL_1, HL_2, and f_2. The particular least-squares algorithm also provides a realistic estimate of the statistical uncertainties of the estimated model parameters (joint confidence intervals). Asymptotic standard errors are not utilized as statistical uncertainties of the estimated parameters because they do not include a contribution for the large covariance of the estimated parameters. These parameter values and uncertainties are used to evaluate the secretory profile and its statistical significance. Thus, this procedure provides a deconvolution of the concentration time series into the secretion time series and the elimination time series.

It is interesting to note that we estimated the natural logarithm of the amplitude of the secretory event, $\ln H_i$, rather than simply the amplitude, H_i, of the secretory event. The purpose in this is to force the amplitude of the secretory event to always be positive. Any real value of $\ln H_i$ will always map into a positive value for H_i. Thus, by including the $\ln H_i$ term in the exponential portion of Eq. (7) we have defined a system of parameters where negative values of H_i cannot exist. The net result is to force the least-squares procedure to provide only physically meaningful positive values for the amplitudes of the secretory events. The method is thus a constrained least-squares method contrary to what has been reported elsewhere (De Nicolao and Liberati, 1993).

An example of the use of this deconvolution is shown in Fig. 2 and Table I. The data points presented in the top graph of Fig. 2 are a typical example of the temporal variation of luteinizing hormone over a 24-hr interval. The particular example is for a normal healthy young adult in the midluteal phase (Sollenberger *et al.*, 1990). The solid curved line shown in the top graph of

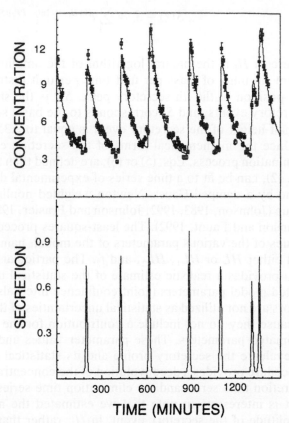

FIG. 2   A typical hormone pulsatility analysis utilizing the DECONV algorithm and computer program. The top graph presents experimental observations, namely, of the temporal variation of serum luteinizing hormone concentrations over a 24-hour interval. Concentration units are mIU/ml. The data are from a normal healthy young adult in the midluteal phase (Sollenberger *et al.*, 1990). The continuous curved line shown in the top graph corresponds to the best fit of Eqs. (2), (5), and (7) to the data. The resulting parameter values are shown in Table I. Seven secretory events and a single-compartment elimination were assumed. The bottom graph presents the deconvolved (predicted) secretory rate as a function of time. The secretory units are mIU/ml/min.

TABLE I   Parameter Values from Analysis Shown
in Fig. 2

Parameter	Value	±1 SD joint confidence interval	
		Lower bound	Upper bound
S_0	0.042	0.030	0.054
C_0	7.69	6.11	9.28
SD	4.55	3.14	6.00
HL	50.4	42.4	58.2
PP_1	197.0	193.4	201.4
H_1	0.74	0.54	1.00
PP_2	424.5	420.8	428.2
H_2	0.85	0.62	1.16
PP_3	616.2	612.6	620.8
H_3	1.03	0.76	1.41
PP_4	888.0	884.7	892.2
H_4	0.94	0.69	1.27
PP_5	1100.9	1097.3	1105.3
H_5	0.80	0.58	1.09
PP_6	1309.5	1304.8	1315.2
H_6	0.60	0.44	0.82
PP_7	1359.4	1349.1	1366.3
H_7	0.76	0.55	1.05
SSR	597.78		

Fig. 2 is the calculated concentration curve based on a fit of Eqs. (2), (5), and (7) to the data. For this analysis seven secretory events and a single-compartment elimination is assumed. The bottom graph in Fig. 2 presents the deconvolved secretory rate as a function of time.

The analysis shown in Fig. 2 and Table I is a weighted nonlinear least-squares analysis of 144 data points to simultaneously estimate 18 parameters. These 18 parameters are S_0, C_0, SD, HL, and a pulse position, PP_i, and log amplitude, $\ln H_i$, for each of the seven secretory events. Normally the simultaneous estimation of 18 parameters is considered extremely difficult. However, in this case most of the parameters are not highly correlated (i.e., they are nearly orthogonal) and thus can easily be simultaneously estimated.

The least-squares algorithm provides a quantitative measure of the cross-correlations between each ij pair of estimated parameters. These cross-correlation coefficients, CC_{ij}, are a measure of the difficulty of simultaneously estimating the ith and jth parameters. The range of values for CC_{ij} is ±1. The closer to ±1, the more difficult it is to simultaneously estimate the ith and the jth parameters. When CC_{ij} is ±1 any variation in the ith parameter can be compensated for by a change in the jth parameter such that the variance-of-fit does not change. If the variance-of-fit does not change then the parameters cannot be simultaneously estimated. When CC_{ij} is within the range ± 0.85 many least-squares procedures will usually have little difficulty simultaneously estimating the parameters. The least-squares procedure that we utilize usually works well with values of CC_{ij} within the range of ± 0.95. However when the cross-correlation is not within ± 0.95, but less than ± 1.0, most least-squares procedures will have significantly more difficulty simultaneously estimating the ith and jth parameters.

For the example presented in Fig. 2 and Table I the CC_{ij} between the secretory event standard deviation, SD, and each of the event amplitudes, ln H_i, is approximately -0.9. As a consequence of this the CC_{ij} between each of the event amplitudes is also moderately high, approximately -0.85. These cross-correlations are within the range that our least-squares procedure can easily accommodate and thus should not be of concern. However by a different choice of parameters to be estimated, these moderate cross-correlations can be almost estimated. If Eq. (7) is written as

$$S(z) = S_0 + \sum_{i=1}^{n} e^{\ln A_i - \ln[(2\pi)^{1/2}SD] - (1/2)[(z-PP_i)/SD]^2}, \tag{8}$$

where ln A_i is the total mass of hormone secreted in the ith secretory event. The mass and the height are related by

$$\ln H_i = \ln A_i - \ln[(2\pi)^{1/2}SD]. \tag{9}$$

The cross-correlation between SD and the ln A_i is lowered by this formulation to approximately -0.3. Thus Eq. (8) is the preferred functional form for the secretory profile.

In contrast to the foregoing, the cross-correlation between another pair of estimated parameters exemplifies a serious problem for all pulse analysis algorithms. For the example in Fig. 2 the cross-correlation between the basal secretion, S_0, and the elimination half-life, HL, is 0.981. It is thus very difficult to simultaneously estimate these parameters. The wide confidence intervals for S_0 and HL shown in Table I are a consequence of this high cross-correlation.

It should be noted that this high cross-correlation is not a consequence of the deconvolution algorithm or mathematical models. It is a consequence of the experimental observations. A visual examination of the top graph of Fig. 2 shows that the hormone concentration does not fall to a constant nadir (basal concentration) before the subsequent secretory event occurs. A longer

TABLE II   Parameter Values from Analysis Shown
in Fig. 3

| Parameter | Value | ±1 SD joint confidence interval | |
		Lower bound	Upper bound
S_0	0.023	0.013	0.035
C_0	10.67	8.23	13.26
SD	5.13	3.91	6.38
HL_1	21.0	11.6	32.6
F_2	0.52	0.31	0.70
HL_2	95.5	74.6	117.4
PP_1	197.5	193.9	201.9
H_1	0.76	0.60	0.98
PP_2	424.9	421.2	428.6
H_2	0.86	0.67	1.12
PP_3	616.9	613.1	621.3
H_3	1.00	0.78	1.30
PP_4	888.5	885.3	892.7
H_4	0.94	0.73	1.22
PP_5	1101.6	1097.8	1106.2
H_5	0.78	0.61	1.02
PP_6	1310.0	1305.3	1316.4
H_6	0.59	0.44	0.77
PP_7	1359.2	1349.1	1366.5
H_7	0.75	0.57	0.98
SSR	553.99		

half-life with a lower basal secretion or a shorter half-life with a higher basal secretion can describe the experimental observations with almost equal precision, thus accounting for the wide confidence intervals and high cross-correlation. The difficulty in simultaneously estimating basal secretion and elimination half-life is clearly one of the most intractable problems in evaluating hormone pulsatility.

The results of an analysis of the data shown in Fig. 2 with a two-component elimination function, Eq. (6), are presented in Table II and Fig. 3. The weighted sum-of-squared residuals, SSR, is 8% lower for the analysis with the two-component elimination function [Eq. (6) and Table II] as compared to the one-component elimination function [Eq. (5) and Table I]. This decreased SSR is significantly lower ($P < 0.05$) by a F-test for an additional term (Bevington, 1969). Consequently the two-component elimination function provides a better description of the experimental data than does the one-component model. This can be seen by a careful examination of the top

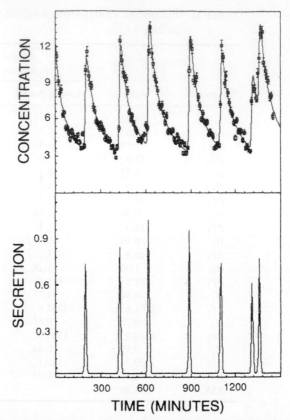

FIG. 3 Another hormone pulsatility analysis utilizing the DECONV algorithm and computer program. The values in the top graph are the same midluteal phase LH data shown in Fig. 2. The solid curved line shown in the top graph corresponds to the best fit of Eqs. (2), (6), and (7) to the data. The resulting parameter values are shown in Table II. Seven secretory events and a two-compartment elimination were assumed. The bottom graph presents the deconvolved secretory rate as a function of time.

graphs in Figs. 2 and 3. The values of the two elimination half-lives for LH shown in Table II are consistent with two-component half-lives reported elsewhere (Veldhuis *et al.*, 1986).

A comparison of the values in Tables I and II indicates that the choice of elimination model does not significantly change the secretory event positions and amplitudes (or mass). The deconvolved secretory patterns shown in the bottom graphs of Figs. 2 and 3 are visually indistinguishable. However the basal secretion is different. This difference is expected from the high cross-correlation between HL and S_0. It is also interesting to note that the cross-correlation for the two-component elimination is improved: the cross-correlation between S_0 and HL_1 was -0.22; the cross-correlation between S_0 and f_2 was 0.47; and the cross-correlation between S_0 and HL_2 was -0.88. The lower cross-correlation coefficients indicate that the least-squares procedure should perform better for this particular data set with the two-compartment elimination model compared to the one-compartment model.

One major disadvantage of the DECONV model and algorithm is that the user is required to specify the number and the approximate locations of secretory pulsations. The algorithm does not provide an automated method to test for an additional secretory pulse or to determine when an additional secretory event might have occurred. Thus, it is difficult to decide if the deviations between the fitted line and the data points in the top graphs of Figs. 2 and 3 at approximately 150, 590, and 710 min are actually significant unassigned secretory events or just random fluctuations with the data. The larger the number of consecutive positive residuals the greater the likelihood is that a pulse has been overlooked.

A second disadvantage is that the parameter estimation procedure is non-linear in the parameters. The consequence of this nonlinearity is that the estimates of the standard errors, and consequently the confidence intervals, of the determined model parameters are only approximate. An exact solution for the standard errors of estimated parameters with nonlinear fitting equations does not exist (Johnson, 1983, 1992; Johnson and Frasier, 1985; Straume *et al.*, 1991; Johnson and Faunt, 1992). It is thus more difficult to draw precise conclusions about the significance of secretory events due to the approximate nature of parameter confidence intervals.

A third disadvantage is the assumption that the secretory event is Gaussian shaped. The Gaussian shape was employed for specific reasons. An individual secretory cell usually contains multiple secretory organelles containing hormone. When a signal is received to secrete the hormone, the hormone release from individual organelles will be distributed in time. The spatial distribution of the secretory cells within the secretory organ will result in a distribution of capillary length and thus an additional time distribution. The spatial distribution will also generate a different time distribution for when the individual cells are activated. Some mixing is expected during the transport of the blood from the secretory organ to the sampling point, for example, the arm of a patient. This mixing will result in yet another distribution function. Each of the distribution functions will be somewhat Gaussian in shape and the combination of all of the distributions can be closely approximated by a Gaussian distribution. Nevertheless the observed secretory distribution might not actually be Gaussian in shape.

PULSE: *Waveform-Independent Deconvolution Method*

Our waveform-independent deconvolution method, PULSE, was specifically designed to eliminate the assumption of the DECONV method that the secretory event is Gaussian in shape. It also eliminates the requirement to specify the number and approximate positions of the secretory events. The major difference between PULSE and the DECONV algorithm is the assumed form of the secretory pulses [i.e., Eq. (7) for the DECONV algorithm and Eq. (10) for the PULSE algorithm]. The secretory pulses are assumed to have a general form that increases from a nadir to a peak and then decreases back to a nadir. Any waveform of this type can be described as a superposition of a series of closely spaced Gaussian curves:

$$S(t) = \int_{-\infty}^{\infty} e^{\ln I_t}\, e^{-(1/2)[(t-z)/F_{sd}]^2}\, dz = [e^{\ln I_t}] \otimes [e^{-(1/2)(t/F_{sd})^2}] = I(t) \otimes F(t), \quad (10)$$

where I_t is a unique impulse function at each time t and F_{sd} is a filter standard deviation for smoothing the function. There is a unique $\ln I_t$ for every experimental observation time; that is, if there are 144 data points then there are 144 different values of $\ln I_t$. In this case (as above), the exponential and natural logarithms are included to force the $I(t)$ function to always be positive. This is analogous to the amplitude terms in the DECONV method. Thus there is a unique positive amount of secretion corresponding to every experimental observation. The F_{sd} term is included so that each of these secretions is actually a small Gaussian shaped impulse. Typically, F_{sd} is set to one-half of the interval between the experimental observations so that there is a small, but important, amount of overlap between the observations. Note that this is significantly different from smoothing the experimental observations. This operation preserves the integrity of the experimental data and yet forces a relatively smooth $S(t)$ function. By substituting the $S(t)$ function into Eq. (2), the concentration as a function of time is expressed as

$$C(t) = I(t) \otimes F(t) \otimes [H(t)E(t)]. \tag{11}$$

The integral is evaluated as a discrete summation over the data points. The summation is illustrated in Fig. 4. The top graph presents a series of discrete values of secretion as a function of time, $S(t)$. The middle graph presents a typical elimination function, $H(t)E(t)$. The amplitudes of the elimination functions presented in the middle graph are scaled by the secretion function of the upper graph. The lower graph presents the resulting concentration as a function of time, $C(t)$, and is simply the sum of all of the scaled elimination functions presented in the middle graph.

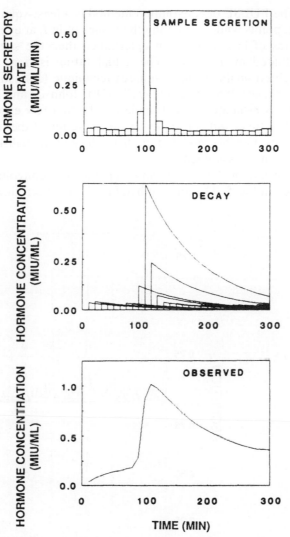

FIG. 4   Schematic description of the waveform-independent deconvolution method (PULSE). The observed serum hormone concentrations (bottom graph) arise from the combined effects of discrete sample secretion (top graph) and the assumed elimination kinetics (middle graph). Reprinted from Veldhuis and Johnson (1992) with permission.

The PULSE deconvolution method is a least-squares fit of Eq. (11) to the data points while evaluating the values of $\ln I_t$ at every data point. Once the values of $\ln I_t$ have been determined then the secretion function can be evaluated by Eq. (10). The actual fitting is done by a steepest-descent weighted nonlinear least-squares technique (Johnson, 1983, 1992; Johnson and Frasier, 1985; Straume *et al.*, 1991; Johnson and Faunt, 1992). The value of C_0 is estimated by extrapolating the first few experimental observations to zero time, and consequently the secretion at the first observation may be inexact. For this model, note that values of the half-lives for the elimination function are assumed.

The use of the waveform-independent deconvolution algorithm is illustrated in Fig. 5. The top graph in Fig. 5 is the secretion pattern corresponding

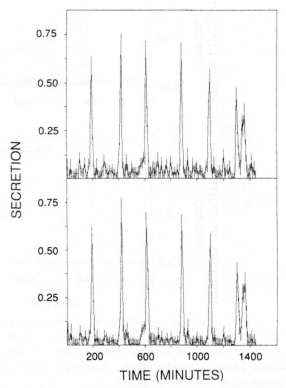

FIG. 5   Example of the use of the PULSE algorithm to obtain the secretory profiles from the experimental observations presented in Figs. 2 and 3. The top graph assumed a single-exponential elimination function and $HL = 50.4$ min as given in Table I. The bottom panel assumed a two-exponential elimination function with the half-lives as given in Table II. The error bars are approximate.

to the experimental data in Figs. 2 and 3, assuming the single-exponential elimination model as described in Table I. The bottom graph in Fig. 5 is the secretion pattern corresponding to the experimental data in Figs. 2 and 3, assuming the double-exponential elimination model as described in Table II. It is clear that the deconvolved secretion pattern is not very dependent on the assumed form of the elimination function. It is also clear the major secretory events are approximately Gaussian in shape, for this example.

The secretory profile produced by the PULSE algorithm is relatively insensitive to the assumed values of the elimination half-lives. The top graph of Fig. 6 presents the same experimental data as in Figs. 2 and 3, and the "best-fit" calculated concentration curve assuming a single-elimination half-life of 95.5 min. Clearly the elimination half-life is too long, as is shown by the poor quality of the resulting "best fit." This general pattern of underestimating the concentration peaks and overestimating the concentration valleys is diagnostic for an assumed elimination half-life that is too long. However, even with a poor choice of elimination half-life the resulting secretory pattern, as shown in the bottom graph of Fig. 6, is qualitatively similar to the patterns shown in Fig. 5.

One advantage of the PULSE method compared to the DECONV method is that the user need not specify the number of secretory events or their approximate positions. A second advantage is that no assumption is made about the shape of the secretory event or the amount of basal secretion.

One disadvantage of this method is that it does not provide a clear way to identify the locations of small secretory events. A second disadvantage is that a large number of tenuous assumptions must be imposed to propagate the uncertainties of the experimental observations into the uncertainties of the secretory profile. The amplitudes and locations of the seven large secretory events in the present example are easily defined, and thus the PULSE method may be of some use in defining the initial parameters required for the DECONV method. However, the PULSE method does not provide an answer as to whether there are additional small but significant secretory events at approximately 150, 590, and 710 min.

PULSE2

The major difficulty with the PULSE method is the inability to quantify accurately the secretion during the nadirs. This results from the fact that the algorithm evaluates a separate and independent value of the secretory rate at every observation. This is clearly required for the duration of a secretory event, but it is preferable to evaluate a single basal secretion S_0 that applies to all of the observations not included in the secretory events. To accomplish

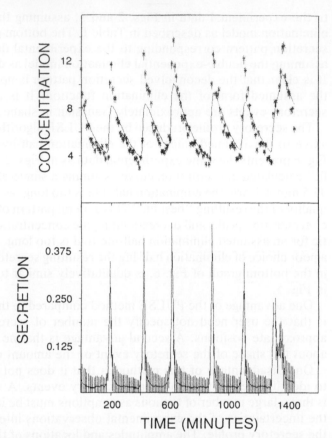

FIG. 6   Second example of the use of the PULSE algorithm to obtain the secretory profiles from the experimental observations presented in Figs. 2 and 3. For this analysis a single-component half-life of 95.5 min was assumed. The top graph presents the actual experimental observations and the calculated concentration profile from this analysis. The bottom graph presents the corresponding secretion profile. The error bars are approximate.

this Eq. (10) is modified to include a term for basal secretion, S_0. The resulting concentration as a function of time is then given by Eq. (12):

$$C(t) = [S_0 + I(t)] \otimes F(t) \otimes [H(t)E(t)]. \qquad (12)$$

A second modification from the PULSE algorithm is that only values of $I(t)$ that are statistically significant and positive are estimated. Those that are not significant and positive are set to 0 and not estimated. The algorithm initially starts with all individual values of $I(t)$ equal to 0 and assumed values for the half-lives in the elimination function. A least-squares fit is performed to evaluate C_0 and S_0. On the basis of this current fit, a secretory event is assigned to the time point that will yield the largest decrease in the variance-of-fit. The fit is repeated to evaluate C_0, S_0, and the amplitudes of all the currently identified secretory events. On the basis of this current fit, another secretory event is assigned to the previously unassigned time point that will yield the largest additional decrease in the variance-of-fit, and the fit is repeated. This process is repeated until the addition of a secretory event does not provide a statistically significant decrease in the variance-of-fit by an F-test for an additional term (Bevington, 1969). The last nonsignificant secretory event is then removed and the process is halted.

The method for deciding where to add an additional secretory event is similar to the algorithm utilized by Munson and Rodbard for one pulse detection algorithm (1989). First, the derivative of the variance-of-fit with respect to the amplitude of a presumptive additional secretory event is evaluated at every possible time point. This derivative will be 0 for those points that have already been assigned a secretory event. These derivatives are subsequently scaled by a quantity that is proportional to the estimated amplitude of a secretory event at each of the data points. This scaling factor is approximated as the absolute value of the difference between the observed concentration and the calculated concentration based on the secretory events at the previous iteration:

$$\Delta \text{ variance}_i \sim \frac{\partial \text{ variance}}{\partial \text{ event}_i} |\text{Obs}_i - \text{Calc}_i| . \tag{13}$$

The absolute value is included to make the scaling factor always positive. The data point with the largest negative predicted change in variance, as per Eq. (13), is taken as the location for the assignment of the next secretory event. That is the point where the assignment of a secretory event with a positive amplitude will provide the largest decrease in the variance-of-fit.

The use of the PULSE2 algorithm for the location of presumptive secretory events is shown in Fig. 7. The data used for Fig. 7 are from a normal healthy young adult in the midluteal phase (Sollenberger et al., 1990). The solid line in the top graph of Fig. 7 corresponds to the best fit of these data to a model with three identified secretory events, an unknown S_0, an unknown C_0, and an elimination half-life of 50.4 min. The middle graph is proportional to the expected change in variance-of-fit for the addition of a secretory event individually at each time point, as given by Eq. (13). The lowest negative value in the middle graph indicates that the next presumptive secretory

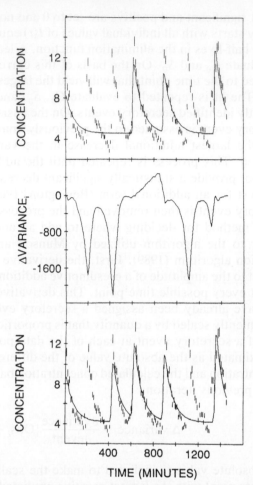

FIG. 7 Example of the secretory event location strategy utilized by the PULSE2 algorithm. The data in the top and bottom graphs are from a normal healthy young woman in the midluteal phase of the menstrual cycle (Sollenberger *et al.*, 1990). The solid line in the top graph corresponds to the best fit of the data to a model with three identified secretory events, an unknown S_0, an unknown C_0, and an elimination half-life of 50.4 min. The middle graph is proportional to the expected change in variance-of-fit for the addition of a secretory event individually at each time point. The solid line in the bottom graph corresponds to the best fit of the data to a model with four identified secretory events, an unknown S_0, an unknown C_0, and an elimination half-life of 50.4 min.

event occurs at approximately 430 min. The solid line in the lower graph corresponds to the best fit of the data after the addition of a presumptive secretory event at 430 min.

This algorithm for locating the positions of presumptive secretory events will normally only provide the positions of positive secretory events. However, it is possible that a particular secretory event that was positive at one step will become negative as additional secretory events are added. If this occurs, the negative secretory event is set to 0 and not considered as a secretory event for the current cycle. This does not preclude a later step from considering a secretory event at this particular position in time.

When the PULSE2 algorithm is correctly applied the differences between the experimental observations and the resulting calculated concentration profile (i.e., the residuals) should reflect only the random observational uncertainties. Thus the validity of the deconvolution process can be tested by applying goodness-of-fit tests such as the runs test or the Kolmogrov–Smirnov statistic (Straume and Johnson, 1992b).

The PULSE2 algorithm has distinct advantages over the PULSE algorithm. The initial identification of the secretory events is automated. Thus the PULSE2 algorithm is a deconvolution algorithm and a pulse detection algorithm. A second advantage is that the least-squares fits employed by the PULSE2 algorithm are all linear in the parameters being estimated. Exact solutions are used to evaluate the precision of estimated parameters and the significance of adding an additional secretory event for linear least-squares parameter estimations. The corresponding exact solutions do not exist for the nonlinear least-squares parameter estimation algorithms utilized by the DECONV and PULSE algorithms. Thus, fewer approximations are required for the linear PULSE2 algorithm.

Note that every position of a presumptive secretory event as identified by the PULSE2 algorithm given above cannot be considered as a unique secretory event. It is possible, for example, for the algorithm to place two presumptive secretory events at adjacent time points. Clearly, presumptive secretory events at adjacent time points should be considered as a single secretory event. Furthermore the secretory events may not be Gaussian in shape. The PULSE2 algorithm can easily accommodate non-Gaussian secretory events as the sum of the presumptive secretory events at adjacent time points. The details of identifying unique secretory events from the presumptive secretory events as identified above are described in the next section.

A Post Hoc Pulse Detection Method

The first step of the PULSE2 algorithm is to deconvolve the concentration profile into its constituent parts, a secretory profile and an elimination function. The secretory pattern from this deconvolution process will be referred to as a series of presumptive secretory events for the remainder of this discussion. The second step is to locate unique secretory events, or pulses, within the secretory profile. The first step does not require any assumptions about the true shape of the secretory event. However, to accomplish the second step a definition of the expected shape of a secretory event is required. Thus the second portion of the PULSE2 algorithm is somewhat arbitrary. This section outlines a realistic definition of a unique secretory event that we have found to be useful.

We expect a secretory event, or pulse, to consist of a smooth rise of secretion from a basal level to a maximum followed by a smooth fall back to basal. Clearly, if the secretory pattern is multimodal (i.e., multiple distinct maxima of secretion), then that portion of the pattern must contain at least as many secretory events as maxima. It is also clear that if two presumptive secretory events occur at sequential data points then there are not sufficient experimental observations to resolve these into separate events. On the basis of these concepts we have developed the following protocol to resolve the presumptive secretory events into unique secretory events.

First, if no minima exist in the calculated secretion pattern, $[S_0 + I(t)] \otimes F(t)$ of Eq. (12), between any two presumptive secretory events, then those two presumptive secretory events, and all of the intervening points, are assumed to be part of the same secretory event. This, of course, implies that any two adjacent presumptive secretory events will be part of the same unique secretory event.

Consider the situation where three presumptive secretory events occur in a row and the middle event corresponds to a minimum in the calculated secretion pattern. Because there is not a minimum between the first two of the presumptive secretory events, the previous step would group these as a single event. Similarly, because no minimum exists between the second and third presumptive secretory events, again they would be grouped together. Because the second presumptive event is common to both of these groups, all three would be grouped into a single event. However, this grouping of all three presumptive secretory events would incorrectly incorporate a minima in the calculated secretory pattern. Therefore, the grouped presumptive secretory events are scanned for the existence of secretion minima that are internal to the groups. If such a minimum exists, the corresponding group is split at the minimum point into two unique secretory events.

If a secretory event, or pulse, is to consist of a smooth rise of secretion from a basal level to a maximum followed by a smooth fall back to basal, then the first derivative of this profile will increase from 0 to a maximum then decrease to a negative minima and then return to 0. Thus a distinctive feature of a unique secretory event is that the derivative of the calculated secretion pattern must contain one maxima followed by one minima. If two unique secretory events are nearly superimposed then the calculated secretory pattern may contain a pronounced shoulder but not be completely resolved into two distinct maxima of secretion. In the presence of pronounced shoulder peaks, the first derivative of the calculated secretion pattern will contain a maxima followed by a minima followed by a second maxima and a second minima. Thus if this pattern occurs, the secretory event is split into two unique secretory events at a point halfway between the first minimum in the derivative and the second maximum in the derivative.

Once the presumptive secretory events have been grouped into these unique secretory events each group must be tested for statistical significance. This test is again performed by an F-test for an additional term (Bevington, 1969). This test compares the overall variance-of-fit, s_i^2, with the variance-of-fit for a new least-squares fit with the ith group of presumptive secretory events removed, s_i^2:

$$\frac{s_i^2}{s_m^2} = 1 + \frac{(ndf_i - ndf_m)}{ndf_m} F(ndf_i - ndf_m, ndf_m, 1 - PROB), \qquad (14)$$

where ndf_m is the number of degrees of freedom corresponding to s_m^2, ndf_i is the number of degrees of freedom corresponding to s_i^2, PROB is the significance probability, and F is the F-statistic. The definition of the ndf terms is the number of data points minus the number of parameters being estimated by the fit. It is important to remember that two degrees of freedom are used for each of the presumptive secretory events, one for the amplitude and one for the position.

The results of the completed analysis of the same set of experimental data are presented in Fig. 8. For this analysis a single elimination half-life of 50.4 min and a filter standard deviation of 5 min (one-half the interval between the data points) are assumed. It is interesting that small but significant pulsatile secretory events occur at 590 and 710 min. The PULSE algorithm indicated that a significant amount of secretion occurred at 590 min but not conspicuously at 710 min (see Fig. 5). A significant amount of basal secretion occurs between the secretory events. This small amount of basal secretion accounts for approximately 50% of the total secretion.

Limitations of PULSE2 are its sensitivity to the choice of the SD of the smoothing function. Simulation studies show increased type I (false-positive) errors in peak identification, if the SD chosen is too small, and higher type II (false-negative) errors in pulse detection, if the assigned SD is too large (J. D. Veldhuis and M. L. Johnson, unpublished, 1995). In contrast, there is considerably less sensitivity of burst detection to variations in the assumed hormone half-life.

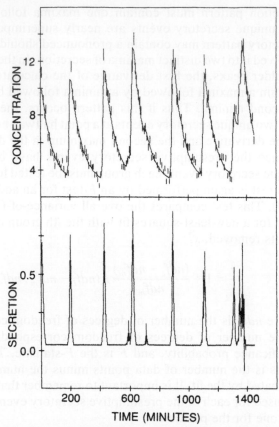

FIG. 8   Example of the use of the PULSE2 algorithm to obtain the secretory profiles from the experimental observations presented in Figs. 2 and 3. The top graph assumed a single-exponential elimination function and HL = 50.4 min as given in Table I. Significant secretory events occur at 200, 430, 620, 890, 1110, 1310, 1350, and 1370–1380 ($P < 0.001$) min. Less significant events also occur at 590 ($P < 0.025$) and 710 ($P < 0.05$) min. The bottom graph shows the corresponding secretory profile.

Summary

This chapter describes several deconvolution approaches that we have developed. These were specifically designed to interpret pulsatile hormone time series observations. Such endocrine data are unique in that there is a limited number of observations with a large amount of experimental uncertainty. These unique features preclude the use of most earlier deconvolution techniques that require large numbers of data points and negligible experimental uncertainties. The PULSE2 program is available for 80386-80387, or faster microprocessors, for IBM-compatible computers on written request from the authors.

Acknowledgments

The authors acknowledge the support of the National Science Foundation Science and Technology Center for Biological Timing at the University of Virginia, the Diabetes Endocrinology Research Center at the University of Virginia (NIH DK-38942), the University of Maryland at Baltimore Center for Fluorescence Spectroscopy (NIH RR-08119; M.L.J.), National Institutes of Health Grant GM-35154 (M.L.J.), and RCDA1K04 HD00634 from National Institute of Child Health and Development, (NIH) (J.D.V.), and Baxter Healthcare Corp. (Round Lake, IL).

References

P. R. Bevington, *in* "Data Reduction and Error Analysis for the Physical Sciences," p. 200. McGraw-Hill, New York, 1969.

C. Desjardins, *Biol. Reprod.* **24**, 1–21 (1981).

W. S. Evans, M. J. Sollenberger, R. A. Booth, A. D. Rogol, R. J. Urban, E. C. Carlsen, M. L. Johnson, and J. D. Veldhuis, *Endocr. Rev.* **13**, 81–104 (1992).

P. A. Jansson, "Deconvolution with Applications in Spectroscopy." Academic Press, New York, 1984.

M. L. Johnson, *Biophys. J.* **44**, 101–106 (1983).

M. L. Johnson, *Anal. Biochem.* **206**, 215–225 (1992).

M. L. Johnson and L. M. Faunt, *in* "Methods in Enzymology" (L. Brand and M. L. Johnson, eds.), Vol. 210, pp. 1–37. Academic Press, San Diego, 1992.

M. L. Johnson and S. G. Frasier, *in* "Methods in Enzymology" (C. H. W. Hirs and S. N. Timasheff, eds.), Vol. 117, pp. 301–342. Academic Press, San Diego, 1985.

P. J. Munson and D. Rodbard, Proceedings of the Statistical Computing Section of the American Statistical Association, Alexandria, Virginia, pp. 295–300. 1989.

G. De Nicolao and D. Liberati, *IEEE Trans. Biomed. Eng.* **40**, 440–455 (1993).

M. J. Sollenberger, E. C. Carlsen, M. L. Johnson, J. D. Veldhuis, and W. S. Evans, *J. Neuroendocrinol.* **2**, 845–852 (1990).

M. Straume and M. L. Johnson, *in* "Methods in Enzymology" (L. Brand and M. L. Johnson, eds.), Vol. 210, pp. 117–129. Academic Press, San Diego, 1992a.

M. Straume and M. L. Johnson, *in* "Methods in Enzymology" (L. Brand and M. L. Johnson, eds.), Vol. 210, pp. 87–105. Academic Press, San Diego, 1992b.

M. Straume, S. G. Frasier-Cadoret, and M. L. Johnson, *in* "Topics in Fluorescence Spectroscopy" (J. R. Lakowicz, ed.), Vol. 2, pp. 117–239. Plenum, New York, 1991.

R. J. Urban, W. S. Evans, A. D. Rogol, D. L. Kaiser, M. L. Johnson, and J. D. Veldhuis, *Endo. Rev.* **9**, 3–37 (1988a).

R. J. Urban, D. L. Kaiser, E. van Cauter, M. L. Johnson, and J. D. Veldhuis, *Am. J. Physiol.* **254,** E113–E119 (1988b).

J. D. Veldhuis and M. L. Johnson, *Am. J. Physiol.* **250,** E486–E493 (1986).

J. D. Veldhuis and M. L. Johnson, *in* "Methods in Enzymology" (L. Brand and M. L. Johnson, eds.), Vol. 210, pp. 539–575. Academic Press, San Diego, 1992.

J. D. Veldhuis, F. Fraioli, A. D. Rogol, and M. L. Dufau, *J. Clin. Invest.* **77,** 1122–1128 (1986).

J. D. Veldhuis, M. L. Carlson, and M. L. Johnson, *Proc. Natl. Acad. Sci. U.S.A.* **84,** 7686–7690 (1987).

[2] Specific Methodological Approaches to Selected Contemporary Issues in Deconvolution Analysis of Pulsatile Neuroendocrine Data

Johannes D. Veldhuis and Michael L. Johnson

Introduction

Neuroendocrine glands and ensembles typically signal remote target cells via an intermittent burstlike release of effector molecules (1–28). Accordingly, serial monitoring of effector concentrations or their release rates over time discloses pulsatile profiles that consist of punctuated and delimited episodes of accelerated secretion. Special technical problems surround the valid, reliable, specific, and sensitive detection of these burstlike release episodes (1, 9, 19, 29–52). Consequently, various analytical tools have been developed to address the unique challenges of quantitatively appraising neuroendocrine data.

Among such strategies are discrete peak detection methods, which are designed to identify significant increases and decreases in serial neurohormone concentrations or successive estimates of neurohormone release rates obtained by direct or indirect monitoring of the neuroendocrine gland (1, 9, 49, 51, 53–56). Techniques of deconvolution analysis have been adapted from the physical sciences, revised, and in some cases designed in novel forms in order to estimate underlying secretory rates from the measured serial neurohormone concentrations over time (19, 29–33, 36, 43, 56–63). For example, one of our recent deconvolution techniques has been termed multiparameter deconvolution analysis, because it attempts to estimate the number, amplitude, mass, and duration as well as temporal location of individual neurohormone secretory bursts, while simultaneously calculating the apparent half-life of elimination of the neuroendocrine effector molecules (31, 32). This particular approach assumes a nominal waveform (Gaussian, skewed distribution, etc.), with or without *a priori* knowledge of neurohormone half-life. Complementary methodologies are waveform-independent, and typically require assumed knowledge of neurohormone half-life (32). The latter techniques may be especially useful when attempting to estimate basal and pulsatile neurohormone release simultaneously (32). However,

Methods in Neurosciences, Volume 28

these mathematically sophisticated technologies are not devoid of their own special technical problems, issues, and requirements for enhancement.

In this chapter, we examine the tools of deconvolution analysis as a methodological achievement in quantitative neuroendocrinology. We specifically identify selected problems inherent in deconvolution analysis of pulsatile neurohormone data. Our intent is to highlight issues of difficulty and offer suggested solutions in the form of specific technical and methodological notes on the individual issues.

Issues

Experimental Uncertainty within Data

The measurement of serial concentrations or release rates of neuroendocrine effector molecules *in vitro* or *in vivo* inevitably entails experimental uncertainty. The sources of experimental uncertainty include first the host biological system under examination (e.g., the intact animal with sampling ports placed in relevant tissue compartments; *in vitro* perifusion of one or more secretory units within or of an entire neuroendocrine ensemble such as the hypothalamo-pituitary unit). This kind of variability arises from unique characteristics of the neuroendocrine observational system(s) selected. Second, the collection technique for withdrawal of blood or other fluid samples contributes additional variability to the final measurement. Third, processing of the samples (e.g., centrifugation, clot removal, freezing, storage, thawing, and dividing into aliquots) introduces further, presumably random uncertainty in the final data. Last, the measurement technique itself (e.g., an immunologically based neurohormone assay, a target-tissue bioassay) creates further random within-sample and between-sample variability.

In many reports, within-assay variability is approximated by the intraassay coefficients of variation, which represent ratios at various neurohormone concentrations (or doses) of the corresponding standard deviations (*SD*) to the series means (1, 32). The coefficients of variation may exhibit dose dependence. They are determined by many factors, such as the number of replicates utilized in the estimation, the choice of the replicated control samples and their similarity to the experimentally unknown samples, the methodology employed for data reduction (see Chapter 9 by Straume *et al.*, this volume), and the nature of the assay system utilized (e.g., a high-precision, high-count fluorometry or photometry-based approach, as in immunofluorometric or chemiluminescence assays, versus a low disintegration per unit time radionuclide assay). In short, a host of factors, some identified and some less evident, contributes in some manner to the resultant overall imprecision inherent in the final measurements of the neurohormone concentrations serially over time. These collective sources of experimental uncertainty pose special problems in the analysis of pulsatile neuroendocrine data, given that true physiological events must exhibit amplitudes substantially higher than those anticipated from the (interactive) sum of all inherent experimental variability in serial observations (1, 32, 51).

The preassay sources of noise are not well defined, but include experimental variations in the system and the collection and processing of the samples for assay. As one approach to the magnitude of this variability, we have measured serial plasma concentrations of stable analytes, such as total serum protein and total serum calcium (1). These end points were chosen because of the high degree of physical precision inherent in their respective assays. Consequently, the majority of experimental uncertainty in the serial serum protein and calcium measurements over time should reflect variability in the observed system and the sample collection and processing steps. As shown in Fig. 1, serial serum total protein and calcium measurements vary over time, but this variability is remarkably less than that of luteinizing hormone (LH) present in the same samples but assayed by immunological techniques. Moreover, the total protein and calcium measurements show no systematic trends over time, such as baseline drift or progressively increasing or declining variance. We estimate from such limited observations that the preassay variability is in the range of 1–5% for the collection of blood samples every 5 min in healthy young men studied on a research unit (1).

Within-assay variability can also be characterized quantitatively. One approach to estimating intraassay noise is the replication of measurements on a single pool of blood, serum, or plasma (1, 32). For example, blood may be collected from one or a dozen or more individuals, pooled, and then aliquoted repeatedly from the single pool and subjected to conventional assay. The aliquots may be stored frozen, thawed, and presented to the technician as individually labeled pseudosamples. In the complete absence of experimental uncertainty, the measurements of several hundred replicates from the single pool of serum or plasma would be identical. Therefore, the *ad seriatim* measurement of such aliquots from a single pool offers an estimation of within-assay experimental uncertainty. As shown in Fig. 2, the resultant distribution of LH and growth hormone (GH) measurements derived from within-assay replication of a single pool of a body fluid is approximately Gaussian (32).

We caution, however, that assays with poor threshold sensitivities (i.e., relatively high lower bounds of detectable concentrations) may exhibit considerable skew, particularly when the mean measured sample concentration approaches the detection limit. Indeed, the assay results in essence are truncated by the detection threshold in some circumstances. Consequently,

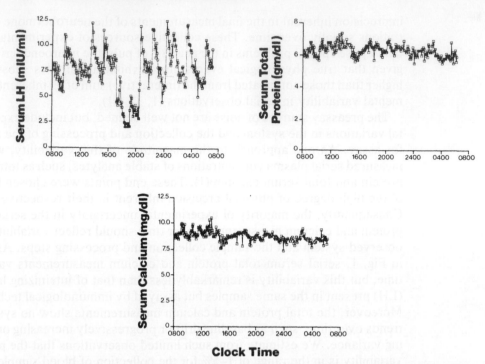

FIG. 1 Serial serum calcium and total protein measurements in blood samples collected at 5-min intervals over 24 hr in one healthy young man compared with serum immunoreactive LH concentrations assayed in triplicate in the same samples. Note that the prominent fluctuations in serum LH concentrations (pulses) do not coincide with any correspondingly large variations in the stable analytes (calcium and total protein). Adapted with permission from Urban *et al.* (1), The paradigm of the lutenizing hormone pulse signal in men, *Endocrine Reviews*, **9**, 3, 1988 © The Endocrine Society.

considerable care and attention must be directed to appropriate representation of the lower assay detection limit. This entails at the very least a precise knowledge of the background measurement level in appropriately prepared tubes and buffers containing zero neurohormone, and in corresponding tubes containing very small amounts of neurohormone approaching the detection limit. A plausible and valid technical approach to data reduction must also be utilized, so that the total experimental variabilities in both zero-dose and lower-end samples are incorporated correctly into the final estimate of within-sample error (see Chapter 9 by Straume *et al.*, this volume).

In brief, within-assay experimental uncertainty must be characterized for each particular measurement system, and each neurohormone effector mole-

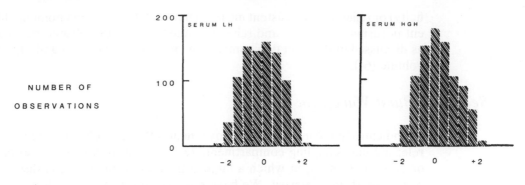

FIG. 2 Frequency histogram of serum GH measurements conducted as multiple replicates from a single pool of normal human blood. For comparison the LH histogram is also shown to illustrate the within-assay error distribution. Adapted with permission from Veldhuis and Johnson (32).

cule. Experimental variability for many analytical tools is assumed to be independently and identically distributed zero-mean random error conforming to a Gaussian distribution. However, assay detection limits, degradation of neurohormone within the sample, peculiarities of the measurement devices, etc., may introduce skew or truncation of the measurement distribution. Accordingly, we recommend wherever possible that a collection of pooled samples be utilized and that 30–100 replicates of at least low, intermediate, and high pools be prepared and analyzed to characterize the distribution of experimental variation in the particular assay of interest. These replicate measurements should conform in every possible respect to the technical processing and assay conditions expected for the experimentally unknown samples. Importantly, conditions of the assay should be optimized so as to place all experimentally unknown samples within the detectable range, definitively exceeding the detection limit by 2 *SD* or more. Moreover, where substantial within-assay variability exists, multiple dose-specific replicates should be carried out, the assay precision enhanced, and/or alternative assay schemes considered. Knowledge of within-assay variability is important analytically, because many curve fitting and other analytical tools utilize one or more measures of intrasample experimental uncertainty to weight the relative contribution of that sample mean to the fit of the data.

We would emphasize that within-assay precision does not in any way guarantee assay validity (measuring the end point of interest with high specificity) or reliability (consistent measurements of the same end point on different occasions). Validity and reliability must also be established rigorously, as discussed in detail in two companion chapters (Chapters 3 and 4), in this volume (64).

Secretory Burst Waveform

As indicated in the introduction above, neuroeffector molecules are typically released into sampling compartments (e.g., blood) in a highly intermittent or episodic fashion, in which a large collection of molecules is discharged over a short time interval. We have referred to this as a burstlike mode of secretion (18, 31, 32, 60, 65–68). The corresponding secretory burst waveform is defined by the plot of secretory rate over time. In general, the exact secretory waveforms of many neurohormones over a range of *in vivo* conditions of experimental relevance are not known at all or known only poorly. However, recent blood sampling at 30-sec intervals in pituitary venous effluent discloses bursts of release of LH and follicle-stimulating hormone (FSH) in the horse (69) (Fig. 3). This *in vivo* complex waveform could be approximated at the point of exit from the pituitary gland into the sampled blood by secretory episodes (Gaussian bursts) with some basal release. A similar waveform is observed also in the ovariectomized sheep, in which the blood draining the anterior pituitary gland exhibits serially changing LH concentrations over time, the plot of which resembles a slightly skewed Gaussian, characterized by a very rapid increase in secretory rates to a finite

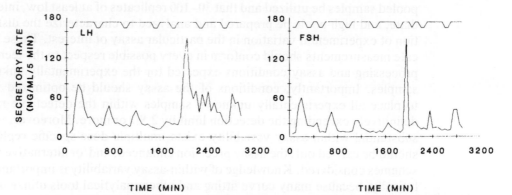

FIG. 3   Equine LH and FSH concentrations in pituitary effluent blood collected at 30-sec intervals. The plots constitute profiles of *in vivo* pituitary LH and FSH secretion rates versus time. Adapted with permission from Veldhuis and Johnson (32). Data were provided by Susan Alexander and Clifford Irvine, Christchurch, New Zealand.

maximal value, followed by a slightly more gradual decline toward baseline (I. Clarke and J. D. Veldhuis, 1994, unpublished). *In vitro* perifusion experiments also permit indirect inferences regarding the waveform of neurohormone release at or near the site of the secretory cells (70). However, *in vitro* manipulations do not necessarily faithfully reflect *in vivo* secretory behavior. Consequently, wherever possible, direct *in vivo* monitoring should be made to clarify the time course of neurohormone secretion from the monitored secretory gland (13, 32, 60, 71).

As the secretory waveform is acted on by fluid turbulence, convection, streaming, admixture, loss, and distribution, the native waveform may undergo significant structural modification. For example, following the secretion of LH into pituitary venous effluent blood, LH molecules are distributed within the far larger plasma volume of the host animal and subjected to intensive admixture as blood from multiple tributaries is combined and reallocated; furthermore, they undergo extensive diffusion and turbulence within large chambers such as those of the heart and pulmonary vasculature. Indeed, as an approximation, one can estimate that the resultant secretory waveform as observed in peripheral blood samples would resemble a Gaussian or a skewed distribution of secretory rates over time (32, 66–68). Consequently, in many deconvolution applications, we have assumed that the underlying neurohormone secretory waveform as inferred indirectly in peripherally sampled blood remote from the site of glandular discharge can be approximated by a Gaussian or skewed distribution of release rates over time (10, 31, 32, 38, 41, 42, 48, 60, 72).

The impact of an assumption of a symmetric Gaussian burstlike secretory event compared to a skewed secretory burst is illustrated in Fig. 4. Here we show that the convolution integral (integrated algebraic product of the functions) of a Gaussian secretory burst and a monoexponential elimination function is a typical plasma neurohormone concentration pulse over time (31, 32, 66–68). Given the same mass of hormone discharged within a skewed versus a Gaussian secretory burst, and identical monoexponential kinetics of elimination, the two corresponding predicted plasma neurohormone effector concentration time courses are only minimally discrepant. Simulation studies indicate that the impact of skewness within the secretory event on the overall observed output (time course of the plasma neurohormone concentration) depends on the mass of hormone secreted per unit distribution volume, the amplitude and relative half-width of the secretion burst, and the neurohormone half-life (42). Thus, under some hypothetical conditions, such as a large mass of hormone secretion in association with a very short half-life, the waveform of the primary secretory event may influence to a greater degree the expected time course of the resultant neurohormone concentrations. Therefore, we recommend that the foregoing information regarding primary

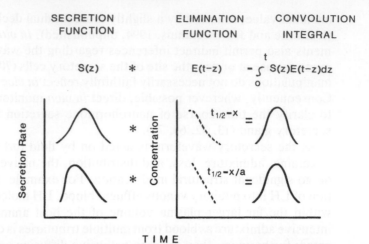

FIG. 4   Schematized illustration of the predicted effect of assuming Gaussian (top) versus skewed (bottom) secretory bursts on the half-life predicted from the resultant plasma hormone concentration profile for a given hormone pulse in the circulation, as acted on by a monoexponential elimination function. Adapted with permission from Veldhuis *et al.* (42).

and peripheral (dissipative) secretory waveforms be obtained from direct measurements wherever possible, and that the deconvolution model of the secretory burst closely emulate the directly monitored waveform at the appropriate point of sampling (e.g., the peripheral circulation versus central hypothalamo-pituitary blood in the case of certain neurohormones).

Half-Life of Neurohormone Removal

After an effector molecule is secreted or released into a sampling compartment (e.g., rapidly flowing blood), the neurohormone is subject to both compartmental distribution and irreversible metabolic elimination. By distribution we refer to the net movement of molecules into a theoretical space during and following their admixture by convection, turbulence, streaming effects, and diffusion into the available physiological compartment [e.g., the plasma volume in the case of a large protein or glycoprotein hormone such as GH or LH (32, 36, 73–75)]. Each released molecule is also subject to one or more fates: possible binding to a tissue receptor; direct excretion; uptake by nonspecific mechanisms into cells; metabolic interconversion by enzymes located within the plasma or other compartments, on cell surfaces, or within cells; and irreversible degradation by hormone-specific or nonspecific enzymatic and nonenzymatic pathways. Collectively, the last sources of removal of the neurohormone from the sampling compartment result in so-called irreversible metabolic clearance.

Following the abrupt burstlike secretion of molecules within a given pulse into the bloodstream, a combination of rapid distribution and more prolonged irreversible metabolic elimination begins to occur virtually simultaneously. The resultant plot of plasma neurohormone concentrations over time is therefore typically curvilinear, with features of monoexponential or often biexponential kinetics (e.g., Refs. 73 and 76). The precise quantitative kinetics of distribution and irreversible elimination can be approximated by injecting appropriate pure and defined preparations of the neurohormone of interest into the sampling compartment in amounts that approximate the physiological secretory burst, ideally with an injection waveform resembling that produced by the secretory gland, and wherever possible with concomitantly suppressed endogenous release of that neurohormone. (Suppression of endogenous release is necessary to avoid the confounding effects of observing both secreted and injected molecules within the same sample; this latter possibility can also be avoided by appropriate labeling of the injected effector substance, although the labeling procedure may itself damage or modify the neurohormone.) A plasma neurohormone disappearance time course is illustrated in Fig. 5, where highly purified human pituitary LH was injected into a patient who had profound deficiency of this particular gonadotropin. The elimination of purified LH was monitored over time in peripheral blood by rapid and repetitive blood sampling followed by immunoassay and *in vitro* bioassay of the corresponding serum samples (73). Note that a monoexponential or biexponential function will provide a good fit of the observed serial hormone concentrations over time.

Although seemingly self-evident, we emphasize that the determination of neurohormone kinetics be carried out with considerable care, so that the resultant plasma disappearance curves are achieved with (i) highly representative biologically active hormone [including in some cases a multiplicity of isoforms (73, 75)]; (ii) similar mass of hormone injected per unit distribution volume as anticipated with an endogenous secretory pulse; (iii) a physiologically relevant amount of baseline hormone release within the system, since the disappearance rate constant appears to depend in some cases on the apparent filling of the distribution space by prior steady-state infusion; (iv) an appropriate host state for the animal (e.g., a physiological number and affinity of peripheral hormone receptors and intact nonreceptor-mediated uptake and elimination mechanisms); (v) a time course of delivery of the neurohormone into the sampling compartment that emulates physiological secretion; and (vi) a sampling duration to include at least five half-lives for

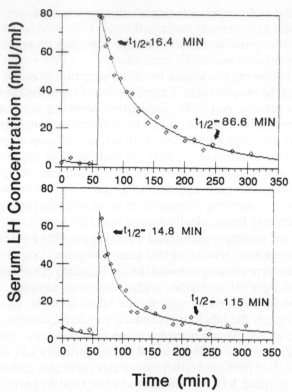

FIG. 5 Exponential curve fitting of serum LH disappearance curves in two hypopituitary men injected with highly purified LH. Adapted with permission from Veldhuis *et al.* (73). Reproduced from the *J. Clin. Invest.* 1986, **77**, 1122 by copyright permission of The American Society for Clinical Investigation.

the decay of the slower (slowest) component in the data, with a sampling frequency initially high enough to capture and resolve the rapid distributional phase half-life, which often approximates 2–4 min (e.g., Refs. 73 and 75). Otherwise, the apparent kinetic values obtained experimentally may differ substantially from those actually operating endogenously within the undisturbed animal (77–100).

In the case of many neurohormones, the apparent half-life of elimination is independent of the mean plasma neurohormone concentration at least over the physiological range (e.g., Ref. 73). However, this is not universally true, and if neurohormone concentrations greatly exceed or fall below the physiological range, significant nonlinearities can be observed more often (73). Moreover, studies using recombinant proteins indicate that even within the physiological range, the apparent half-life of irreversible hormone elimination may vary as a nonlinear function of the plasma hormone concentration. For example, in the case of human recombinant GH infused into healthy individuals made acutely deficient in GH by co-injection of a GH-release-inhibiting peptide, somatostatin, the half-life of GH decay from steady state is proportionate to a power function of the mean plasma GH concentration within the established physiological range. Indeed, the half-life increases exponentially from 8 to 12 min toward a plateau value of 18–22 min as the plasma GH concentration rises through the midportion of the normal physiological range of 1–10 μg/l (101) (Fig. 6). This saturability of GH removal over the physiological range and LH at concentrations exceeding the physiological range may reflect rate-limiting effects of peripheral binding and/or uptake mechanisms via receptor or nonreceptor mediated pathways.

An implication of a significant linear or nonlinear relationship between mean neurohormone half-life and its plasma concentration is the need to consider use of a correspondingly appropriate kinetic function in the convolution integral, which is the algebraic expression relating secretion and clearance over time (31, 32) and hence allowing (by deconvolution) an estimation

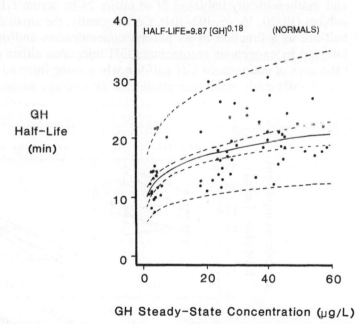

FIG. 6   Nonlinear relationship between GH half-life estimated by steady-state infusion of recombinant human GH and plasma GH concentration in healthy adults. A power-function fit and 95% confidence intervals are given. Adapted with permission from Haffner *et al.* (140).

of one or both of these functions. Alternatively, the deconvolution model can utilize a mean apparent hormone half-life corresponding to the average half-life displayed over the sampling interval and neurohormone concentrations of interest. The latter approach appears to be adequate [e.g., in the case of GH and LH, since 24-hr profiles of 5-min sampled serum GH or LH concentrations can be described by a single monoexponential function with no systematic bias within the 24-hr fit as assessed by the distribution of residuals (differences between the observed serum hormone concentration and the predicted curve) (8, 10, 102–104)].

Studies utilizing recombinant human GH to estimate GH elimination rates in healthy adults also show a statistically consistent difference of several minutes in calculated plasma GH half-life in the morning (shorter) compared to the late evening (longer) (88) (Fig. 7). The finding that the apparent GH half-life following direct infusion of the protein is several minutes longer in the evening is a consistent trend within a group of healthy young men. This numerically small difference (~15%) may not be physiologically evident to target cells and does not appear to create obvious analytical discrepancies; the choice of a single average within-subject GH half-life provides a good and mathematically unbiased fit of entire 24-hr serum GH profiles in that subject (10, 20, 36, 46, 102–104). Consequently, the small differences in GH half-life as a function of its plasma concentration and/or time of day as inferred by exogenous recombinant GH injections either are an artifact of estimates of endogenous GH half-life when using injected hormone and/or are of sufficiently small magnitude that an average subject- and hormone-

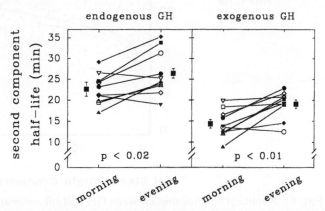

FIG. 7 Estimated GH half-life of endogenous GH and following injections of exogenous recombinant human GH in the morning versus the evening in healthy young men. Adapted with permission from Porksen et al. (88).

specific half-life in most pathophysiological states will adequately represent the pulsatile neurohormone profile for endogenously released GH molecules.

The foregoing issues illustrate the complexity of ostensibly straightforward assumptions regarding neurohormone half-lives of distribution and irreversible elimination from the sampling compartment. Thus, in most circumstances, we recommend injecting a releasing factor or other secretagogue immediately following the prolonged sampling session used for basal pulsatility estimates, so as to obtain a prominent, well-defined hormone secretory burst with five half-lives of decay available for quantitating endogenous half-life. This approach allows one to compute a hormone-specific, subject-specific, and session-specific hormone half-life that is related to the endogenous effector molecules (32, 76). This is preferable in many circumstances in which the endogenous hormone undergoes one or more *in vivo* modifications (e.g., posttranslational carbohydrate addition), which may render its half-life different from that of injected pure recombinant or extracted (and hence variably modified) hormone.

Basal versus Burstlike Secretion

A very difficult contemporary issue in deconvolution analysis is the relative quantitative allocation of neurohormone secretion to basal (time-invariant) versus burstlike release. For example, in the case of GH, conventional radioimmunoassays (RIAs) with detection limits of approximately 0.1 ng/ml have revealed exclusively a burstlike pattern of GH secretion in healthy humans (10, 32, 36, 46, 62, 102–105). However, as illustrated in Fig. 8, the use of an enhanced-sensitivity chemiluminescence assay for GH discloses low rates of underlying basal secretion on which are superimposed distinct bursts of GH release (106). As illustrated in Fig. 8, the conventional RIA sensitivity threshold of 0.1 ng/ml eliminates from detectability the low basal serum GH concentrations, which the ultrasensitive GH assay discloses to be values approximating 15–25 pg/ml. Basal GH release presumably underlies these nearly constant low serum GH concentrations, although a reservoir of GH in plasma (e.g., GH bound to a slowly dissociating binding protein present in low concentrations) and/or release of GH from a slowly turning over pool (e.g., GH released into plasma by steady-state shedding of cytoplasmic GH receptors containing GH with high affinity binding) could also produce this profile. Direct physiological experiments are required to distinguish these and other possibilities. However, assuming that a low rate of basal release can result in pseudo-steady-state serum GH concentrations, and that episodic bursts of GH secretion can produce in distinct superimposed

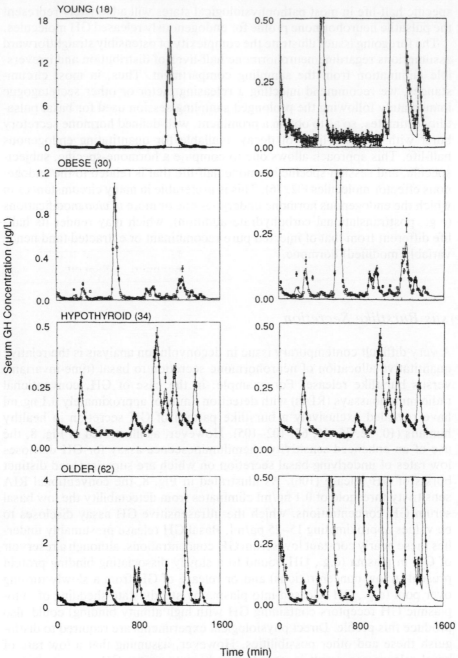

pulses, then an analytical challenge emerges of distinguishing basal from burstlike neurohormone secretion.

Several specific issues arise in distinguishing basal from burstlike neurohormone secretion. First, one must question whether basal secretion is a zero-order function of time (i.e., a constant value over 24 hr), a sinusoidal function of time, a simple linear function of time, and so on. Second, one must define how best to estimate basal secretion as distinct from pulsatile hormone release in any given profile, with or without prior knowledge of the hormone half-life. And, third, given a cascade of immediately successive increases and decreases in neurohormone concentrations over time (e.g., a volleylike pattern), one must decide how to distinguish a collection of secretory bursts that are partially superimposed (or fused) from a combined increase in secretory burst frequency (possibly with no change in amplitude) and a rise in or emergence of some underlying basal secretion.

To evaluate the 24-hr time course of putatively admixed pulsatile and basal hormone secretion, we recommend that direct experimental methods be applied *in vivo*. Such methods should entail techniques to specifically suppress or eliminate pulsatile hormone release, while preserving the basal component or vice versa. This is not a simple matter, because the two pathways or modes of secretion may be mechanistically interdependent within the neurosecretory unit. For example, basal hormone release may reflect so-called constitutive release or leakage of hormone that may be either related to or independent of the processing of hormone into specific secretory granules, their margination within the cell, and their availability for subsequent exocytosis. If basal hormone release is coupled physiologically to granular exocytosis, then these processes will be difficult to unravel experimentally in the whole animal. However, where possible, the time course of basal release over the sampling interval (e.g., over 24 hr) should be defined independently and this information utilized in a relevantly posed model of deconvolution analysis.

Determining basal and pulsatile hormone release rates simultaneously by deconvolution analysis is not a trivial undertaking. The challenge arises

---

FIG. 8   Serum GH concentration profiles over time filled by deconvolution analysis in various individual men sampled every 10 min assessed in an ultrasensitive chemiluminescence assay illustrating basal hormone release with superimposed pulsatility. The right-hand graphs give an expanded view. Ages are given in parentheses. No basal release could be observed by radioimmunoassay (RIA) or immunoradiometric assay (IRMA), for which nominal sensitivities are 0.5 and 0.1 μg/liter. Note expanded scale of right-hand graphs. Adapted with permission from Iranmanesh *et al.* (106), Low basal and persistent pulsatile growth hormone secretion are revealed in normal and hyposomatotripic men, *J. Clin. Endocrinol. Metabol.* **78**, 84, 1994. © The Endocrine Society.

because both processes contribute jointly to the resultant overall plasma hormone concentration profile, and because parameters of the two corresponding functions are highly correlated. This correlation may or may not be linear but could, for example, be sigmoidal (e.g., if basal and pulsatile hormone release are physiologically coupled through intracellular mechanisms, a rise in pulsatile hormone secretion might be accompanied by a linear or an exponential increase or decrease in basal release over a corresponding time course). If one assumes provisionally that basal and pulsatile hormone release are mathematically independent, and that their waveforms are known, then their corresponding secretory parameters could be estimated given half-life and appropriate hormone-concentration data. The data must contain significant regions of basal secretion alone, because our simulation studies show (as anticipated intuitively) that continued pulsatile hormone release without any intervals of observable baseline (unvarying serial neurohormone concentrations over time) would allow the estimated basal secretion rate to produce serum hormone concentrations that lie somewhere between zero and a maximal value not exceeding the nadirs of successive pulses. Indeed, the latter value may be considerably above true basal secretion, because even in a purely pulsatile model it represents the summation of all previously secreted molecules that have not yet been eliminated.

Given a combined profile of basal and pulsatile hormone release with relatively prolonged intervals of apparent basal secretion intervening, then estimates of both maximal basal and minimal pulsatile hormone release could be obtained from an appropriately formulated deconvolution equation that assumes some knowledge of the algebraic form of the basal and pulsatile secretion functions that underlie the data (32). This is illustrated by the fit of the data in Fig. 8, where both basal and pulsatile GH release appear to occur in a healthy human (106). However, in the absence of periods of sustained observable unvarying serum neurohormone concentrations over time, basal release could only be estimated to lie between zero and some maximal rate sufficient to account for the lowest nadirs represented consistently within the plasma neurohormone concentration data. Thus, in essence only an upper (and lower, i.e., zero) bound for the basal secretion estimate can be inferred readily.

A special problem of combined basal and pulsatile hormone release exists in profiles that consist of rising plasma neurohormone concentrations, which could result from partial summation of rapidly occurring secretory bursts (i.e., burst fusion) and/or frequent secretory bursts with a rising (or even falling) baseline. Analytically distinguishing between these two modes of secretion giving rise to any particular data set is extremely difficult, unless some provisos, prior knowledge, or assumptions are held concerning the waveforms of both pulsatile and basal secretion within the overall episode (e.g., an assumption that basal secretion is unchanged, rises exponentially, or conforms to a cosine function). Alternatively, in some contexts, the form of basal secretion could be estimated, if the waveform, amplitude, and/or frequency of the pulsatile component were assumed, known, and/or fixed. In short, constraining assumptions concerning either the mode of basal or the nature of pulsatile release, or both, are required because these variables are highly correlated statistically and potentially coupled physiologically. For example, when a given profile is represented by combined basal and pulsatile secretion, any inferred increase in the basal estimate will necessarily imply a reciprocal decrease in the pulsatile component in order to represent the same data. This strongly negative correlation makes the estimation of basal and pulsatile hormone secretion as independent parameters extraordinarily difficult, if not in some cases impossible, without *a priori* assumptions concerning or independent knowledge of features of basal and pulsatile release within the system under study.

By way of example, if one assumes that basal hormone release is zero or time-invariant, then the proestrous LH surge in the rat can be modeled as a cascade of partially summating LH secretory bursts of increasing amplitude, frequency, and duration, with possibly a prolonged LH half-life as well (48) (Fig. 9). The same inference applies, if one estimates basal secretion as some value between zero and the largest amount of basal release that would produce the lowest nadirs observed in the data. More particular estimates of the exact time course of true basal secretion often are not possible without further information or additional assumptions.

Influence of Binding Proteins on Apparent Neurohormone Secretory Pulse Waveform

Many neurohormone effector molecules associate with high affinity to one or more binding proteins present in plasma and/or on target tissues (107). Tissue binding sites are represented typically by specific hormone receptors that mediate cellular actions of the hormone, and which in some cases can be proteolytically cleaved and released into the bloodstream where continued hormone binding can occur at high affinity (108–111). For example, in the case of GH, a high-affinity GH binding protein circulates in plasma in significant concentrations under physiological conditions. In the human this binding protein is the extracellular domain of the tissue GH receptor, which is released into the bloodstream after enzymatic scission (112–115). Alternatively, in the case of sex-steroid hormone binding proteins, at least two important species are present in blood, namely, the low-affinity, high-capacity, binding protein albumin and a high-affinity, low-capacity binding protein,

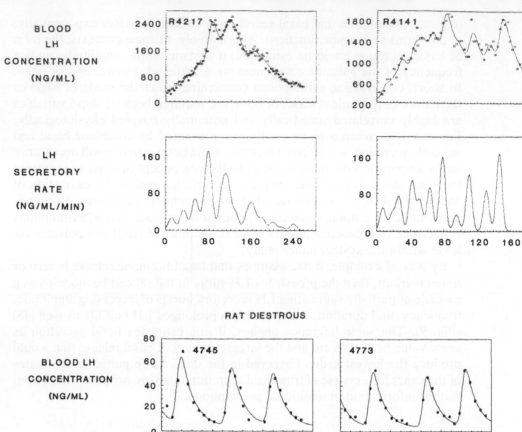

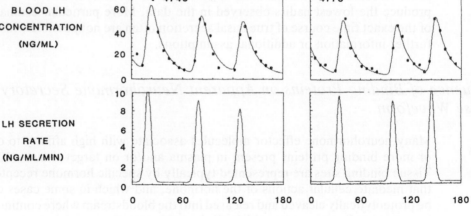

TIME (MIN)

FIG. 9 Spontaneous proestrous LH surge (top subpanel fitted data) subjected to deconvolution analysis (bottom subpanel plots) to reveal a succession of LH secretory bursts partially superimposed but with increasing mass, amplitude, and duration. For comparison, LH release at diestrus is also shown. Adapted with permission from Veldhuis *et al.* (48).

sex hormone-binding globulin (107, 116). Indeed, the latter consists of a collection of posttranslationally modified glycoprotein isoforms.

As might be predicted intuitively, the existence of one or more binding proteins in plasma imparts an additional complexity to the apparent waveform of the bound, free, and total concentrations of the effector molecule (46, 117). For example, Fig. 10 illustrates several features: the apparent time course of free GH secretion into the bloodstream assuming a Gaussian burst of release; the corresponding time courses of free, bound, and total GH in plasma given a particular distribution volume in the presence of a high-affinity

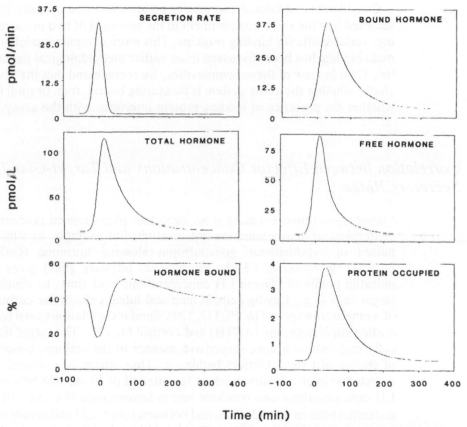

FIG. 10 Profiles of the theoretical GH secretory rate within a single Gaussian burst, the resultant serum total, bound and free GH concentrations, the percentage of GH bound, and the percentage of binding protein occupied in plasma. Data assume a single burst of GH secretion centered at time zero. Adapted with permission from Veldhuis *et al.* (46).

and finite-capacity specific GH binding protein in plasma; the temporally sequential changes in percentage occupancy of the GH binding protein; and the time-dependent variations in the percentage of GH bound at any moment to its cognate binding protein (46). Extensive simulation studies indicate that the time courses of total, free, and bound hormone depend on the quantitative features of the secretory impulse itself, the amount and affinity of the binding protein, the rate of elimination of the unbound hormone, the presence of other relevant ligands if any, etc. (46). Most notably, even a relative brief burst of hormone secretion results in an extended plasma hormone concentration peak if a high-affinity binding protein is present in the circulation in significant amounts and remains unsaturated during the release episode.

Considerable additional complexity arises when two or more ligands are secreted into the same compartment in the presence of two or more competing, variable-affinity binding proteins. This more complex model of multidomain binding has been discussed in an earlier methodological paper in detail (66, 118). In view of these complexities, we recommend that the investigator clarify whether the assay system is measuring bound, free, or total hormone, whether the presence of binding protein interferes with the assay, etc.

Correlation between Effector Concentrations and Target-Gland Secretory Rates

Stimulus–secretion coupling is an important physiological process that is a centerpiece of neuroendocrine organization. For example, as stimulated by pulses of hypothalamic gonadotropin-releasing hormone (GnRH), the burstlike secretion of LH by the anterior pituitary gland gives rise to a pulsatile profile of plasma LH concentrations over time, to which gonadal target cells (e.g., Leydig cells in men and luteal cells in the corpus luteum of women) are exposed (6, 59, 119, 120). Similar correlations exist for adrenocorticotropic hormone (ACTH) and cortisol (4, 121). The target tissue typically responds in a dose-responsive manner to the ambient concentrations of the agonist (or inhibitor) bathing it. Thus, there is a strongly positive correlation over a relatively short lag interval of 10–40 min between serum LH concentrations and resultant serum testosterone (Leydig cell product) concentrations in men (3, 122), and between serum LH and serum progesterone concentrations in women (8, 11). Although the serum concentrations of the two hormones are strongly positively correlated, in principle the physiological linkage of greater interest is the serum effector concentration compared to the resultant release or secretory rate of the product by the target tissue; for example, serum LH concentrations should be strongly correlated in the male with testicular testosterone secretion rates, and in the female with the progesterone secretory rates of the corpus luteum.

Although not formally addressed in the literature to date, we believe that deconvolution analysis is a practical technique by which to assess and quantitate the foregoing concordance between plasma neurohormone concentrations and the secretory rate of the target tissue. This is illustrated in Fig. 11, which depicts coupling between a particular pair of relevant hormones, namely, pituitary-derived ACTH and adrenally derived cortisol. The plot shows the plasma concentrations of ACTH over time, which are paralleled closely by the calculated cortisol secretory rates over time. The latter could be derived by deconvolution analysis assuming a Gaussian waveform for underlying cortisol secretory bursts (18, 31), or waveform-independent deconvolution analysis using literature-based estimates of the two-component cortisol half-life (32, 103). These mathematically independent techniques yield similar plots of secretory rates. Indeed, we recommend that cross-correlation analysis be applied to relate the serial plasma hormone (ACTH) concentrations to the calculated target-tissue (cortisol) secretory rates in the same samples and at various lags (times in minutes separating the two series of interest) (32, 42, 66–68, 123). This proposed correlation utilizes the circulating concentrations of the stimulating hormone and calculated secretory rates of the target gland, and therefore assesses the relevant biological associations and eliminates the confounding effects of different plasma half-lives for the two distinct hormones (effector and product hormone). In this context, we are less interested in deconvolving the effector hormone concentrations over time (e.g., ACTH), as physiological precepts dictate that some correlate of the time-integrated plasma ACTH concentration (rather than its secretory rate) conveys the stimulus to the target tissue (adrenal cortex).

One may further assess concordance of individual pulsatile events (rather than cross-correlations between paired serial hormone concentrations, above) using discrete probability analysis (42, 71, 124). More specifically, either with computer simulations or with assumed models of coincidence, the degree of expected random concordance between discrete pulsatile events in two hormone pulse series can be estimated by calculating a presumptive mean, variance, and probability density function that is based on purely chance associations between bursts within the two pulse trains (42, 71, 124). For example, in the case of ACTH and cortisol, individual spontaneous or pharmacologically stimulated ACTH pulses or release episodes often occur in close temporal proximity to cortisol release episodes (as detected by a discrete peak-detection methodology, for example), and this concordance of events is significantly nonrandom (4, 123). Typically, such comparisons of event concordance have utilized the number and locations of plasma hormone concentration pulses, rather than the numbers and locations of

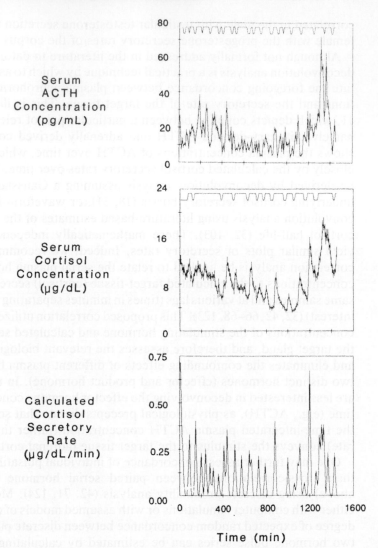

FIG. 11 Illustrative relationships among plasma ACTH and cortisol concentration profiles (top and center) over time and deconvolved cortisol secretory rates (bottom). Adapted with permission from Veldhuis *et al.* (18), Amplitude, but not frequency, modulation of ACTH secretory bursts give rise to nyctohemeral rhythm of the corticotropic axis in man, *J. Clin. Endocrin. Metabol.* **71,** 624, 1990; © The Endocrine Society.

calculated hormone secretory bursts per se (3, 4, 8, 11, 18, 121, 122). We suggested earlier that another strategy for approaching the issue of hormone pulse concordance is to deconvolve the output of the target gland, so as to determine the number and location of resultant secretory bursts generated by the target tissue (31, 32, 42, 71, 124). The number and location of secretory bursts generated in the target tissue should then be related to the number and locations of effector hormone concentration pulses (peaks in concentration values of the stimulatory hormone bathing the responsive cells). This is also illustrated in Fig. 11, where one can directly assess concordance between plasma ACTH concentration peaks and calculated cortisol secretory bursts. Such concordance analysis should describe more appropriately the anticipated physiological coupling between two functionally linked pulsatile neurohormone time series.

When the question arises of cosecretion of two hormones from the same gland or from two putatively coregulated glands, we suggest that deconvolution analysis be applied to both hormone time series. For example, one might ask if LH and FSH are cosecreted, because these two glycoproteins are contained to substantial degree within the same gonadotrope cell population (125). Similarly, one might determine whether the protein hormones ACTH and β-endorphin are cosecreted, because these two peptide hormones are colocalized within the same secretory granules of the anterior pituitary gland (123). To assess this type of question, we suggest that correlations be carried out of the calculated ACTH and β-endorphin secretory rates, rather than merely their corresponding plasma concentrations. The limitation in correlating only plasma hormone concentrations when the question of cosecretion is posed lies in the potential for significantly different elimination rates of the two hormones. If the hormone elimination rates are substantially different, or if they are quite long, then considerable spurious cross-correlation can be observed (i.e., cross-correlations are expected for hormones with long half-lives, which result in significant autocorrelation in each series), or correlation could be artificially underestimated because of a disproportionately short half-life for one hormone and a long half-life for the other.

Our strategy for assessing possible cosecretion of two hormones is illustrated in Fig. 12, where we show deconvolution estimates of serum LH and FSH secretory rates calculated by waveform-independent deconvolution analysis of LH and FSH concentrations measured in blood collected at 10-min intervals over 24 hr. We suggest that a waveform-independent deconvolution technique (e.g., Refs. 32 and 103) be employed, in order to minimize any additionally spurious cross-correlations otherwise introduced by model-specific (e.g., symmetric Gaussian) hormone secretory bursts.

When two secretory profiles are generated by deconvolution analysis of the relevant serum hormone concentration pairs (e.g., LH and FSH, ACTH and β-endorphin, ADH and oxytocin), the physiological queries that are posed may suggest that the resultant (secretory) profiles be analyzed as described in detail elsewhere for possible coincidence of events and/or cosecretion (3, 8, 11, 42, 66, 67, 71, 122, 124). Specifically, the technique of cross-correlation analysis can be applied at different lags to assess parallel

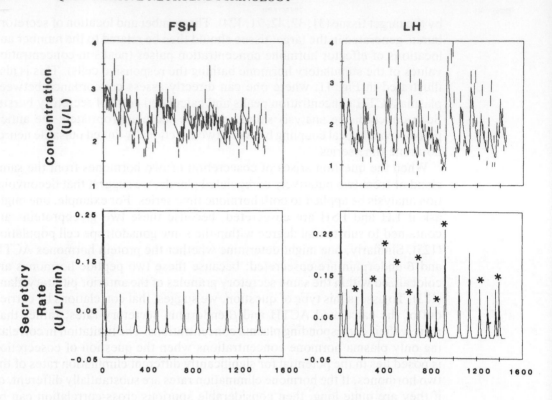

FIG. 12   Serum immunoreactive LH and FSH concentration profiles (top) and the corresponding calculated secretion patterns (bottom) in a single male estimated by deconvolution analysis using PULSE2 and literature values for LH (60 min) and FSH (240 min) half-lives (73, 75). These hormones are ostensibly cosecreted. Coincident secretory pulses are marked by asterisks in the computed LH secretory profile. Data from J. D. Veldhuis and M. L. Johnson, 1994, unpublished.

changes in calculated sample secretion rates for the two hormones (positive cross-correlation) and/or divergent secretion rates that systematically differ (e.g., reciprocal coupling, in which high secretory rates of a given effector are accompanied by reduced secretory rates of the companion neurohormone before or after some lag). Second, discrete probability coincidence analysis may be applied after defining the number and locations of the significant secretory bursts in the two series. Typically the resultant secretory data are discretized; that is, the location of a peak is assigned to a single sample corresponding to the peak maximum, or the first occurrence of a significant increase in secretion, or via some other reasonable *a priori* rules for creating dichotomous data. These principles have been reviewed in detail in earlier methodological papers (42, 66, 67).

Use of Convolution Integral as Pulse Simulator

Several investigators have recognized the merit of using the convolution integral to generate simulated neuroendocrine pulse profiles (1, 9, 32, 38, 41, 42, 44–47, 49, 52, 66, 67). Such simulations may be of particular utility in testing the validity and reliability of neuroendocrine pulse detection methodologies (see Chapters 3 and 4 by Mulligan *et al.*, this volume), deconvolution techniques, coincidence equations, etc. For example, validity of pulse identification can be explored systematically via simulated neuroendocrine pulse trains, in which the number, location, amplitude, duration, and mass of all underlying simulated hormone secretory bursts are known *a priori*, and the concurrent apparent half-life of neurohormone elimination is also defined explicitly (32, 66–68). The profiles can be generated by imposing reasonable parameter values on the secretion and elimination functions, for example, a plausible mean hormone burst frequency, amplitude and duration, and half-life (e.g., Refs. 1, 9, 32, 38, 41, 42, 46, 49, and 52). The resultant numerical profiles are then corrupted by the addition of noise (experimental uncertainty) of different distributions and amounts in order to emulate observable physiological pulsatile neurohormone profiles closely.

A companion chapter (Chapter 9 by Straume *et al.*, this volume) discusses the special issue of pulse simulators designed to mimic more complex physiological processes, such as episodic release of GH in humans in which serial interpulse intervals show negative autocorrelation (i.e., short intervals tend to be followed by long intervals, and long intervals by short intervals), and serial GH secretory burst amplitudes tend to be positively autocorrelated (large bursts tend to be followed by large events and vice versa) (10) (see Fig. 13). Conversely, in the case of the LH axis, two pulse generators may exist at least in the luteal phase of the normal menstrual cycle, because low- and high-amplitude LH secretory bursts appear to coexist as two distinct populations with different intrinsic mean frequencies (8, 42). Multiple pulse generators may be required to explain such data series, or to emulate their physiological properties closely (42). Moreover, in the case of pulsatile LH release in women in the midluteal phase of the menstrual cycle (8), but not at other times, or in men (6, 120), there is a negative autocorrelation of successive LH interpulse intervals; that is, long intervals tend to be followed

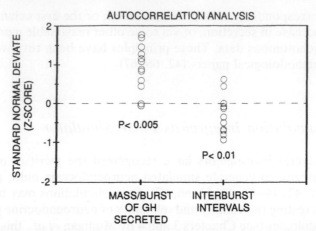

FIG. 13 Autocorrelation analyses of serial GH secretory burst mass and interpulse intervals in healthy young men as assessed by multiparameter deconvolution analysis (31). Adapted with permission from Hartman *et al.* (10).

by shorter ones, and vice versa. Pulse simulators that impose purely stochastic properties without such correlative or even deterministic relationships between and among parameters may not encompass accurately the full range of physiological variability anticipated in the data. Thus, classic stochastic modeling with normal (Gaussian) randomly and independently controlled pulse generator frequency, amplitude, and duration is probably an oversimplification for certain axes under some pathophysiological conditions.

An example of apparently nonrandom physiological variability over 24 hr is the hormone interburst interval for GH in normal men (10). Note that there is significant 24-hr variability in the detectable GH interpulse interval over 24 hr in healthy young men. Although such observations were obtained in an immunoradiometric assay (IRMA) that could not detect GH in all daytime blood samples, our reanalysis of this problem using a chemiluminescence GH assay illustrates a substantial decrease in the GH interburst interval at nighttime compared to during the day, which may be due to comparing GH secretion stimulated by sleep and fasting (nighttime) versus GH release suppressed by the subject's being awake and fed (daytime) (10, 20, 36, 62, 89, 102, 103, 106). Consequently, appropriate pulse simulation for the GH axis requires stochastic variability with deterministic trends, namely, random GH interburst intervals with a systematic trend of 24-hr variation in the mean value of the interpulse interval. These approaches to more realistic physiological simulations of pulsatile neuroendocrine axis are discussed in further detail with specific application to the GH axis in a companion chapter (Straume *et al.*, Chapter 9, this volume).

We emphasize that there are other implications of deterministic trends within neuroendocrine axes, and therefore in the corresponding biomathematical models for them. For example, significant 24-hr variability in intersecretory burst interval (and hence reciprocally in hormone burst frequency) is important to recall when evaluating daytime versus nighttime GH pulse concordance, since the expected random associations with GH pulses will differ over 24 hr as a function of the mean burst frequencies in the two series (42, 71, 124). In addition, significant diurnal variations in interpulse interval may affect the sensitivity and specificity of peak detection applied to simulated series during the apparent daytime versus nighttime sampling intervals, if significantly discriminable features of the data vary substantially; for example, the signal shape, amplitude, or frequency, the ratio of peak amplitude to interpulse nadir, and the relative ratio of peak amplitude to basal secretion. We believe that methodological advances in convolution modeling will require further comprehension of and attention to deterministic elements contained within the otherwise stochastic models simulating neuroendocrine pulsatility (e.g., Keenan, Chapter 12, this volume).

Assumptions of Nonlinear Curve Fitting Methods Used in Deconvolution Analysis

Our particular waveform-defined multiparameter approach to deconvolution analysis (31, 32) essentially entails iterative curve fitting of a nonlinear multiparameter equation to the measured pulsatile neurohormone concentrations over time. Other waveform-independent deconvolution methods may involve linear estimation models such as PULSE2 (see Chapter 1, this volume, by Johnson and Veldhuis). Inherent in curve-fitting applications are assumptions regarding nonlinear methodologies. In brief, these include the following: the random zero-mean independent and identical Gaussian distribution of experimental uncertainty solely in measures of the dependent (y axis) variable; the statistical independence of serial observations; the absence of experimental uncertainty in the independent (x axis) variable; the lack of any systematic bias within the data; and the structural relevance of the model to the observed measurements (126).

In the section on experimental uncertainty within data (above), we discussed the distribution of experimental uncertainty in the y axis measurements. The second assumption of independence of serial measurements is a condition not uniformly satisfied in neuroendocrine studies. For example, the sample experimental uncertainty which should be statistically independent might show successive autocorrelation. In addition, many hormones are monitored under conditions in which successive measurements show a strong artifactual autocorrelation; that is, any given or measured value at time t can be used to predict a value at time $t - n$, or $t + n$, where n is a finite time increment. Because hormone elimination occurs over seconds, minutes, or hours, any particular hormone concentration predicts its predecessor as well as its successor because of the decay function. In general, valid assays in themselves produce virtually no autocorrelation, in that the sample values of successive measurements are almost completely independent technically. Indeed, systematic drift within an assay is a cause for serious concern and corrective action. However, based on the expected autocorrelation of successive hormone measurements *in vivo*, we recommend that in appropriate studies autoregressive modeling be utilized to reduce autocorrelation when carrying out cross-correlation analysis (e.g., Refs. 3, 122, and 127). Otherwise, cross-correlation will disclose apparent (but spurious) cross-correlations between two series, even though such cross-correlation is attributed to the serial autocorrelation present in each series itself. However, autoregressive modeling eliminates the elimination content of the data series. Alternatively to an autoregressive model, one might perform waveform-independent deconvolution analysis to remove the autocorrelative influence of hormone specific half-life in the data and then cross-correlate the resultant sample secretory profiles (see section on correlation between effector concentrations and target-gland secretory rates, above).

Error is assumed to exist exclusively in the dependent variable measurement (y axis data). However, in the case of blood sampling, etc., some experimental uncertainty does exist in the independent variable (x axis) as well. The extent of this variability is typically not known precisely to the investigator. However, in evaluating the timing of collection of blood samples, we observed a skewed distribution of only positive incremental times, in which the increment was the time difference between the actual and the scheduled collection of the blood sample (1). No blood samples were collected early, but some blood samples were collected late, resulting in a right-skewed distribution. Thus, the apparent distribution of experimental error in the x axis had a slightly nonzero mean and was non-Gaussian. The exact impact of this violation of an integral assumption of nonlinear methods of curve fitting will vary with the extent of the deviations.

In the case just illustrated, the error presumably is relatively small, inasmuch as the delay in sample acquisition in our particular research unit was usually less than 10% of the sampling interval. For any hormone with a half-life severalfold longer than the sampling interval, this sampling delay would contribute on average only minimally to a biased (artificially low) serum hormone concentration estimate in that sample. Even so, we recommend that this issue be explored further and that, where the experimental uncertainties in the x and y axes are approximately known and Gaussian in nature, two-dimensional error propagation be utilized in the analysis of bivariate data. This concept is illustrated in Fig. 14, where simple linear regression is used to relate sample immunoreactive LH concentrations (and their approximately Gaussian error) to sample LH measurements in an *in vitro* Leydig cell bioassay (also with approximately Gaussian experimental uncertainty). The measurements of LH in the two assays results in an elliptical two-

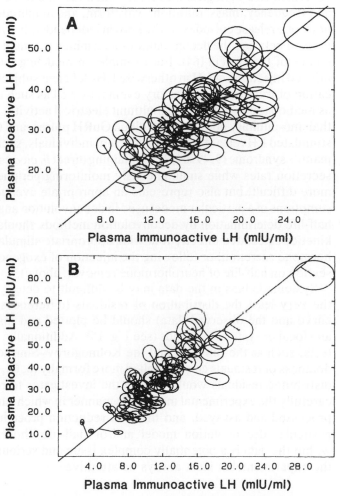

FIG. 14 Two-dimensional error propagation for the curvilinear regression relating immunoreactive LH and bioactive LH concentrations in 10 minutely sampled blood in 2 young men (A and B). Adapted with permission from Veldhuis *et al.* (119), Preferential release of bioactive lutenizing hormone in response to endogenous and low-dose exogenous gonadotrophin releasing hormone (GnRH) pulses in man, *J. Clin. Endocrinol. Metabol.*, **64,** 1275, 1987. © The Endocrine Society.

dimensional error space contributed by the joint uncertainties in the two measurement systems (119). Regression analysis should then minimize the sum of squares of residuals in both the *x* axis and *y* axis dimensions simultaneously, as discussed elsewhere in greater detail (126).

Determining model relevance or validity also is a crucial aspect of valid deconvolution analysis. To this end, we suggest that the investigator define several *in vivo* biological models in which the rate of basal secretion (if any) and frequency, mass, duration, waveform, and amplitude of pulsatile neurohormone release episodes can be known independently *a priori*, allowing comparison between the deconvolution-determined estimate and the true performance of the system (64). For example, in modeling the LH axis, we have utilized injections of LH in otherwise LH-deficient subjects (6) for the identification of true-positive secretory events, as well as such independent criteria as mediobasal hypothalamic multiunit electrical activity, simultaneous hypothalamo-pituitary portal venous blood GnRH pulses, and (exogenous) GnRH-stimulated LH release in GnRH-deficient individuals, such as those with Kallmann's syndrome (9, 49, 64, 71). Obtaining direct *in vivo* estimates of glandular secretion rates while simultaneously monitoring peripheral blood has been more difficult, but also represents an appropriate avenue for testing the assumptions of a particular model used in deconvolution analysis. Similarly, the half-life determination by deconvolution methods should be compared with kinetics calculated directly following appropriate stimulation of endogenous hormone secretion, or following the injection of exogenous effector (see the section on half-life of neurohormone removal, above).

Systematic bias in the data may be difficult to detect. We believe that at the very least the distribution of residuals (differences between the fitted curve and the observed data) should be plotted and inspected visually for any local or systematic trends (see Fig. 15). Additionally, objective statistical tests, such as the runs test and the Kolmogorov–Smirnov statistic, for randomness of residuals can be applied more formally (6, 68, 128). Nonrandomly distributed residuals would require the investigator to review and appraise carefully the experimental model, the manner in which samples are collected, processed and assayed, and the data reduction procedures, as well as the particular deconvolution model and/or half-life chosen. Systematic bias within the data is a potentially complex issue and serious concern, to which the experimentalist must always be attentive.

Statistical Assessment of Deconvolution Measures

A typical physiological or pathological study in neuroendocrinology requires comparisons of specific measures of hormone secretion or half-life under two or more conditions. For example, pulsatile prolactin secretion may be

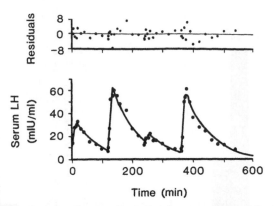

FIG. 15   Deconvolution fit of pulsatile LH release profile with plot of (randomly distributed) weighted residuals (differences between predicted fit and observed serum hormone concentrations). Data from J. D. Veldhuis and M. L. Johnson, 1991, unpublished.

evaluated in women during postpartum lactation versus during the normal menstrual cycle. Specific measures of prolactin secretion are typically of interest, such as the number, amplitude, duration, or mass of the calculated secretory bursts, and/or the half-life of the hormone in question (129). Unfortunately, few studies have addressed the statistical distributions of these important measures. Moreover, in general, relatively few measurement values are available to the investigator, making a reliable assessment of statistical distribution even more difficult. As shown in Fig. 16 in the case of the GH axis, in any given subject the calculated mass of GH secreted per burst may not conform to a simple Gaussian distribution (Fig. 16A). Similarly, the mean mass of GH secreted per burst in a group of N subjects considered together may or may not exhibit a normal distribution (Fig. 16B). Because parametric statistical tests assume an underlying Gaussian distribution of measured estimates of the dependent variables, significant departures from this assumption could make parametric estimates less meaningful. Accordingly, the median and range of a deconvolution measure are important complementary statistics to report in addition to the mean and its standard error, as both the central tendency and the dispersion of deconvolution measures are not necessarily well represented by parametric statistics. In addition, where normality cannot be adequately determined, we would recommend nonparametric comparisons of the two distributions of the measures; for example, a Wilcoxon paired or unpaired test of the population medians, or the Kolmogorov–Smirnov statistic to compare their cumulative probability distributions (130). Whenever possible, we suggest that the central tendency and the dispersion of the deconvolution measure be assessed by frequency histograms, as well as by appropriate statistical tests of normality, skewness,

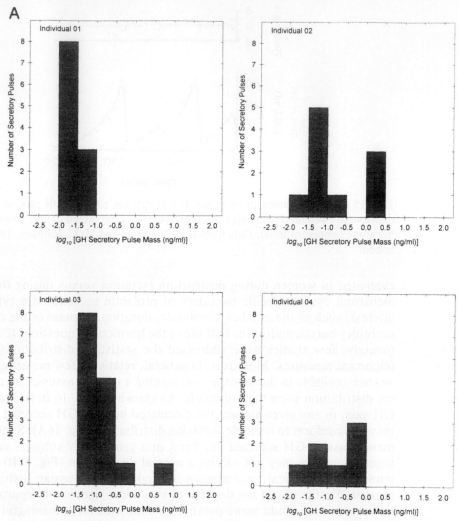

FIG. 16   (A) Frequency histograms of the deconvolution-estimated (31) mass of GH secreted per burst in 4 young adult males. (B) Frequency histogram of mass of GH secreted per burst in a group of 17 men. Data from J. D. Veldhuis, M. Straume, and M. L. Johnson, 1995, unpublished.

etc. Where appropriate, the measures may be subjected to mathematical transformation (e.g., the logarithms or square roots taken of the values) in order to normalize the distribution. One should then document normality of the resultant new distribution.

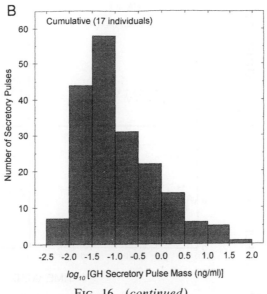

FIG. 16   (*continued*)

Within- and Between-Subject Variability

Experiments in which the same animal or subject is monitored serially with or without intervention are typically assumed to yield stable and reproducible within-subject physiological measures over the corresponding intervals. However, comparisons of the pulsatile profiles of serum LH concentrations using Cluster analysis for discrete peak detection (130a), and the deconvolution-calculated measures of LH secretory burst frequency, mass, and half-life (64, 131) obtained during two or three consecutive sessions a few weeks apart, show significant within-subject variability over time when blood is collected at 10-min intervals over 24-hr sessions. Although intrasubject variability of physiological measures of pulsatile hormone secretion is evident on serial evaluation, the mean hormone concentration, its half-life, and the 24-hr production rate are all strongly conserved across successive measurement sessions. Thus, serial within-subject physiological variability must be considered in the analysis of statistical power, and in the interpretation of physiological data. For example, as shown in Fig. 17, serum LH concentration profiles in the same individual collected approximately 2 weeks apart without other intervention may appear to be quite similar or remarkably different. The exact mechanistic basis behind such physiological stability versus variability in a healthy subject over time is not evident, but it may reflect the particular

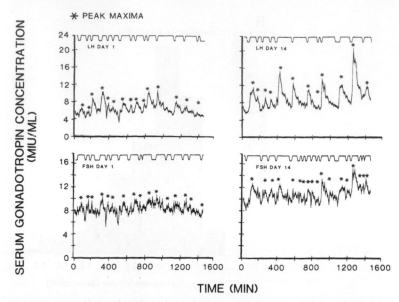

FIG. 17 Physiological variability in serum immunoreactive LH and FSH profiles over 24 hr in a healthy young man studied twice 14 days apart without intervention. Adapted with permission from Veldhuis *et al.* (124).

stochastic properties of the individual pulse generator and/or the nonlinear feedback loops involved in systems control (see Chapter 9 by Straume *et al.*, this volume, for an example in the GH axis). Indeed, irregularity and regularity of patterns within neurohormone time series may be subtle in the presence of strongly stochastic forces (132). Moreover, if large numbers of observations (e.g., 10^5 data values) were to be made, we expect that nonlinear dynamic features consistent with deterministic chaos may appear to operate within neuroendocrine time series, and produce apparent irregularity and apparent nonuniformity despite minimal differences in starting conditions.

Intersubject variability is also a prominent feature of normal neuroendocrine physiology (e.g., for LH axis see Refs. 1 and 53). This is evident in Fig. 17, which compares two different subjects who underwent identical blood sampling and subsequent assay of serum samples for immunoreactive LH concentrations. A similar inference is illustrated in Fig. 18, where the calculated daily GH production rates are given for a number of healthy individuals. Note that in the second circumstance, the deconvolution measure (i.e., daily GH secretion rate per unit distribution volume) is strongly determined by physical age, by a measure of relative body fatness (body mass index, BMI), as well as by the interaction between these variables (89).

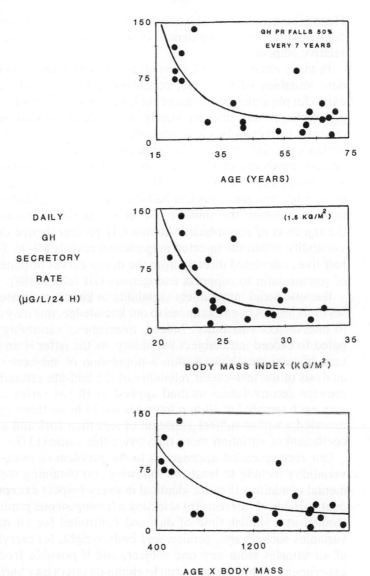

FIG. 18   Dispersion of endogenous GH production rates in healthy men, and the significant variation of this estimate with age (top) and body mass index (center), or their interaction (bottom). Adapted with permission from Iranmanesh *et al.* (89).

Hence, experiments evaluating this particular deconvolution measure must control for prominent influences of the physiological parameters of age and relative adiposity.

In many circumstances, even after controlling for known relevant dependent variables such as age, considerable between-subject variability still exists for physiological measures of hormone secretion or half-life. For example, earlier studies utilizing steady-state infusion of purified pituitary GH to estimate its metabolic clearance rate show a 3- to 4-fold range of estimates within a population of healthy adults (reviewed in Refs. 46, 76, and 88). Our studies consisting of the injection of a GH-releasing peptide followed by somatostatin to block further GH secretion also show a severalfold variation in calculated endogenous GH half-life values among healthy men (76). Similarly, monitoring the time course of serum GH concentrations following the injection of recombinant human GH protein (hence devoid of intrinsic variability within the injected preparation) reveals a 2- to 3-fold range in GH half-lives calculated directly from the decay curves obtained in the presence of somatostatin to suppress endogenous GH release (88).

Because serial intrasubject variability in kinetic estimates across separate sessions has not been examined to our knowledge, this may contribute further to intersubject variability. Indeed, intersubject variability would be anticipated to exceed intrasubject variability, as the latter is an expected component of total variability within a population of subjects. For example, an analysis of the test–retest reliability of the half-life estimate of our multiparameter deconvolution method applied to 10 LH series in as many young men each sampled at 10-min intervals for 24 hr on three separate occasions revealed a within-subject variation of less than 10% and a between-subject coefficient of variation more than twice this value (131).

Our recommended approaches to the problem of intra- and intersubject variability include at least the following: (a) obtaining well-defined experimental conditions that are identical in every respect except for the independent variable of interest; (b) selecting a homogeneous population of subjects studied at a similar time of day and controlled for all major determining variables such as age, gender, and body weight; (c) carrying out the assay of all samples from any one subject, and if possible from any one entire experiment, within the same run to eliminate interassay variability; (d) identical handling, processing, and automated aliquoting of the blood samples; (e) performing preliminary experiments to estimate the variance of a particular deconvolution measure, and hence allow some calculation of statistical power (where power = $1-\beta$ error, and β is the probability of falsely accepting a null hypothesis of no treatment effect) (133); (f) utilizing each subject as his own control wherever possible; (g) selecting primary parameters that are relatively stable within and across subjects, such as the mean hormone concentration and calculated half-life as the basis for principle conclusions; and (h) protecting the choice of p (or α, i.e., the probability of falsely rejecting the null hypothesis of no difference) values, especially when multiple measures are being compared between two groups. For example, one may wish to evaluate differences between two study sessions in relation to neurohormone pulse frequency, amplitude, duration, mass and half-life, total production rate, and basal secretion, which therefore involves a total of at least seven individual statistical contrasts. A protected p value might be chosen at $p = 0.01$ instead of $p = 0.05$, in order to allow for such multiplicity of comparisons. Moreover, only the primary deconvolution measures should be compared first, for example, secretory burst number, mass, and hormone half-life, as other measures, such as total daily GH secretion rate, secretory burst amplitude, and duration, are all strongly correlated with or numerically dependent on these primary measures.

Statistical Validity versus Biological Validity

An implicit assumption in analytical methods is that application of a mathematically and statistically valid tool produces a biologically valid result. This is by no means necessarily correct. For example, withdrawing blood samples at very infrequent intervals may result in a marked underestimate of the frequency of secretory events mediating the biological process, even though the actual analysis of the reduced time series is carried out in a statistically valid manner. This is illustrated in Fig. 19. In particular, as the number of samples obtained per 24 hr at equally spaced intervals decreases (i.e., as sampling density falls), the sensitivity and specificity (133, 134) of deconvolution analysis can be expected to decrease correspondingly (64). Decreases in sensitivity (inversely proportional to number of false-negative events) and specificity (inversely proportional to number of false-positive events) are anticipated biologically, since infrequent blood sampling will predispose to an inadvertent underrepresentation of higher-frequency release episodes that are actually occurring physiologically.

In short, even the statistically reasonable selection of p values (both α and β errors), the appropriate propagation of within-sample error, and the appropriate use of joint parameter statistical confidence intervals, will not yield substantially more information than can be offered by the quality of the data (e.g., as influenced by sampling interval). Accordingly, the investigator must identify and construct conditions of the experiment so as to optimize the extraction of useful biological inferences from the study. This will involve optimization of not only sampling frequency, but also many other factors, such as the techniques for sample withdrawal, processing, handling, and

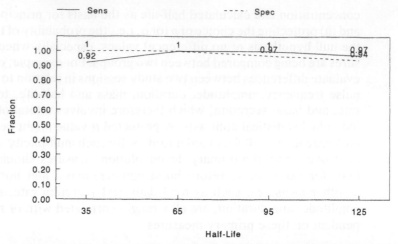

FIG. 19   Relation of sensitivity and specificity of multiparameter deconvolution (12) for detection of hormone secretory bursts in relation to hormone half-life as assessed in simulated LH time series. Adapted with permission from Mulligan *et al.* (64).

assay and data reduction. Indeed, the primary design elements in the experiment are also critical (e.g., time of day, type of animal studied, condition of the animal, and nature of the treatment). We emphasize that there is no single methodological approach that will enhance the sensitivity, specificity, reliability, and precision of estimating deconvolution measures when the scientific quality or information content of the data has become rate limiting.

Given optimal experimental quality in the design and conduct of the experiment, we strongly endorse the view that deconvolution analysis should be validated rigorously with respect to its sensitivity, specificity, precision, and reliability. The issue of validity testing and reliability estimation are discussed in accompanying Chapters (see Chapters 3 and 4 by Mulligan *et al.*, this volume) and elsewhere (32, 60, 64, 66–68). The precision of estimating deconvolution parameters is strongly controlled by the number of observations available from which each deconvolution measure is estimated (e.g., sampling duration and intensity and the number of samples obtained per estimated half-life) as well as the joint intrinsic experimental uncertainty contributed by all the procedures and measurements considered collectively. Thus, the importance of the intrasample coefficient of variation must also be emphasized.

Although some deconvolution methods yield increasingly variable estimates of secretion at high data densities (due to an ill-posed nature of the deconvolution problem, e.g., with discrete deconvolution, see below), the multiparameter methodology produces estimates of secretion and half-life that are increasingly precise as the sampling intensity increases and the intraassay variability decreases. These points are illustrated in Fig. 20. Note that if hormone measurements are imprecise (high assay variability) or made in blood collected at 20-min versus 1-min intervals, the estimate of half-life is less precise. An important consideration in addition to the density of data collection (number of samples per unit time) is the total duration of observation. For example, a half-life cannot be estimated well when the total duration of sampling approaches or is less than the value of the half-life itself. Accordingly, we would recommend empirically that at least five half-lives of observation time be employed to estimate any given half-life with some precision. We have verified this thesis for stimulated data, for example, where a half-life of 60 min is estimated from approximately 250 data points

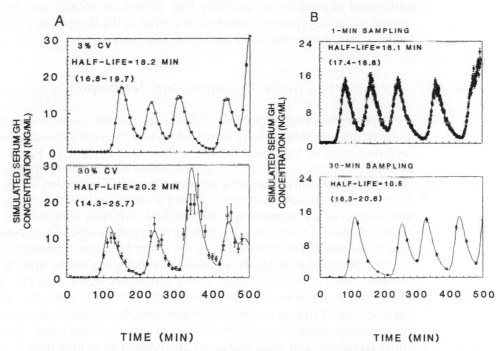

FIG. 20 (A) Comparison of 3 versus 30% intraassay coefficients of variation (cv) for 10-min sampling protocols to estimate a hormone half-life in simulated pulse profiles. (B) Comparison of 30-min and 1-min sampling for hormone half-life estimates by multiparameter deconvolution analysis of a simulated pulse profile extending over 500 min. Parentheses give 95% confidence intervals for the estimates. The true "synthetic" half-life was 18 min. Adapted with permission from J. D. Veldhuis and M. L. Johnson (141).

collected at 1-min intervals, compared to only the first 25 of those data points also collected at 1-min intervals. The precision (coefficient of variation) and the reliability of the half-life estimate are predictably better when the sampling duration extends over multiple half-lives. Thus, adequate data precision, density, and duration are all important in order to achieve desired precision and reliability when estimating deconvolution measures.

In summary, we recommend that appropriate *in vivo* biological models with known mechanisms of generating true-positive secretory events and/ or clearance mechanics be utilized to establish the validity of the biomathematical model, delineate its sensitivity and specificity under different biological conditions, and assess its reliability both for a single observer (serial reliability of applying the same test to the same data on different occasions) and among multiple individuals (degree of concordance among independent investigators applying the same analytical tool to the same data). Moreover, anticipated physiological variability (see section on within- and between-subject variability, above) must be considered in the design, measurement, analysis, and interpretation of deconvolution analysis.

The Ill-Posed Nature of Some Deconvolution Techniques

As discussed in an earlier volume of this series (67, 68), some deconvolution methods produce substantially increased variability in the estimate of sample secretion rates at high data densities. This is illustrated in Fig. 21. For example, if sample secretion rates are calculated from the sample derivatives or first differences in consecutive serum hormone concentration values [e.g., discrete deconvolution analysis (56)], then at sampling densities that are high with respect to the hormone half-life, marked variability (ringing) can occur in the calculated instantaneous secretory rate (ISR) profile. Thus, collecting data at 1-min intervals and analyzing the serum hormone concentrations by discrete deconvolution yields a pattern of rapidly oscillating apparent LH secretory rates with some values above and others below zero (1, 29, 32). The negative secretion rates cannot be realized physiologically, and the phenomenon of ringing is also presumably nonphysiological. In contrast, a model-specific methodology, here illustrated as the multiparameter deconvolution technique, can yield increased precision of fit at high data densities. Consequently, an important caveat in formulating the deconvolution integral is a statement of the secretion and clearance relationship so as to minimize the ill-posed nature of the deconvolution problem. Waveform-independent deconvolution analysis can also be formulated, so as to minimize the ill-posed issue (32) (Fig. 21).

We also emphasize that although a variety of smoothing and filtering techniques can be employed to reduce the serial variability in calculated secretion rates that can occur when the deconvolution problem is particularly ill-posed (68), such smoothing techniques often produce autocorrelation when applied to the output, and thereby violate the assumption of statistical independence of consecutive measures that ordinarily underlies most curve-fitting approaches in current use (see the section on assumptions of nonlinear curve-fitting methods used in deconvolution analysis). Accordingly, we recommend smoothing the waveform function, rather than filtering the original data or the predicted output (fit of the original data). This smoothing issue is discussed further in relation to a new waveform-independent deconvolution method developed by Johnson and Veldhuis (PULSE2, see Chapter 1, this volume).

Evaluating Equimolar Secretion of Two or More Neuroendocrine Substances

An interesting problem in deconvolution analysis entails calculating a common secretion rate for two or more hormones secreted in equimolar or proportionate amounts, but which exhibit different half-lives in the bloodstream or another sampled compartment. A well-defined example of this circumstance is the cosecretion of insulin and C-peptide, which are released from common granular stores by pancreatic β cells in equimolar amounts following the cleavage of the C-peptide moiety from the proinsulin precursor molecule. Because the half-lives of insulin and C-peptide in the plasma are considerably different, approaching 2.5–3.5 min for a monoexponential estimate of insulin half-life and 18–25 min for C-peptide in men, substantially different plasma concentration profiles of the two molecules are achieved despite their cosecretion. Of further interest, C-peptide is not removed substantially by hepatic extraction, whereas insulin concentrations are reduced by 30–50% during passage through the liver. These circumstances provide several interesting solutions to the estimate of insulin and C-peptide secretion (30, 58, 61). Indeed, earlier approaches in the literature took advantage of the lack of significant extraction of C-peptide through the liver, and utilized peripheral C-peptide concentration profiles and half-lives to estimate by deconvolution analysis prehepatic C-peptide secretion, which by definition if correctly determined would equal on a molar basis prehepatic insulin release.

We have considered an adaptation of our multiparameter deconvolution technique to approach this problem somewhat differently. The modified program, DECONV2, reads both a serum C-peptide concentration versus time file and a concurrent insulin concentration versus time file, and performs curve fitting of both profiles simultaneously. The fit of the insulin profile is accorded an insulin-specific half-life, whereas that of C-peptide is furnished

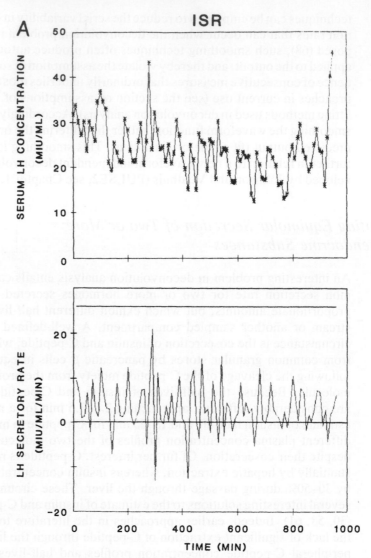

FIG. 21   Comparison of a discrete deconvolution [instantaneous secretory rate, ISR (40)] estimate of 10-min sampled bioactive LH data (A) compared to (B) multiparameter deconvolution analysis (31, DECONV), or (C) a waveform-independent methodology [PULSE, (32) see text Chapter 1]. Adapted with permission from Veldhuis and Johnson (29).

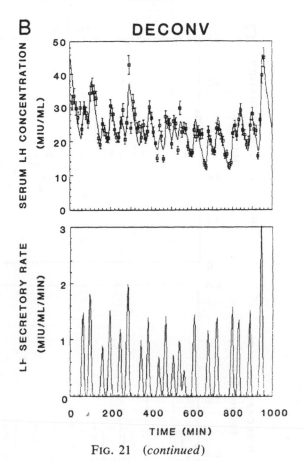

FIG. 21 (*continued*)

with its relevant (different) half-life. If the half-lives of these two peptides are not well known in the condition under study, then these values can be estimated by the deconvolution algorithm. Putative secretory bursts with or without basal release giving rise to the individual serum insulin and C-peptide concentration profiles are estimated simultaneously from both series, assuming that the locations and durations of secretory bursts are identical, and that their amplitude (maximal secretory rate per pulse) is different by only a linear factor (accounting for any differences in distribution volumes, molecular weights, and the fraction of insulin extracted via the liver). Unknown parameters then include the locations of all secretory bursts, their amplitudes, the half-duration of the secretion events, a fraction relating the C-peptide and insulin secretion rates as estimated posthepatically, basal secretion if

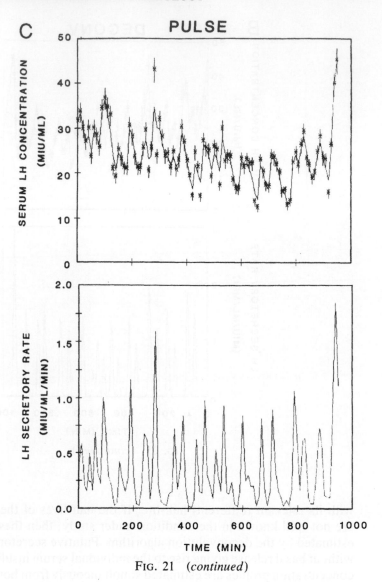

FIG. 21 (continued)

any, and if desired the individual half-lives of insulin as well as C-peptide. This strategy is illustrated in Fig. 22 for human insulin and C-peptide release in a healthy young man. Note that the individual plasma insulin and C-peptide profiles are fit adequately by the preceding assumptions, and that the resultant

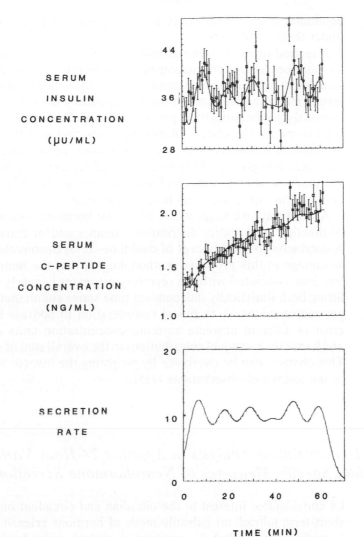

FIG. 22   (Top) Plasma insulin (half-life 2.5 min) and (center) C-peptide (half-life 20 min) concentration profiles with deconvolution-predicted curves, and (bottom) calculated equimolar secretion rate following glucose in a healthy young man sampled every 1 min using a modified multiparameter deconvolution technique (DECONV2). The common secretory plot was estimated by considering both the insulin and C-peptide secretory profiles simultaneously assuming that they were related by a molarity-scaling factor. Data from P. Butler, N. Porksen, and J. D. Veldhuis, unpublished.

equimolar secretion rate of insulin/C-peptide can be estimated numerically under these conditions.

A second application of multiparameter deconvolution analysis to equimolar hormone secretion encompasses measurements of the same hormone sampled at two different anatomical sites within the circulation, for example, insulin measurements carried out prehepatically in the portal vein and simultaneously posthepatically in peripheral blood. Such profiles should be identical except for the effects of hepatic extraction and the unequal absolute distribution volumes in the portal vessels and the peripheral venous circulation. Accordingly, the foregoing model of equimolar secretion but unequal secretory burst amplitudes and unequal apparent hormone half-lives should apply to the two concurrent insulin time series.

Accordingly, we suggest that equimolar hormone secretion with different elimination rates and/or distribution volumes and/or extraction coefficients is conducive to this concept of dual-time-series deconvolution analysis. An advantage of this proposed methodology is that the number of degrees of freedom associated with the secretory estimates is nearly doubled by curve fitting both statistically independent time series simultaneously for common parameter values. A technical consideration to address is the unbalanced error in different absolute hormone concentration units in the two series, and hence their unequal contributions to the overall sum of squared residuals. This obstacle can be overcome by weighting the inverse absolute variances of the individual observations (135).

Using Deconvolution Analysis to Appraise 24-Hour Variations in Individual Specific Measures of Neurohormone Secretion

Of considerable interest to the ultradian and circadian biologist is how the short-term (ultradian) pulsatile mode of hormone release varies over 24-hr (circadian periodicity). Deconvolution analysis provides a unique tool with which to address the corresponding question: Is the 24-hr rhythm in serum hormone concentrations developed by 24-hr variations in secretory burst amplitude and/or interpulse interval and/or basal secretion? To address this physiological query, we recommend a two-step approach (50, 103). First, the 24-hr serum hormone concentration profile is submitted to either model-specific or waveform-independent deconvolution analysis. Such analysis will identify the number and location of significant hormone secretory bursts, and quantitate their amplitude and mass. Second, two plots of interest are created and fit with a cosine or other relevant function to represent 24-hr periodicity. One plot relates each individual secretory burst mass to the location in time of its center, and thus represents a plot of burst mass versus time over 24 hr. The second plot relates the numerical value of the interburst interval (minutes) between any two successive secretory pulses to the midpoint in time between the centers of the same two calculated secretory pulses. Thus, the plot represents a graph of interpulse interval duration against 24 hr time.

Such plots are illustrated for the multiparameter deconvolution estimates of ACTH secretory burst mass and ACTH interburst interval over 24 hr in a group of healthy young men in Fig. 23. Note that this particular hormone exhibits a significant 2- to 3-fold variation in the mass of individual (ACTH) secretory bursts over 24 hr, but that there is no significant nyctohemeral variation over time for the ACTH interburst interval. Accordingly, whereas ACTH burst frequency is essentially constant throughout the day and night, the amount of hormone secreted per event is strongly controlled by 24-hr rhythmicity (18). In the case of cortisol, secretory burst mass also predominates in its magnitude of diurnal rhythmicity compared to burst frequency (interburst interval) (136).

In Fig. 23, we have plotted collective data from the group of subjects, each of whom generated 15–25 individual bursts of ACTH release. A single cosine function was plotted for all the data considered together. An alternative analytical strategy is to plot a corresponding profile for each man, and then utilize cosine curve fitting to fit simultaneously all 6–8 data sets assuming a common 24-hr periodicity, individually distinct acrophases and mesors (mean value about which the 24-hr rhythm varies) for each of the profiles, and a common fitted mean amplitude. This would answer the experimental question: Given individually distinct acrophases in plots of ACTH secretory burst mass and/or frequency over 24-hr in different individual subjects, is there a nonzero (significant) 24-hr rhythmicity in the ACTH secretory burst mass and/or interpulse interval for the group of subjects considered as a whole? The statistical confidence intervals defining the estimate of the group mean amplitude should be nonzero at 95% in order to affirm this hypothesis.

How Should Deconvolution Analysis Deal with Possible Secretory Events Contained within the Leading or Trailing Edges of an Observed Time Series?

A common analytical problem is the occurrence of incomplete (possible) secretory events presumptively identified at the leading edge or trailing end

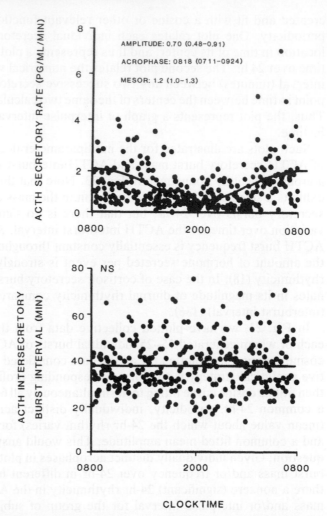

FIG. 23 Significant diurnal variation in ACTH secretory burst mass (top) but not interpulse interval (bottom) in a group of 8 healthy young men studied via 10-min sampling and multiparameter deconvolution analysis (31). Fitted cosine parameters (±67% confidence intervals) are noted. Adapted with permission from Veldhuis *et al.* (18), Amplitude, but not frequency, modulation of ACTH secretory bursts gives rise to the nyctohemeral rhythm of corticotropic axis in man, *J. Clin. Endocrinol. Metabol.* **71**, 811, 1990. © The Endocrine Society.

of a hormone time series. This problem is illustrated in Fig. 24. For example, the data series might begin with a collection of serially declining hormone concentrations followed by well-defined peaks each flanked by nadirs. The series might end analogously with a progressive decline or an unidirectional

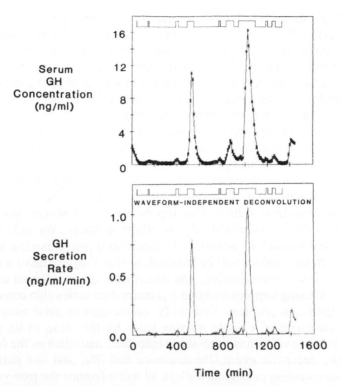

FIG. 24   Illustrative problems of leading and trailing edge artifacts in a GH time series. (Top) Sampled GH data and (bottom) calculated sample secretion rates (PULSE). The exact amplitude, mass, and duration of the first and last partial peaks cannot be determined readily. Data from J. D. Veldhuis and M. L. Johnson, 1993, unpublished. See text for further discussion.

upstroke. To our knowledge, there is no mathematically perfect way to deal with incomplete observations. However, we would recommend that if the data at the outset of a series show only unidirectional decline this segment be fit as simple exponential decay, as the precise amplitude and location of any prior secretory event (beginning before time zero of sampling) cannot be known with certainty. If, however, one or more points are observed on the ascending limb, and a maximal value is achieved before the decline, then a peak may be tentatively located by deconvolution analysis. If the position of this peak cannot be established without infinite uncertainty (i.e., the lower bound of the temporal location of its center is not distinguishable from minus infinity), then the putative secretory event should be regarded as indeterminate and deleted, and the data fit for decaying values only. A similar logic could be applied arbitrarily to the trailing end of the hormone time series; that is, one could argue to regard as indeterminate or delete the data values within a terminal unidirectional upstroke or downstroke if (i) no inflection or maximum exists and (ii) the putative peak position remains indeterminate, whether or not the resultant fit of the data in the absence of adding an edge peak produces consecutively positive or negative residuals within that region (differences between the fitted curve and the observed data). Five consecutive positive or negative residuals would be improbable on the basis of chance alone at $p < 0.033$ by the binomial probability distribution. This would imply that a peak was overlooked, but if it remains indeterminate because of incomplete data, we would suggest that the remaining data be fit without it, thus eliminating unrealistic constraints imposed by these irreconcilable values. This represents one of several possible arbitrary approaches to so-called edge or fringe artifacts. We believe the preeminent rule to which to adhere is to establish *a priori* explicit guidelines for how such data values will be handled, so that they are dealt with uniformly in an unbiased, reproducible, and describable manner for all data sets.

Missing samples within the primary data series also constitute a potentially significant problem. Ordinarily, convention in pulse analysis allows the replacement of a single missing value by the mean of its nearest neighbors. This convention is necessarily incorrect, inasmuch as the function underlying the secretion event, the hormone half-life, and the particular location of surrounding peaks and valleys all will influence the true value of the missing sample. Accordingly, in most circumstances, we believe that it is preferable to let the missing sample remain missing. In our multiparameter deconvolution method, as well as in our new waveform-independent procedure PULSE2 (see Chapter 1, this volume, by Johnson and Veldhuis), the data may be irregularly spaced and/or contain missing sample values. Curve fitting the remaining data will project a curve through the region of the missing data point, and this curve will be defined appropriately by the combined influences of the other observed samples in this region of the series and not merely the two nearest flanking values. If multiple successive data values are missing and cannot be reassayed or retrieved, the investigator must evaluate how much significant information is lost on this account. For example, half-life estimates from the data series may not be greatly influenced by several missing samples even in succession, if large stretches of data are still available showing five or more half-lives of decay. However, the absolute enumeration of peak frequency and amplitude will be vitiated if one or more peaks are omitted because of missing data. Hence, peak frequency and amplitude estimates would be expected to show deteriorating reliability when stretches of samples are omitted over a span that could contain one or more secretory bursts. We recommend great caution about accepting inferences based on such data series. An *a priori* rule should be stated and complied with to address this difficulty before beginning the analysis.

Reliability of Deconvolution Estimates

When pulsatile neurohormone time series are subjected to curve fitting by deconvolution analysis, the resultant mean parameter estimates will exhibit a range of precision, reliability, and probability. Precision is the inverse of the quantitative error associated with the parameter estimte; that is, a half-life value of 20 with the precision of ±1 min (*SD*) denotes that there is approximately a 67% probability that the true half-life value lies between 19 and 21 min, and a 95% probability that it falls approximately between 18 and 22 min (assuming a symmetrical error space for the purposes of this example). The test–retest reliability of an estimate for a given data series designates the test–retest correlation when the same phenomenon is studied twice; interobserver reliability is the correlation between results obtained by two independent observers applying the same tool to the same data set; and physiological reliability denotes the consistency of the physiological measure over repeated observations. These various facets of reliability are discussed in Chapter 3 by Mulligan *et al.*, this volume.

Any particular set of parameter estimates is ideally accompanied by a probability estimate; for example, the pulse frequency is estimated at a particular α level, or p value, which is the probability of falsely rejecting the null hypothesis that this number of pulses is not present. Simulation experiments and analysis of *in vivo* biological data indicate that the probability values for true-positive peak identification must be not only conditioned by the mathematical and statistical qualifications inherent in the analytical methodology, but also adjusted for the experimental conditions under which the measurements were made. For example, Fig. 25 compares estimates of LH secretory burst amplitude over a range of apparent sampling intensities utilizing computer-simulated LH pulse profiles (64). Note that for any particular data series emulating physiological LH pulsatility in normal young men, a prolongation of the sampling interval beyond 20 min produces a progressive overestimate of the mean LH secretory burst amplitude despite equivalent statistical confidence interval testing. The change in parameter reliability is not a function of the analytical tool per se, but represents an inevitable consequence of curtailing data density and thereby eliminating observable information that otherwise would denote the presence of additional peaks. Thus, parameter reliability must be understood not only in relation to the statistical features of the analytical tool, but also in conjunction with the physical properties of the neuroendocrine pulse system, the sampling condi-

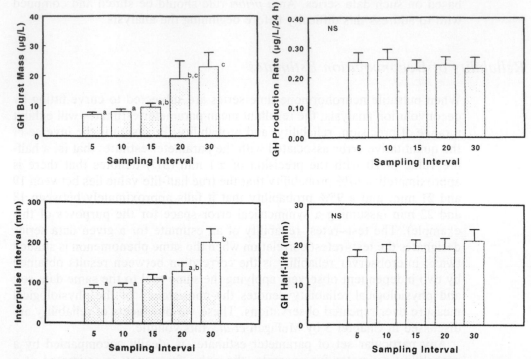

FIG. 25   Spurious increases in apparent GH secretory burst mass and interpulse interval compared with stable half-life and production-rate estimates, as sampling interval is extended from every 5 min to every 30 min in young men. Data adapted with permission from Hartman *et al.* (10). Different superscripts above the bars denote statistically different means.

tions under which the observations were made, the physical limitations of the assay system, etc. Accordingly, we recommend that all analytical tools including those of deconvolution analysis be submitted to reliability testing under simulated and observed *in vivo* biological conditions that most closely resemble the state expected during the actual investigation proposed.

As demonstrated in several earlier papers (9, 41, 42, 44, 45, 49, 64), the errors that could be introduced by failing to evaluate the impact of an experimental condition such as sampling intensity can be large for some measures and minimal for other measures. For example, in GH deconvolution analysis of time series collected at 5-min intervals for 24 hr, we observed that collecting data at 10-, 20-, or 30-min intervals still allowed a reliable estimate of GH half-life (assuming that the 5-min data yielded the closest approximation to true *in vivo* half-life). The calculated 24-hr production rate of GH was also well estimated from even 20- or 30-min data (10). In contrast, apparent

GH secretory burst amplitude was overestimated as was burst duration and mass when sampling intensity declined from every 5 to every 15, 20, or 30 min. Collectively, our computer simulation studies show that the exact effects of sampling intensity on parameter reliability depend on the associated noise and the underlying hormone half-life, as well as secretory burst characteristics including expected pulse frequency, duration, and amplitude (64).

Estimating Endogenous Hormone Half-Life

As alluded to briefly above (see section on half-life of neurohormone removal), where we discuss determining whether the half-life of a neurohormone is concentration dependent, endogenous hormone half-lives are not necessarily equivalent numerically to those estimated by infusions of one or more similar or identical molecules, even if the experimental conditions of sampling and measurement are identical. Consequently, estimates of endogenous hormone half-life are of merit. One approach to estimating hormone half-life and secretion simultaneously is discussed in the work by Munson and Rodbard (137), and our strategy using multiparameter deconvolution analysis has been reviewed earlier (31, 32, 67). The Munson methodology requires zero basal secretion, as the existence of basal hormone release would mathematically confound interpretation of the natural logarithmic transformation utilized to linearize the estimate of a monoexponential half-life. Indeed, the methodology was designed to compute a single-component model of elimination. Our multiparameter deconvolution technique also estimates monoexponential hormone half-life. This also is more difficult to accomplish in the presence of basal secretion, unless basal secretion is known *a priori* and/or unless fixed basal secretion can be estimated from intervals of unchanging serum hormone concentrations separating regions of burstlike hormone release (see above in section on half-life of neurohormone removal).

However, neuroendocrine time series resulting from variable or indeterminate basal secretion combined with a variable pulsatile component provide a unique challenge to investigators wishing to estimate pulsatile hormone secretion, hormone half-life, and basal secretion simultaneously. Indeed, when evaluated as putative solutions to any particular neurohormone time series, these three primary sets of parameters (basal secretion, pulsatile release, and half-life) are highly interdependent statistically and functionally, allowing for multiple possible solutions rather than a unique fit of the data. One approach to addressing this difficulty is to administer a secretagogue to each subject at the end of the sampling session, with the intent of stimulating a solitary peak or a volley of peaks with putatively negligible basal secretion compared to the magnitude of the release episode. If possible, neurohormone elimination then should be observed over five half-lives in order to obtain reliable and precise estimates of subject-, session-, and hormone-specific half-life. We recommend that rather than simply subjecting the decay portion of the induced pulse to curve fitting, as sometimes recommended (40), deconvolution analysis of the release episode should be carried out so as to fit simultaneously underlying secretion and neurohormone half-life (31, 32, 68). Ideally, the waveform of the secretory event should be known *a priori* so that a relevant approximate waveform can be utilized in the multiparameter deconvolution analysis.

As illustrated in Fig. 26 (top), a secretogogue-stimulated hormone secretory burst of very brief duration compared to the hormone half-life can be estimated reasonably by simply curve fitting an exponential function to the decay data. However, the lower graphs show that a prolonged hormone secretory burst associated with a relatively short half-life of elimination results in a plasma neurohormone concentration profile that cannot be treated reliably as exclusively a decay-mediated event. Indeed, simply curve fitting the declining serum hormone concentrations to an exponential function could greatly overestimate the hormone half-life, when unaccounted secretion contributes to maintaining the declining hormone concentrations and thereby causes an apparent delay in hormone removal. Consequently, significant artifacts can be introduced into the estimate of hormone half-life if the induced hormone-concentration peak is merely subjected to exponential curve fitting of the decay limb (1, 31, 32). The degree of artifact engendered will depend on the hormone half-life, the duration and amplitude of the secretion signal, and the amount of concurrent basal hormone release.

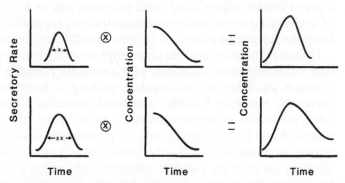

FIG. 26   Schematic comparison of the predicted effects of a brief (top) versus prolonged (bottom) hormone secretory burst and a given monoexponential half-life of decay on the decay limb of the concentration curve. Data from J. D. Veldhuis and M. L. Johnson, 1992, unpublished.

Preliminary Curve Fitting for Estimating Deconvolution Parameters

When multiple individual parameters of hormone secretion and/or half-life must be estimated simultaneously, the multiparameter fit proceeds more efficiently and rapidly if reasonable nominal or preliminary parameter estimates can be obtained. We suggest that a prefitting algorithm be prepared to accomplish this. We have used either of two waveform-independent techniques in this context. Both of these strategies are illustrated in Fig. 27.

Historically, we first carried out provisional peak identification with a program called Cluster (51) and preliminary curve fitting with an algorithm designated as PULSE (32, 103). Thus, the initial strategy that we adopted was to apply a discrete peak detection method that has been validated with *in vivo* and computer-simulated data, such as Cluster analysis (9, 49). For example, Cluster analysis (51) may be utilized to identify the apparent number and locations of significant release episodes in a simulated or actual biological time series. This algorithm will identify plausible peak positions from the serial serum hormone concentration measurements, but such positions of course reflect the maximal serum concentration of the hormone that is attained, rather than the underlying secretory pulse maximum (a maximal secretory *rate*) that occurs earlier (31, 32). With the use of a scaling factor that is a function of the hormone half-life and burst duration, and a plausible estimate of the hormone half life, knowledge of the approximate peak positions provided by Cluster analysis can offer a useful starting point for multiparameter curve fitting.

The PULSE methodology has been presented in detail earlier (32), but, in brief, it consists of estimating by nonlinear least-squares methods all sample secretory rates associated with each plasma hormone concentration given a particular known or assumed biexponential hormone half-life. Because this program was not designed initially to provide prefitting parameters, the definitions of the sample secretion rate amplitude in PULSE and the Gaussian burst amplitude in the multiparameter deconvolution algorithm (DECONV) are dissimilar. This results in systematic underestimation of the burst amplitude for DECONV when using the value for the PULSE sample amplitude (Fig. 27, top). However, the algorithm is still useful to identify plausible peak numbers and positions. This is followed by a rapid single-step fitting of amplitudes in DECONV to achieve a good approximate fit of the data. Further parameter fitting is carried out thereafter with DECONV so as to eliminate low-probability events by statistical confidence-interval testing and more nearly estimate the correct endogenous hormone half-life, and secretory burst duration, locations, and amplitudes (32, 64).

A third-generation deconvolution technique is described elsewhere in this volume, which we have designated PULSE2 (Chapter 1). This approach

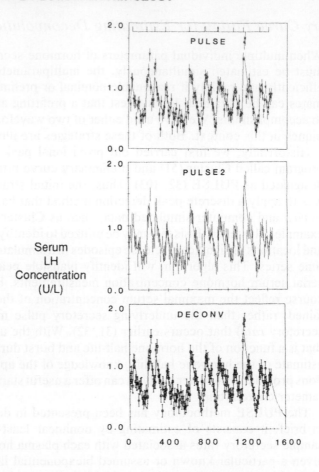

Serum
LH
Concentration
(U/L)

Time (min)

FIG. 27   Comparison of preliminary fits of a 10-min serum LH concentration profile by PULSE (top) and PULSE2 (center), with the final fit (bottom) by DECONV. The different analytical tools tested here as prefitting devices for multiparameter deconvolution (DECONV) are described in the text. Data from J. D. Veldhuis and M. L. Johnson, 1994, unpublished.

identifies in a waveform-independent manner significant secretory bursts scattered throughout the time series given an assumed hormone half-life and/ or basal secretion value. The algorithm will estimate pulsatile secretion alone, or estimate pulsatile and basal secretion simultaneously. Figure 27, center, illustrates the prefitting utility of this waveform-independent technique, which provides excellent starting guesses for the multiparameter deconvolution approach. Typically following a prefitting estimation by PULSE2, a 24-hr time course of serum hormone concentrations obtained by sampling blood at intervals of 5–10 min can be curve fit in 5–20 min by first refitting the burst amplitudes, then estimating all the amplitudes and positions, followed by calculating the amplitudes, positions, and burst duration, and lastly computing all secretion and half-life parameters simultaneously. This final fit is shown in Fig. 27, bottom.

In summary, when multiple secretory (and half-life) parameters must be estimated simultaneously and are interdependent, and if some physiological knowledge exists concerning the system (e.g., plausible hormone half-life and burst duration), we recommend the development or application of a reasonable, rapid prefitting deconvolution methodology such as PULSE2 (above), or some other technique (below), so as to obtain low-stringency estimates of peak number, amplitude, location, and/or duration with or without an inference regarding possible basal secretion.

Use of a Weighting Function in Nonlinear Curve Fitting

As discussed in some detail in an earlier review (1) and in Chapter 9, this volume, hormone measurements are confounded by dose-related within-assay experimental uncertainty, which can be designated by dose-dependent standard deviations, absolute variances, or coefficients of variation. Many nonlinear curve-fitting techniques utilize the degree of experimental uncertainty inherent in any given sample hormone concentration to inversely weight the iterative fit of the function (curve) of interest. For example, our multiparameter deconvolution algorithm utilizes the square of the sample deviation (and, hence, the sample variance) as an inverse weighting term in curve fitting the convolution integral to the pulsatile hormone concentration versus time series (31, 32, 67, 68). This allows the fitted curve to approximate data points that have high precision more closely, and to approach data points known with lesser precision less closely. This is illustrated in Fig. 28, where we compare weighted and nonweighted deconvolution fits. For the analysis of many neurohormone time series, the use of an appropriately chosen weighting function is appropriate, and can be shown to be mathematically suitable in obtaining maximal likelihood estimates of fitted parameters (126).

The correct assignment of experimental uncertainty to any given measurement is not a trivial undertaking. For example, many neuroendocrine assays are carried out in duplicate, so that only two measures are available from which to estimate the variability inherent in any particular sample measure-

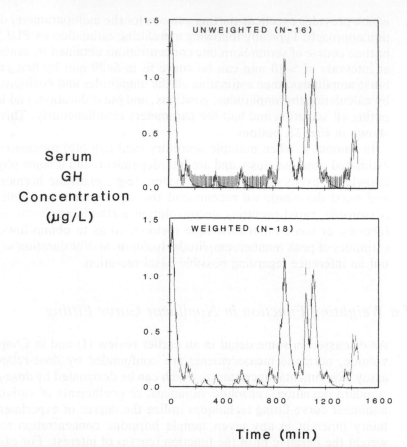

**Serum
GH
Concentration
(µg/L)**

Time (min)

FIG. 28   Fit by PULSE2 of a serum GH concentration profile with variable (weighted, bottom) or constant (unweighted, top) intrasample standard deviations (12). Data are serum GH concentrations determined by chemiluminescence assay of blood samples collected at 10-min intervals. The profile pulse frequencies were estimated: $N = 16$, unweighted fit and $N = 18$ weighted. Data from J. D. Veldhuis and M. L. Johnson, 1995, unpublished.

ment. Because the standard deviation of duplicates is equal to their range, very limited information is obtained from duplicates. Somewhat greater insight into experimental variability is achieved by assaying multiple replicates, but the number of such replicates desired is typically greater than can be practically applied to extended biological data series because of either limiting blood volume and/or limiting assay costs. An alternative approach is to estimate error from more extensive quality control samples contained within the same assay and/or its standard curve. As discussed elsewhere in this volume in greater detail (Chapter 9), the valid estimation of within-sample error requires appropriate fitting of the standard curve with respect to the zero-dose tube, a correct choice of the algebraic function to describe the standard curve, an appropriate iterative curve fitting approach utilizing within-sample variance to weight the fit of the standard curve, and calculation of joint experimental uncertainties in the final within-sample variance estimates that derive from not only the variability among sample replicates (standard curve and quality control samples) but also the experimental uncertainty in the estimated parameters of the standard curve. The resultant joint statistical confidence limits for a sample hormone concentration value should then closely approximate the theoretical value obtained by assaying multiple replicates, even though the unknown samples are assayed only in singlet or duplicate (138).

The exact choice of the weighting function (i.e., within-sample variance) is also not a trivial consideration. The arguments related to the choice of the weighting term are reviewed elsewhere in considerable detail (135).

Resolving Pulses within Pulses

As intimated earlier (see Fig. 9), some physiological and pathophysiological secretory events may result from the summation of multiple successive release episodes and/or the fusion of incompletely fulfilled secretory bursts. The proestrous LH surge is a plausible example, in which 3- to 4-min blood sampling in the rat shows continuously elevated serum LH concentrations (48). Visually the profile suggests either basal hormone release with superimposed secretory bursts, or a succession of incompletely resolved (partially fused) secretory impulses. These two models are difficult to distinguish in most physiological data, although the mathematical forms of the descriptions are distinct. An example of this discriminative challenge is illustrated for TSH (Fig. 29A) and prolactin release (Fig. 29B). In these graphs, we show 24-hour serum TSH and prolactin concentration profiles generated by IRMA or RIA of blood collected at 10-min intervals in healthy young men. The multiparameter deconvolution model (DECONV) was applied assuming purely pulsatile hormone release (31). As shown by the calculated (deconvolved) secretory profiles below the observed data, a succession of partially overlapping TSH or prolactin secretory bursts emerges from this analysis with this assumption. Of considerable interest, if a waveform-independent deconvolution methodology, such as PULSE or PULSE2 (31, 32), is applied to the same data, then a similar overall secretion profile is resolved. However, the last two models interpret the origins of the profiles differently, from DECONV namely, that they are composed of a substantial basal (time-

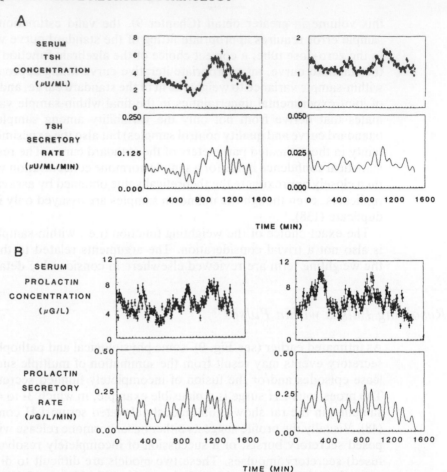

FIG. 29   Deconvolution analysis of 10-min serum TSH and prolactin profiles collected over 24 hr and submitted to multiparameter deconvolution analysis (31). The top graphs contain the fitted curves and the serially observed serum TSH (A) or prolactin (B) concentrations over time, whereas the bottom graphs depict the calculated secretion profile assuming a model of delimited punctuated secretory bursts without intervening tonic hormone release. Fusion of rapidly successive hormone secretory events in this model leads to apparent basal hormone release, as typified for the two pituitary hormones. Waveform-independent deconvolution analysis identifies putative time-invariant basal hormone secretion with superimposed lower-amplitude release episodes (103). The two models (fused secretory bursts versus pulses superimposed on basal release) are difficult to distinguish analytically (see Discussion text). Data from J. D. Veldhuis and M. L. Johnson, 1993, unpublished.

posed of a substantial basal (time-invariant) secretory component with super-imposed lower amplitude secretory bursts (103).

In general, the information content of available neurohormone time series collected under conventional conditions *in vivo* does not permit one to resolve definitively the two different models of release, namely, fused higher amplitude secretory pulses versus basal secretion with superimposed smaller pulses. Efforts to discriminate between these two models we believe should include the following: (a) sampling the neuroendocrine ensemble at or near the point of hormone release, for example, pituitary venous effluent; (b) performing high-density observations to determine if secretion does fall to or near zero between successive events; (c) demonstrating selective abolition of the pulsatile component, for example, following administration of a relevant antagonist of endogenous secretagogues; (d) specifically altering the apparent basal component to show its existence and susceptibility to regulation; (e) selectively modifying the pulsatile component, for example, by withdrawal or enhancement of negative feedback; and/or (f) slowing the frequency of the pulsatile component to observe whether secretion declines to zero between events and/or basal secretion continues. Other strategies may also be required, but this challenge remains a difficult one technically and analytically.

In attempting to unravel admixed basal and pulsatile neurohormone release in humans, we have observed the greatest success to date in studying the LH axis. As shown in Fig. 30, we find that administration of a potent and selective GnRH antagonist (Nal-Glu GnRH) preferentially abolishes the pulsatile component of immunoreactive and bioactive LH release in postmeno-pausal women, while leaving residual largely time-invariant LH concentrations (139). This strongly suggests to us that LH secretion in the postmenopausal (feedback-withdrawn) environment consists of both pulsatile and basal hormone secretion, and that the pulsatile component is significantly GnRH dependent. The basal component may or may not be partially GnRH dependent, as it continues despite abolition of effective GnRH action (confirmed in other studies by the administration of synthetic GnRH in the presence of the antagonist). Because the degree of putative basal LH release prior to administration of the antagonist is not known, but could be any value between zero and that corresponding to the interpulse nadir serum LH concentrations, we do not know definitively whether the GnRH antagonist lowered (or even increased) basal release. Although this experiment cannot establish whether the decapeptide antagonist decreased or had no effect on basal LH secretion, we may infer that at the doses chosen it did not eliminate basal (time-invariant, and virtually pulseless) LH release. Consequently, this illustrates the complexity of examining basal and pulsatile hormone release *in vivo* when these two modes of secretion are mingled.

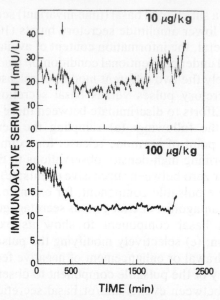

FIG. 30   Dissection of pulsatile versus basal LH release in a postmenopausal woman treated with 10 or 100 μg/kg Nal-Glu-GnRH (injected at the time indicated by the arrow), which is a potent selective antagonist of gonadotropin-releasing hormone (GnRH). Note the virtual abolition of discernible pulsatile LH secretion following injection of the higher dose of GnRH antagonist (bottom), which discloses time-invariant (basal) LH release. Adapted with permission from Urban *et al.* (139).

Summary

This chapter has been designed to address important technical queries that arise in the course of carrying out deconvolution analysis of neuroendocrine data. We have stated a variety of contemporary technical controversies, issues, and limitations and in some cases offered possible solutions. Solutions were rarely if ever singular, and rarely if ever absolutely definitive. Indeed, the evolving complexities of deconvolution analysis applied to *in vivo* neuroendocrine data raise many novel and interesting challenges, while beginning to answer only a few questions. Nonetheless, we believe that the systematic and explicit appraisal of technical hurdles is an essential first step toward methodological advances. Consequently, we hope that the foregoing technical essay will help to clarify, stimulate, and in some cases address contemporary issues and obstacles in the timely and progressive domain of deconvolution analysis of neuroendocrine data.

Acknowledgments

We thank Patsy Craig for skillful preparation of the manuscript and Paula P. Azimi for the artwork. This work was supported in part by National Institutes of Health Grant RR 00847 to the Clinical Research Center of the University of Virginia, NICHD RCDA 1 KO4 HD00634 (J.D.V.), NIH Grants GM-35154 and RR-08119 (M.L.J.), the Baxter Healthcare Corporation, Roudlake, IL (J.D.V.), the Diabetes and Endocrinology Research Center Grant NIH DK-38942, the NIH-supported Clinfo Data Reduction Systems, the Pratt Foundation, the University of Virginia Academic Enhancement Fund, and the National Science Foundation Center for Biological Timing (NSF Grant DIR89-20162). This work was also supported in part by an NIH P-30 Reproduction Center (C. Desjardins, Director) #NIH P-30 HD-28934.

References

1. R. J. Urban, W. S. Evans, A. D. Rogol, D. L. Kaiser, M. L. Johnson, and J. D. Veldhuis, *Endocr. Rev.* **9,** 3 (1988).
2. W. V. R. Vieweg, J. D. Veldhuis, and R. M. Carey, *Am. J. Physiol.* **262,** F871 (1992).
3. J. D. Veldhuis, J. C. King, R. J. Urban, A. D. Rogol, W. S. Evans, L. A. Kolp, and M. L. Johnson, *J. Clin. Endocrinol. Metab.* **65,** 929 (1987).
4. J. D. Veldhuis, M. L. Johnson, E. Seneta, and A. Iranmanesh, *Acta Endocrinol.* **126,** 193 (1992).
5. M. H. Samuels, J. D. Veldhuis, C. Cawley, R. J. Urban, M. Luther, R. Bauer, and G. Mundy, *J. Clin. Endocrinol. Metab.* **76,** 399 (1993).
6. J. D. Veldhuis, M. L. Johnson, and M. L. Dufau, *Am. J. Physiol.* **256,** E199 (1989).
7. J. D. Veldhuis and S. J. Winters, *J. Androl.* **10,** 248 (1989).
8. M. L. Sollenberger, E. C. Carlson, M. L. Johnson, J. D. Veldhuis, and W. S. Evans, *J. Neuroendocrinol.* **2**(6), 845 (1990).
9. R. J. Urban, M. L. Johnson, and J. D. Veldhuis, *Endocrinology (Baltimore)* **124,** 2541 (1989).
10. M. L. Hartman, A. C. S. Faria, M. L. Vance, M. L. Johnson, M. O. Thorner, and J. D. Veldhuis, *Am. J. Physiol.* **260,** E101 (1991).
11. W. G. Rossmanith, G. A. Laughlin, J. F. Mortola, M. L. Johnson, J. D. Veldhuis, and S. S. C. Yen, *J. Clin. Endocrinol. Metab.* **70,** 990 (1990).
12. J. D. Veldhuis, W. S. Evans, L. A. Kolp, A. D. Rogol, and M. L. Johnson, *J. Clin. Endocrinol. Metab.* **66,** 414 (1988).
13. S. J. Winters and P. E. Troen, *J. Clin. Invest.* **78,** 870 (1986).
14. J. Isgaard, L. Carlsson, O. G. P. Isaksson, and J. O. Jansson, *Endocrinol (Baltimore)* **123,** 2605 (1988).
15. S. L. Davis, D. L. Ohlson, J. Klindt, and M. S. Anfinson, *Am. J. Physiol.* **233,** E519 (1977).
16. G. S. Tannenbaum and J. B. Martin, *Endocrinology (Baltimore)* **98,** 562 (1976).

17. D. J. Dierschke, A. N. Bhattaracharya, L. E. Atkinson, and E. Knobil, *Endocrinology (Baltimore)* **87,** 850 (1970).
18. J. D. Veldhuis, A. Iranmanesh, M. L. Johnson, and G. Lizarralde, *J. Clin. Endocrinol. Metab.* **71,** 452 (1990).
19. R. P. McIntosh, J. E. A. McIntosh, and L. Lazarus, *J. Endocrinol.* **118,** 339 (1988).
20. J. D. Veldhuis, R. M. Blizzard, A. D. Rogol, P. M. Martha, Jr., J. L. Kirkland, B. M. Sherman, and Genentech Collaborative Group, *J. Clin. Endocrinol. Metab.* **74,** 766 (1992).
21. L. M. Winer, M. A. Shaw, and G. Baumann, *J. Clin. Endocrinol. Metab.* **70,** 1678 (1990).
22. H. Hiruma, M. Nishihara, and F. Kimura, *Brain Res.* **582,** 119 (1992).
23. D. S. Weigle, W. D. Koerker, and C. J. Goodner, *Am. J. Physiol.* **247,** E564 (1984).
24. C. Monet-Kuntz and M. Terqui, *Int. J. Androl.* **8,** 129 (1985).
25. T. F. Gallagher, K. Yoshida, H. D. Roffwarg, D. K. Fukushida, E. D. Weitzman, and L. Hellman, *J. Clin. Endocrinol. Metab.* **36,** 1058 (1973).
26. D. C. Parker, L. G. Rossman, and E. F. Vanderlaan, *J. Clin. Endocrinol. Metab.* **36,** 1119 (1973).
27. G. W. Randolph and A. R. Fuchs, *Am. J. Perinatol.* **6,** 159 (1989).
28. P. M. Plotsky and W. Vale, *Science* **230,** 461 (1985).
29. J. D. Veldhuis and M. L. Johnson, *J. Neuroendocrinol.* **2,** 755 (1991).
30. E. Van Cauter, *Am. J. Physiol.* **237,** E255 (1979).
31. J. D. Veldhuis, M. L. Carlson, and M. L. Johnson, *Proc. Natl. Acad. Sci. U.S.A.* **84,** 7686 (1987).
32. J. D. Veldhuis and M. L. Johnson, *in* "Methods in Enzymology" (L. Brand and M. L. Johnson, eds.), Vol. 210, p. 539. Academic Press, San Diego, 1992.
33. R. Rebar, D. Perlman, F. Naftolin, and S. S. C. Yen, *J. Clin. Endocrinol. Metab.* **37,** 917 (1973).
34. W. J. Jusko, W. R. Slaunwhite, Jr., and T. Aceto, Jr., *J. Clin. Endocrinol. Metab.* **40,** 278 (1975).
35. F. O'Sullivan and J. O'Sullivan, *Biometrics* **44,** 339 (1988).
36. K. Albertsson-Wikland, S. Rosberg, E. Libre, L. O. Lundberg, and T. Groth, *Am. J. Physiol.* **257,** E809 (1989).
37. R. J. Henery, B. A. Turnbull, M. Kirkland, J. W. McArthur, I. Gilbert, G. M. Besser, L. H. Rees, and D. S. Tunstall Pedoe, *Chronobiol. Int.* **6,** 259 (1989).
38. J. D. Veldhuis and M. L. Johnson, *Am. J. Physiol.* **255,** E749 (1988).
39. C. M. Swartz, V. S. Wahby, and R. Vacha, *Acta Endocrinol.* **112,** 43 (1986).
40. K. E. Oerter, V. Guardabasso, and D. Rodbard, *Comput. Biomed. Res.* **19,** 170 (1986).
41. R. J. Urban, M. L. Johnson, and J. D. Veldhuis, *Am. J. Physiol.* **257,** E88 (1989).
42. J. D. Veldhuis, A. B. Lassiter, and M. L. Johnson, *Am. J. Physiol.* **259,** E351 (1990).
43. P. L. Toutain, M. Laurentie, A. Autefage, and M. Alvinerie, *Am. J. Physiol.* **255,** E688 (1988).
44. V. Guardabasso, G. De Nicolao, M. Rocchetti, and D. Rodbard, *Am. J. Physiol.* **255,** E775 (1988).

45. E. Van Cauter, *Am. J. Physiol.* **254,** E786 (1988).
46. J. D. Veldhuis, M. L. Johnson, L. M. Faunt, M. Mercado, and G. Baumann, *J. Clin. Invest.* **91,** 629 (1993).
47. A. D. Genazzani and D. Rodbard, *Acta Endocrinol.* **124,** 295 (1991).
48. J. D. Veldhuis, M. L. Johnson, and R. V. Gallo, *Am. J. Physiol.* **265,** R240 (1993).
49. R. J. Urban, M. L. Johnson, and J. D. Veldhuis, *Endocrinology (Baltimore)* **128,** 2008 (1991).
50. J. D. Veldhuis, M. L. Johnson, A. Iranmanesh, and G. Lizarralde, *J. Biol. Rhythms* **5,** 247 (1990).
51. J. D. Veldhuis and M. L. Johnson, *Am. J. Physiol.* **250,** E486 (1986).
52. V. Guardabasso, A. D. Genazzani, J. D. Veldhuis, and D. Rodbard, *Acta Endocrinol.* **124,** 208 (1991).
53. W. S. Evans, E. Christiansen, R. J. Urban, A. D. Rogol, M. L. Johnson, and J. D. Veldhuis, *Endocr. Rev.* **13,** 81 (1992).
54. R. J. Urban, D. L. Kaiser, E. Van Cauter, M. L. Johnson, and J. D. Veldhuis, *Am. J. Physiol.* **254,** E113 (1988).
55. D. K. Clifton and R. A. Steiner, *Endocrinology (Baltimore)* **112,** 1057 (1983).
56. K. E. Oerter, V. Guardabasso, and D. Rodbard, *Comput. Biomed. Res.* **19,** 170 (1986).
57. R. P. McIntosh and J. E. A. McIntosh, *Endocrinol.* **107,** 231 (1985).
58. A. Pilo, E. Ferrannini, and R. Navalesi, *Am. J. Physiol.* **233,** E500 (1977).
59. J. D. Veldhuis, V. Guardabasso, A. D. Rogol, W. S. Evans, K. Oerter, M. L. Johnson, and D. Rodbard, *Am. J. Physiol.* **252,** E599 (1987).
60. J. D. Veldhuis and M. L. Johnson, *In* "Advances in Neuroendocrine Regulation of Reproduction" (S. S. C. Yen and W. W. Vale, eds.), p. 123. Plenum, Philadelphia, Pennsylvania, 1990.
61. R. C. Turner, J. A. Grayburn, G. B. Newman, and J. D. N. Nabarro, *J. Clin. Endocrinol.* **33,** 279 (1972).
62. J. D. Veldhuis, A. Faria, M. L. Vance, W. S. Evans, M. O. Thorner, and M. L. Johnson, *Acta Paediatr. Scand.* **347,** 63 (1988).
63. P. A. Jansson, "Deconvolution Methods in Spectrometry," p. 99. Academic Press, New York, 1984.
64. T. Mulligan, H. A. Delemarre-van de Waal, M. L. Johnson, and J. D. Veldhuis, *Am. J. Physiol.* **267,** R202 (1994).
65. F. de Zegher, H. Devlieger, and J. D. Veldhuis, *Pediatr. Res.* **32,** 605 (1992).
66. J. D. Veldhuis and M. L. Johnson, *in* "Methods in Enzymology" (M. L. Johnson and L. Brand, eds.), Vol. 240, p. 377. Academic Press, San Diego, 1994.
67. J. D. Veldhuis, J. Moorman, and M. L. Johnson, *Methods Neurosci.* **20,** 279 (1994).
68. J. D. Veldhuis, W. S. Evans, J. P. Butler, and M. L. Johnson, *Methods Neurosci.* **10,** 241 (1992).
69. S. L. Alexander and C. H. G. Irvine, *J. Endocrinol.* **114,** 351 (1987).
70. J. D. Veldhuis and M. L. Johnson, "Frontiers in Neuroendocrinology," Vol. 11, p. 363. Raven, New York, 1991.
71. J. D. Veldhuis, A. Iranmanesh, I. Clarke, D. L. Kaiser, and M. L. Johnson, *J. Neuroendocrinol.* **1,** 185 (1989).

72. W. C. Nunley, R. J. Urban, W. S. Evans, and J. D. Veldhuis, *J. Clin. Endocrinol. Metab.* **73,** 629 (1991).

73. J. D. Veldhuis, F. Fraioli, A. D. Rogol, and M. L. Dufau, *J. Clin. Invest.* **77,** 1122 (1986).

74. E. Christiansen, J. D. Veldhuis, A. D. Rogol, P. Stumpf, and W. S. Evans, *Am. J. Obstet. Gynecol.* **157,** 320 (1987).

75. R. J. Urban, V. Padmanabhan, I. Beitins, and J. D. Veldhuis, *J. Clin Endocrinol. Metab.* **73,** 818 (1991).

76. A. CS. Faria, J. D. Veldhuis, M. O. Thorner, and M. L. Vance, *J. Clin. Endocrinol. Metab.* **68,** 535 (1989).

77. G. T. Campbell, E. D. Balir, G. H. Grossman, A. E. Miller, M. E. Small, and E. M. Bogdanove, *Endocrinology (Baltimore)* **103,** 674 (1978).

78. G. T. Campbell, D. D. Nansel, W. M. Meinzer III, M. S. Aiyer, and E. M. Bogdanove, *Endocrinology (Baltimore)* **103,** 683 (1978).

79. A. M. Akbar, T. M. Nett, and G. D. Niswender, *Endocrinology (Baltimore)* **94,** 1318 (1974).

80. T. M. Badger, W. J. Millard, S. M. Owens, J. Larovere, and D. O'Sullivan, *Endocrinology (Baltimore)* **128,** 1065 (1991).

81. G. Baumann, M. A. Shaw, and T. A. Buchanan, *Metabolism* **38,** 330 (1989).

82. G. Baumann, M. W. Stolar, and T. A. Buchanan, *Endocrinology (Baltimore)* **119,** 1497 (1986).

83. G. M. Besser, D. N. Orth, W. E. Nicholson, R. L. Byyny, K. Abe, and J. P. Woodham, *J. Clin. Endocrinol.* **32,** 595 (1971).

84. A. K. Dubdey, A. Ahanukoglu, B. C. Hansen, and A. A. Kowarski, *J. Clin. Endocrinol. Metab.* **67,** 1064 (1988).

85. R. C. Fry, L. P. Cahill, J. T. Cummins, B. M. Bindon, L. R. Pipper, and I. J. Clarke, *J. Reprod. Fertil.* **81,** 611 (1987).

86. C. M. Hendricks, R. C. Eastman, S. Takeda, K. Asakawa, and P. Gorden, *J. Clin. Endocrinol. Metab.* **60,** 864 (1985).

87. P. C. Hindmarsh, D. R. Matthews, and C. E. Brain, *Clin. Endocrinol. (Oxford)* **30,** 443 (1988).

88. N. Porksen, T. O'Brien, J. Steers, S. Munn, J. D. Veldhuis, and P. C. Butler, *Am. J. Physiol.* in press, 1995.

89. A. Iranmanesh, G. Lizarralde, and J. D. Veldhuis, *J. Clin. Endocrinol. Metab.* **73,** 1081 (1991).

90. F. J. Lopez and A. Negro-Vilar, *Endocrinology (Baltimore)* **123,** 740 (1968).

91. G. W. Montgomery, G. B. Martin, S. F. Crosbie, and J. Pelletier, *in* "Reproduction in Sheep" (D. R. Lindsay and D. T. Pearce, eds.), p. 22. Australian Academic Science, Canberra, 1984.

92. A. G. Morell, G. Gregoriadis, I. H. Scheinberg, J. Hickman, and G. Ashwell, *J. Biol. Chem.* **246,** 1461 (1971).

93. D. Owens, M. C. Strivastave, C. V. Tompkins, J. D. N. Nabarro, and P. H. Sonksen, *Eur. J. Clin. Invest.* **3,** 284 (1973).

94. W. D. Peckham, T. Yamaji, D. J. Dierschke, and E. Knobil, *Endocrinology (Baltimore)* **92,** 1660 (1973).

95. D. M. Robertson, L. M. Foulds, R. C. Fry, J. T. Cummins, and I. Clarke, *Endocrinology (Baltimore)* **129,** 1805 (1991).

96. K. L. Sydnor and G. Sayers, *Proc. Soc. Exp. Biol. Med.* **83,** 729 (1953).

97. A. Vermeulen, L. Verdonck, M. Van der Straeten, and N. Orie, *J. Clin. Endocrinol. Metab.* **29,** 1470 (1969).

98. R. F. Weick, *Endocrinology (Baltimore)* **101,** 157 (1977).

99. L. Wide, *Acta Endocrinol.* **112,** 336 (1986).

100. L. Wide and M. Wide, *J. Clin. Endocrinol. Metab.* **54,** 426 (1984).

101. F. Schaefer, G. Baumann, L. M. Faunt, D. Haffner, M. L. Johnson, M. Mercado, E. Ritz, O. Mehls, and J. D. Veldhuis, The 1994 Annual Endocrine Society Meeting (1994) (abstract).

102. J. D. Veldhuis, A. Iranmanesh, K. K. Y. Ho, G. Lizarralde, M. J. Waters, and M. L. Johnson, *J. Clin. Endocrinol. Metab.* **72,** 51 (1991).

103. J. D. Veldhuis, A. Iranmanesh, M. L. Johnson, and G. Lizarralde, *J. Clin. Endocrinol. Metab.* **71,** 1616 (1990).

104. P. M. Martha, Jr., K. M. Goorman, R. M. Blizzard, A. D. Rogol, and J. D. Veldhuis, *J. Clin. Endocrinol. Metab.* **74,** 336 (1992).

105. N. M. Wright, F. J. Northington, J. D. Miller, J. D. Veldhuis, and A. D. Rogol, *Pediatr. Res.* **32,** 286 (1992).

106. A. Iranmanesh, B. Grisso, and J. D. Veldhuis, *J. Clin. Endocrinol. Metab.* **78,** 526 (1994).

107. W. M. Pardridge and E. M. Landaw, *Am. J. Physiol.* **249,** E534 (1985).

108. D. W. Leung, S. A. Spencer, G. Cachianes, R. G. Hammonds, C. Collins, W. J. Henzel, R. Barnard, M. J. Waters, and W. I. Wood, *Nature (London)* **330,** 537 (1987).

109. G. Baumann and M. A. Shaw, *J. Clin. Endocrinol. Metab.* **70,** 680 (1990).

110. G. Baumann, M. A. Shaw, and R. J. Winter, *J. Clin. Endocrinol. Metab.* **65,** 814 (1987).

111. W. H. Daughaday and B. Trivedi, *Proc. Natl. Acad. Sci. U.S.A.* **84,** 4636 (1987).

112. G. Baumann and M. A. Shaw, *Biochem. Biophys. Res. Commun.* **152,** 573 (1988).

113. A. C. Herington, S. Yemer, and J. Stevenson, *J. Clin. Invest.* **77,** 1817 (1986).

114. B. C. Cunningham, M. Ultsch, A. M. DeVos, M. G. Mulkerrin, K. R. Clauser, and J. A. Wells, *Science* **254,** 821 (1991).

115. S. A. Spencer, R. G. Hammonds, W. J. Henzel, H. Rodriquez, M. J. Waters, and W. I. Wood, *J. Biol. Chem.* **263,** 7862 (1988).

116. R. Horton, J. Shinsako, and P. H. Forsham, *Acta Endocrinol.* **48,** 446 (1965).

117. G. Baumann, M. W. Stolar, and T. A. Buchanan, *J. Clin. Endocrinol. Metab.* **64,** 657 (1987).

118. J. D. Veldhuis, L. M. Faunt, and M. L. Johnson, *in* "Methods in Enzymology" (M. L. Johnson and L. Brand, eds.), Vol. 240, p. 349. Academic Press, San Diego, 1994.

119. J. D. Veldhuis, M. L. Johnson, and M. L. Dufau, *J. Clin. Endocrinol. Metab.* **64,** 1275 (1987).

120. J. P. Butler, D. I. Spratt, L. S. O'Dea, and W. F. Crowley, *Am. J. Physiol.* **250,** E338 (1986).

121. A. Iranmanesh, G. Lizarralde, M. L. Johnson, and J. D. Veldhuis, *J. Clin. Endocrinol. Metab.* **68,** 1019 (1989).

122. J. D. Veldhuis and M. L. Johnson, *J. Clin. Endocrinol. Metab.* **67,** 116 (1988).

123. A. Iranmanesh, G. Lizarralde, and J. D. Veldhuis, *Acta Endocrinol.* **128,** 521 (1993).

124. J. D. Veldhuis, M. L. Johnson, and E. Seneta, *J. Clin. Endocrinol. Metab.* **73,** 569 (1991).

125. M. H. Samuels, J. D. Veldhuis, P. Henry, and E. C. Ridgway, *J. Clin. Endocrinol. Metab.* **71,** 425 (1990).

126. M. L. Johnson and S. G. Frasier, *in* "Methods in Enzymology" (C. H. W. Hirs and S. N. Timasheff, eds.), Vol. 117, p. 301. Academic Press, San Diego, 1985.

127. R. J. Urban, M. R. Davis, A. D. Rogol, M. L. Johnson, and J. D. Veldhuis, *J. Clin. Endocrinol. Metab.* **67,** 1149 (1988).

128. J. S. Bendat and A. G. Piersol, "Random Data." Wiley, New York, 1986.

129. W. C. Nunley, R. J. Urban, J. D. Kitchen, B. G. Bateman, W. S. Evans, and J. D. Veldhuis, *J. Clin. Endocrinol. Metab.* **72,** 287 (1991).

130. P. R. Bevington, T. Bick, M. B. Youdim, and Z. Hochberg, "Data Reduction and Error Analysis for the Physical Sciences" p. 118. McGraw-Hill, New York, 1969.

130a. J. Y. Weltman, J. D. Veldhuis, A. Weltman, J. R. Kerrigan, W. S. Evans, and A. D. Rogol, *J. Clin. Endocrinol. Metab.* **71,** 1646 (1990).

131. C.-J. Partsch, S. Abrahams, N. Herholz, M. Peter, J. D. Veldhuis, and W. G. Sippell, *Acta Endocrinol.* in press (1994).

132. S. M. Pincus, *Proc. Natl. Acad. Sci. U.S.A.* **88,** 2297 (1991).

133. A. R. Feinstein, *Clin. Pharmacol. Ther.* **17,** 104 (1975).

134. B. M. Bennett, *Biometrics* **28,** 793 (1972).

135. E. DiCera, *in* "Methods in Enzymology" (L. Brand and M. L. Johnson, eds.), Vol. 210, p. 68. Academic Press, San Diego, 1992.

136. J. D. Veldhuis, A. Iranmanesh, G. Lizarralde, and M. L. Johnson, *Am. J. Physiol.* **257,** E6 (1989).

137. P. J. Munson and D. Rodbard, "Proceedings of the Statistical Computing Section of the American Statistical Association," p. 295. Washington, D.C., 1989.

138. M. Straume, J. D. Veldhuis, and M. L. Johnson, "Methods in Enzymology" (M. L. Johnson and L. Brand, eds.), Vol. 240, p. 121. Academic Press, San Diego, 1994.

139. R. J. Urban, S. N. Pavlou, J. E. Rivier, W. W. Vale, M. L. Dufau, and J. D. Veldhuis, *Am. J. Obstet. Gynecol.* **162,** 1255 (1990).

140. D. Haffner, F. Schaefer, J. Girard, E. Ritz, and O. Mehls, *J. Clin. Invest.* **93,** 1163 (1994).

141. J. D. Veldhuis, and M. L. Johnson. *Endocrinology (Baltimore)* **127,** 2611 (1990).

[3] Physiological within Subject Variability and Test–Retest Reliability of Deconvolution Analysis of Luteinizing Hormone Release

Thomas Mulligan, Michael L. Johnson, and
Johannes D. Veldhuis

Importance of a Reliable Instrument

Whether building an antique rifle or inventing a new assay for growth hormone (GH), one must ensure that the instrument performs consistently. For example, when the rifle is in a fixed position the projectile should repeatedly hit the same area of the target (Fig. 1). Similarly, if a new growth hormone assay is applied multiple times to a single aliquot of blood, all assay results (replicates) should be similar. One important statistical measure of consistent performance is test reliability. As with validity, an analytical instrument is of unknown value until its reliability is established (1). However, validity and reliability are not necessarily related. For example, a new assay for GH may consistently provide the same answer when repeatedly applied to a single aliquot of blood (high reliability), but the assay result may be consistently incorrect (low validity) (2).

Reliability can be best understood in three domains. First, does the instrument perform consistently when applied multiple times by the same operator to the same data set? This is test–retest reliability. Second, does the instrument perform consistently when applied to the same data set by two different operators? This is the issue of interoperator reliability. Finally, an investigator may want to determine physiological reliability, that is, whether a given phenomenon, such as pulsatile luteinizing hormone (LH) secretion, is stable when assessed repeatedly over time in a given individual. To assess whether the parameters of LH secretion are physiologically consistent over time, the same operator would apply the same analytical instrument to LH data collected from the same subject on different days. Reliability, therefore, complements validity as a discriminative measure of an assessment instrument. An ideal instrument should be both valid (accurately measuring the phenomenon) and reliable (consistently measuring the phenomenon). Here, we introduce specific methods by which to appraise the reliability of an important neuroendocrine analytical tool, namely, deconvolution analysis, and we illustrate its application to

Methods in Neurosciences, Volume 28

pulsatile pituitary LH secretion in healthy young men. Validity testing is presented in Chapter 4 by Mulligan *et al.*, this volume.

Assessing the Reliability of Deconvolution Analysis

Determining the test–retest reliability of deconvolution analysis of endocrine time series requires that the technique be applied multiple times by the same operator to the same data set. Importantly, the operator must use the same rules of analysis (order and criteria for analytical choices) each time he/she analyzes the data. In addition, to avoid subjective bias, the operator must be blinded to the identity of the data sets that are being (re)analyzed.

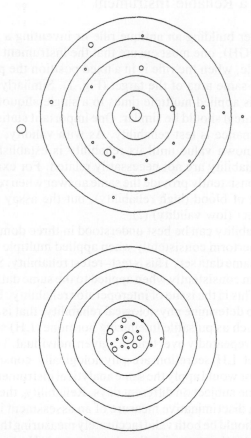

FIG. 1 Comparison of the consistency (reliability) of two antique rifles when each is fired at the target from a fixed position. Results in the top bull's-eye demonstrate low reliability, whereas those in the lower bull's-eye demonstrate high reliability.

Determining the interoperator reliability of deconvolution analysis requires that two or more operators independently apply the analytical technique to the same data sets. Again, to avoid undesired bias, each operator must be blinded to the analytical results obtained by the other.

Finally, to determine the reliability of the phenomenon (e.g., pulsatile LH secretion), the same operator must apply the same technique to data collected from the same subject, under similar conditions, on at least two distinct occasions in time. To minimize biological variation within the subject, study episodes should be spaced far enough apart in time to minimize the effect of blood loss, while close enough in time to avoid the development of seasonal differences, intercurrent illness, or aging. We suggest, for example, that study episodes for gonadotropin analysis in men be at least 2 but no more than 8 weeks apart. In women, the same stage (e.g., early follicular phase) of the menstrual cycle should be studied serially over 2 or more months. Such reliability analysis for a discrete peak-detection application, cluster analysis (3), has been presented for LH time series in men and young women (4).

Data Characteristics

To assess reliability properly, one should employ data that include a spectrum of representative features (e.g., known absent or present secretory events, and secretory events of varying amplitude, shape, mass, duration, and frequency). In addition, the waveform of the underlying events, the half-life, and the experimental uncertainty in the data should embrace a full range of values anticipated to occur in biological data. We have used both computer-simulated time series from a previously described biophysical model (5) and, wherever possible, biological data sets. Biological data must be accompanied by independent markers of true secretory events to ensure valid interpretation and should emulate the anticipated experimental conditions expected as closely as possible. Even so, a combination of computer-simulated and biological data is particularly useful, we believe, as the two strategies are strongly complementary. For example, biological data contain less well-defined experimental uncertainty, which can be made explicit in computer simulations. On the other hand, biological data realistically represent experimental observations, whereas model-based simulations may not uniformly capture all nuances of biological behavior including within- and between-animal variability, secretory waveform, *in vivo* sources of random variations, and obfuscating aspects of sample withdrawal and processing. Thus, we vigorously urge the use of both biological and computer-modeled time series for reliability and validity testing.

In computer simulations reported elsewhere (6), and reanalyzed here, all sample values were created as duplicates spanning 24 hr at simulated collection intervals of 10 min. The hormone time series had varying apparent hormone (LH) half-lives ($t_{1/2}$ of 35, 65, and 95 min), interpulse intervals (45, 60, and 90 min), mean secretory burst amplitudes (0.03, 0.1, 0.3 IU/liter/min), and random experimental variation or noise (defined as intraassay coefficients of variation of 10, 20, and 30%). A constant secretory burst half-duration (the duration of the idealized Gaussian secretory event at half its maximal amplitude) of 10.5 min was used in the simulations. All mean LH interpulse interval and amplitude values within a series were made to vary about their nominal means, assuming a Gaussian distribution with a coefficient of variation of ±30%. These arbitrary parameters were chosen for the present methodological exercise because the resultant computer-simulated data represent a span of realistic human LH time series (7, 8).

We also used authentic human serum LH concentration time series to assess interoperator reliability. The human data sets consisted of six men with isolated hypogonadotropic hypogonadism, who were given intravenous (i.v.) pulses of 7.5, 25, 75, and 250 ng/kg of GnRH in randomized order at a fixed interval (every 120 min) over 10 hr. The LH secretory pulses of varying amplitude followed gonadotropin-releasing hormone (GnRH) injections (9).

Deconvolution Analysis

The steps we follow in multiparameter deconvolution analysis, regardless of which form of reliability we are assessing, are as follows. First, all duplicate sample values are used to estimate dose-dependent intrasample variance as a power function of the sample means (a program called FIX is employed for this purpose). This procedure provides a mean (of duplicates) serum LH value and an interpolated standard deviation (SD) for each sample. The formatted time series (Fig. 2A) are then submitted to preliminary, waveform-independent, pulse estimation [PULSE (10)] to identify statistically significant secretory events at low stringency ($p < 0.10$) assuming some nominal and plausible hormone half-life (e.g., $t_{1/2}$ of 60 min). The putative number of pulses and their amplitudes and locations as identified by this prefitting step are used as initial guesses for the multiparameter deconvolution analysis (DECONV), which fits a calculated reconvolution curve to the concentration data. The model equations employed by DECONV are reviewed elsewhere (10). The definitions of waveform-independent (PULSE) and waveform-specific (DECONV) secretory pulse amplitudes differ. Hence, at this point, the initial parameter guesses underestimate the serum LH levels (Fig. 2B). Indeed, a superior prefitting algorithm (PULSE2) is described elsewhere in this volume (see Chapter 1 by Johnson and Veldhuis).

Iterative curve fitting is performed by DECONV via a modified

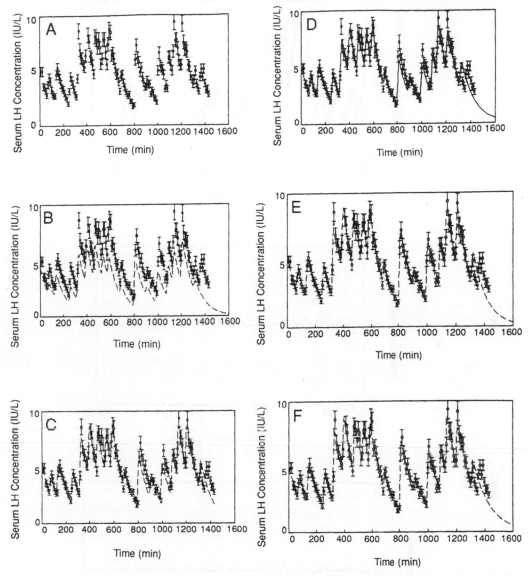

FIG. 2 Sequential parameter estimation in multiparameter deconvolution analysis, illustrated for one particular (arbitrary) data set. In this computer-simulated endocrine time series, peripheral blood was "sampled" at 10-min intervals for 24 hr for the duplicate "measurement" of serial serum LH concentrations (A). The simulated serum LH concentrations and the predicted reconvolution curve after a single prefitting step using PULSE are also shown (B). Note that the preliminary parameter estimates produce a curve that is an imperfect fit of the data. As additional parameters are sequentially estimated by deconvolution analysis (C–F), the reconvolution curve more closely approximates the serum LH concentrations.

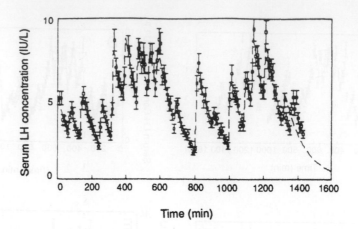

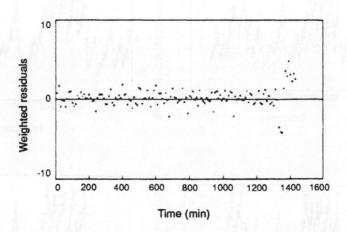

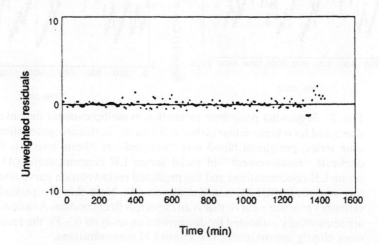

Fig. 5. Sequential parameter estimation in multiparameter deconvolution analysis of data and for a (similar) cyclone. . . . In this coupled parametric-stochastic time series, peripheral blood was sampled at 10-min intervals for 24 hr for the analysis. "Measurement" of serial serum LH concentrations (MA). The simulated serum LH concentrations and the predicted resolved values, together with their peak . . . . probability amplitude samples provide a certain estimate over the median. As indicated, parameter . . are sequentially estimated by deconvolution analysis (0. 5), the reconvolution curve more closely approximates the original LH concentrations.

Gauss–Newton approach to estimate multiple parameters of secretion as well as hormone half-life using a minimum four-step procedure so as to (1) estimate only the starting (T_0) concentration and amplitudes of the Gaussian secretory peaks (Fig. 2C); (2) determine only the positions of the secretory peaks after preliminary amplitude optimization (Fig. 2D); (3) evaluate both the amplitudes and positions of the secretory peaks as fitted parameters (Fig. 2E); and (4) fit the T_0 concentration, hormone $t_{1/2}$, and secretory pulse amplitudes, half-duration, and positions simultaneously (Fig. 2F). For example, when deconvolving an LH time series, we first obtain preliminary parameter guesses with PULSE or PULSE2, and then allow only the T_0 concentration and amplitudes of the secretory peaks to vary while holding all other parameters fixed (e.g., half-life, burst duration, and peak position estimates). The other parameters are progressively varied (steps 2 to 4 above). Notably, during the fourth step, all parameters are fit simultaneously. By this fourth step, the predicted reconvolution curve has begun to approximate closely the measured serum LH concentrations.

After the first series of four fitting steps (above), the serum LH concentration data and the fitted curve are inspected for consecutive positive residuals. We plot the raw data, the fitted curve, and the weighted and unweighted residuals (Fig. 3). A weighted residual is the residual value weighted inversely as the square of the interpolated standard deviations for that sample. The data analyst can estimate a provisional new peak wherever two or more consecutive residuals (differences between the fitted curve and observed data) are positive (Fig. 4). This corrected set of parameter estimates is then reevaluated iteratively using the same four-step fitting procedure described above. Throughout the fitting procedure, tentative hormone pulses whose amplitudes or locations are assigned mathematically indeterminate values [i.e., values that could not be distinguished from -10^{38} or $+10^{38}$ (approaching negative and positive infinity), respectively] are deleted one at a time beginning with the smallest peak, and the data refit (steps 1 to 4 above).

The foregoing tentative "best fit" of the data yields pulse measures (parameters) constrained only by finite and determinable peak amplitudes and positions. These parameters are then used in the calculation of statistical (e.g., 95%) confidence limits. Although various levels of statistical stringency can

FIG. 3 Illustration of weighted (inversely as the within-sample variance) and unweighted residuals (differences between fitted curve and the observed sample values) as a function of sample number for the particular deconvolution fit of the illustrative computer-simulated serum LH concentration time series given in Fig. 2. Note that there are multiple consecutive positive residuals after the last deconvolution-derived secretory peak, suggesting a missed secretory pulse.

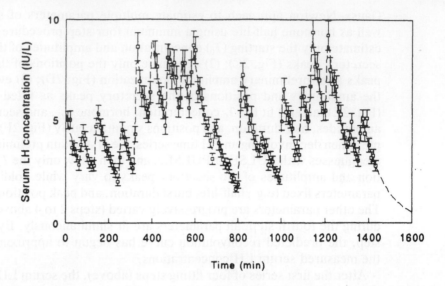

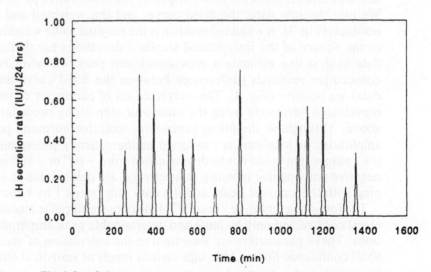

FIG. 4 Final fit of the computer-simulated endocrine time series in Figs. 2 and 3 after application of deconvolution analysis, insertion of a provisional peak based on positive residuals, and the use of statistical confidence limits to test for significant secretory pulses. The last reconvolution curve closely approximates the simulated serum LH concentrations, as can be assessed objectively by the mean residual error, the Aikaike information coefficient (AIC), and the randomness of the residuals. The last feature can be evaluated by the runs test, autocorrelation analysis, or the Kolmogorov–Smirnov statistic, as discussed elsewhere (e.g., Ref. 10).

be applied, we recommend that, for 10-min sampled LH time series in humans, the amplitudes (or mass) of all resolved LH secretory bursts have nonzero lower bounds at 95% joint confidence limits calculated for the secretory burst half-duration, hormone $t_{1/2}$, and all secretory burst amplitudes considered simultaneously in an n-parameter variance space. The choice of statistical stringency depends on a collection of factors, including, but not limited to, sampling intensity, secretory burst signal-to-noise ratio, noise, hormone half-life, the presence or absence of confounding basal hormone secretion, pulse frequency, as well as the experimental question and, hence, the nominally required rates of false-positive and false-negative errors. Secretory pulses not conforming to a given statistical stringency test are deleted one at a time beginning with the smallest or least determinate value. Parameters are then reestimated by iterative fitting (as outlined above).

Reliability Testing

After deconvolution analysis of the endocrine time series, the parameters of secretion and elimination (e.g., secretory burst position, amplitude, half-duration, and hormone $t_{1/2}$) for the two fits are compared to determine the Pearson's parametric correlation coefficient between the two applications of the analytical technique. When sufficient data are available, we recommend using the Wilk–Shapiro or other test of normality to assess the distribution of the secretory (or half-life) measure. If the assumption of normality is violated, a distribution-free test of rank correlation (e.g., Spearman's) should be employed.

We found that when one person (T. Mulligan) reanalyzed computer-simulated data sets ($n = 12$) having a spectrum of representative features (e.g., varying noise, hormone $t_{1/2}$, secretory pulse amplitude, pulse frequency), test–retest reliability was high (Table I). Such reanalysis was carried out blind at a remote interval (9 months). We also found that when two operators analyzed the same data sets (computer-simulated $n = 10$, human $n = 6$), interoperator reliability was high (Table I). The least reliable parameters were secretory burst amplitude and duration, which are highly (negatively) correlated parameters. Their product, secretory burst mass, was reliably determined. Note that these inferences apply to 10-min simulated LH time series, wherein the burst half-duration (10.5 min) approximated the sampling interval and, hence, is not a well-determined parameter. More frequent (higher density) blood sampling and assay would be required in authentic biological series to enhance the determinability of the absolute duration and amplitude of such brief secretory bursts.

TABLE I   Reliability of Deconvolution Analysis of Luteinizing Hormone Time Series[a]

Parameter	Test–retest	Interobserver	Physiological
Half-duration of secretory burst	+0.98	+0.89	−0.41
Hormone $t_{1/2}$	+0.97	+0.95	+0.32
Number of secretory peaks	+0.99	+0.98	+0.39
Interpulse interval	+0.99	+0.99	+0.13
Secretory burst area	+0.99	+0.93	+0.47
Secretory burst amplitude	+0.97	+0.76	−0.14
Hormone production rate	+0.99	+0.97	+0.40

[a] The calculated correlation coefficients are given for the parameters of LH secretion and half-life when deconvolution analysis is performed twice on the same endocrine time series (i.e., the same computer simulation or individual profile). In the test–retest example, the same investigator analyzed 12 endocrine time series on two separate occasions several months apart, while blinded to the results of the first analysis. In the interobserver example, two different investigators each independently analyzed 10 computer-simulated and 6 human endocrine time series, while blinded to one another's results. Physiological reliability was assessed by analyzing two independently collected 10-min serum immunoreactive LH concentration time series over 24 hr in each of 15 young healthy men (T. Mulligan and J. D. Veldhuis, 1992, unpublished).

Intersession Physiological Reliability

In contrast to test–retest or interoperator reliability, the physiological inter-session consistency (test–retest reliability or stability of the phenomenon) of the parameters of LH secretion estimated in a particular radioimmunoassay (RIA) in young men sampled over 24 hr at 10-min intervals is not high. Using the cluster pulse detection algorithm, Weltman et al. (4) found limited day-to-day consistency in the number and frequency of LH concentration pulses among healthy young women. However, intersession consistency was high for mean and integrated serum LH concentrations ($r = +0.79$) (4). Using deconvolution analysis, we found limited day-to-day physiological reliability in the parameters of LH secretion among healthy young men ($n = 15$) but observed significant consistency in the mean (24-hr) serum LH concentration ($r = +0.614$, $p = 0.015$); the apparent LH half-life and LH secretory pulse amplitude, duration, and frequency were nonuniform across sessions in individual men whose serum LH concentrations were determined in duplicate in a dual-label RIA (Table I and Fig. 5). Similar studies have not yet been

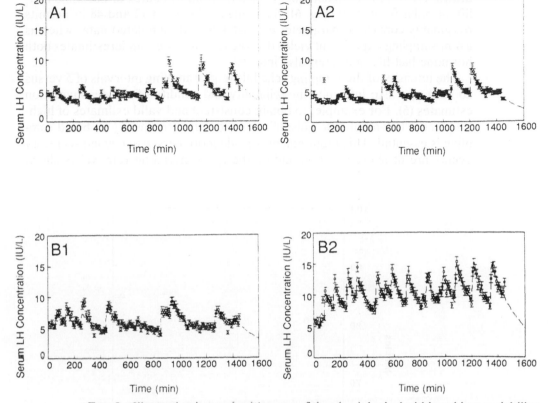

Fig. 5   Illustration in two healthy men of the physiological within-subject variability of immunoreactive (RIA) LH secretion and clearance as estimated by deconvolution analysis. Note that the serum LH concentration profiles of subject A (A1 and A2) are similar despite their being obtained 4 weeks apart. In contrast, the LH concentration profiles depicted for another volunteer (B1 and B2) demonstrate substantial within-subject variability for a similar 4-week interval.

carried out using an LH bioassay, but our deconvolution analyses of serum LH series assessed by two-site immunoradiometric assay in young men each sampled on three independent occasions disclose high physiological reliability for estimates such as LH half-life and daily secretion rate (11).

Sources of Variability in Endocrine Time Series

In addition to physiological variability of the phenomenon, there are various extrinsic sources of variability. For example, in frequent venous sampling experiments, the total duration of the sampling session (e.g., blood withdrawal for 4–48 hr) impacts the precision of the estimates of hormone half-life or pulse frequency (Fig. 6). Sampling for between 12 and 48 hr permits reasonably consistent parameter estimates for our simulated data, whereas a 4-hr sampling experiment yields divergent values (i.e., underestimates both hormone half-life and interpulse interval).

The intensity of the sampling schedule (e.g., sampling intervals of 5 versus 20 min) may also influence the reliability and validity of the deconvolution estimates (6). For example, to obtain consistent and valid estimates of high-frequency, low-amplitude pulsatility, an adequate frequency of blood sampling is essential. This requirement is made more persuasive when secretory events are of short duration and/or the $t_{1/2}$ of hormone removal is short.

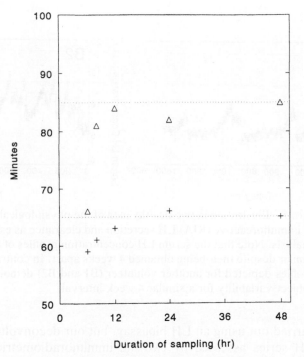

FIG. 6   Illustration of the relationship between the duration of blood sampling (e.g., 4 to 48 hr) and the consistency of the estimates of hormone $t_{1/2}$ and secretory interpulse interval. The computer-simulated endocrine time series ($n = 3$) were generated with a nominal hormone $t_{1/2}$ of 65 min (bottom dotted line) and secretory interpulse interval of 85 min (top dotted line). Peripheral blood was "sampled" at 10-min intervals for 48 hr. The data for the 4-, 6-, 8-, 12-, and 24-hr subsets were derived by simply truncating the data at the time points noted. Note that as the duration of sampling drops below 12 hr, the $t_{1/2}$ (+) and interpulse interval (△) values are underestimated.

Consistent estimates (i.e., high interseries reliability) of LH secretory measures are attained when 2.5- and 5-min blood sampling series are compared, but not when either of these regimens is compared to 30-min subseries (Fig. 7). Similar inferences were made when comparing 5-min and 15-, 20-, or 30-min subseries in the same healthy young men (12). Of interest, consistency of secretory or $t_{1/2}$ measures across different sampling intensities depended on the parameter selected; for example, half-life estimates (but not burst duration) had high reliability when compared over 5-, 10-, 15-, and 20-min sampling rates.

Contributing further to variability is the timing of sample withdrawal. Although the phase of the sampling protocol is not critical (Fig. 8), it is unlikely that the venous samples are obtained at exactly equal intervals (e.g., 10 min). More likely, actual observations will be made at intervals that vary between 10 min and slightly longer times. Early samples were quite unlikely, as we found in single-blind observations of our clinical research staff (13).

In addition, there is variability in sample withdrawal, preparation (e.g., aliquot quantity, defibrination, centrifugation), freezing, storage, and thawing, as well as in the precision of the assay. Error accumulation continues at the level of assay data reduction in such domains as instrumentation, photon or scintillation counting, and RIA curve fitting). These factors can be assessed indirectly, for example, by assaying the time series twice (11).

Finally, the choice of the analytical tool that is applied to the endocrine time series (e.g., cluster or deconvolution analysis) and the selection of statistical constraints can impart differences to the final details reported for the analysis. Quantitating such total error is difficult, but all sources of variability must be kept in mind and minimized by the investigator.

As predicted above, we found that there could be substantial disagreement between operators if statistical confidence limits were not applied. For example, discrepant fits could occur if one but not the other operator was overly zealous at inserting putative peaks to fit the data, especially at low signal-to-noise ratios (e.g., reduced burst amplitude with high within-assay variability). However, the deconvolution algorithm corrected fully for overinterpretation of initial parameter estimates when statistical confidence limits were imposed [e.g., when 95% confidence limits were applied simultaneously to all parameters, there was remarkable interoperator agreement (6)].

Summary

Deconvolution analysis is a powerful analytical tool, which can provide valuable information regarding hormone secretion. However, as with all analytical tools, it must be properly applied to maximize reliability. The

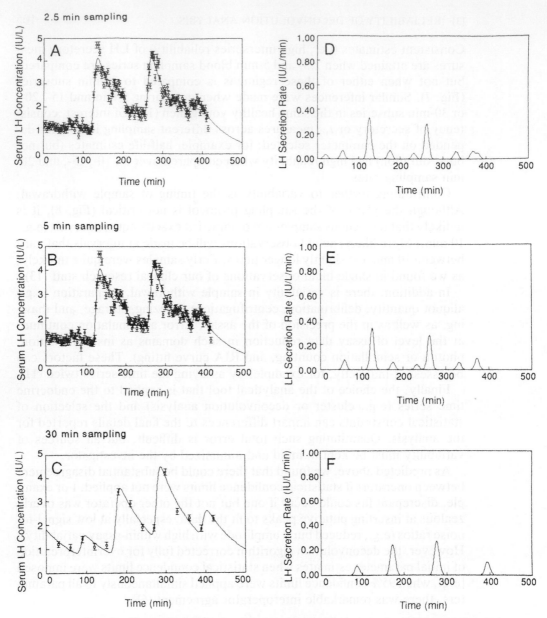

FIG. 7 Illustration of interseries consistency between 2.5- and 5-min sampling paradigms but not 2.5- and 30-min sampling schedules. Infrequent observation intervals degrade the reliability of interseries comparisons, here shown for a single healthy male sampled at 2.5-min intervals overnight. One of the constituent 5-min and 30-min LH series are shown for interseries comparison. The serum LH concentrations and the fitted reconvolution curve are depicted (A–C), along with the associated LH secretory bursts (D–F).

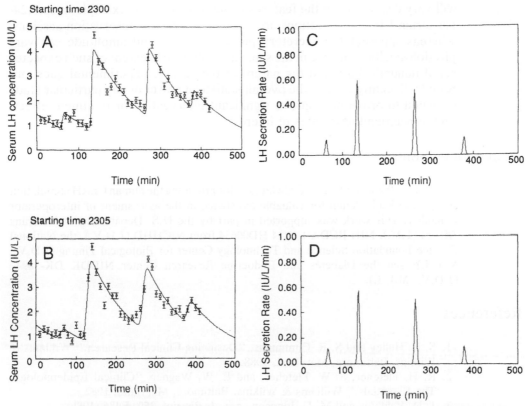

FIG. 8   Illustration of the lack of impact of sample collection phase (time when sampling session started) on the consistency of deconvolution fits of 10-min LH subseries in a normal male volunteer who underwent blood sampling at 2.5-min intervals overnight. The parent 2.5-min LH series were edited to yield two (of four possible) constituent 10-min series, one beginning at 2300 hr and the other at 2305 hr. The serum LH concentrations and the fitted curves for the two constituent series are shown (A and B), one of which has a phase shift of 5 min. The associated LH secretory bursts are also depicted (C and D).

investigator should use a standardized approach including the assessment of statistical confidence limits. Optimal experimental conditions must be utilized in collecting the data in order not only to maximize sensitivity and specificity of the deconvolution method, but also to ensure the reliability and validity of individual parameter estimates. Finally, biological variability of a system output (e.g., pulsatile LH secretion) over time must be recognized. Indeed, estimates of the intrinsic physiological reliability of a system will vary depending on the features being observed; for example, mean (24-hr) serum LH concentrations are quite stable across observation sessions, whereas apparent LH secretory pulse frequency and amplitude can vary physiologically over time in the same individual. The extent of such expected serial nonuniformity and the nature of the specific experimental questions posed will jointly control the overall statistical power of any particular study designed to prove or disprove significant intervention or treatment effects on the neuroendocrine axis of interest.

Acknowledgments

The authors thank William F. Crowley, Jr., for providing the human GnRH stimulation series, and Paula Azimi for valuable assistance in the assessment of interoperator reliability. This work was supported in part by the U.S. Department of Veterans Affairs (T.M.), NIH RCDA 1 KO4 HD00634 from NICHHD (J.D.V.), the National Science Foundation Science and Technology Center for Biological Timing (J.D.V., M.L.J.), and the Diabetes and Endocrine Research Center NIDDK DK-38942 (J.D.V., M.L.J.).

References

1. S. B. Hulley and S. R. Cummings, "Designing Clinical Research." Williams & Wilkins, Baltimore, Maryland, 1988.
2. R. H. Fletcher, S. W. Fletcher, and E. W. Wagner, "Clinical Epidemiology: The Essentials." Williams & Wilkins, Baltimore, Maryland, 1982.
3. J. D. Veldhuis and M. L. Johnson, *Am. J. Physiol.* **250,** E486 (1986).
4. J. Y. Weltman, J. D. Veldhuis, A. Weltman, J. R. Kerrigan, W. S. Evans, and A. D. Rogol, *J. Clin. Endocrinol. Metab.* **71,** 1646 (1990).
5. J. D. Veldhuis and M. L. Johnson, *Am. J. Physiol.* **255,** E749 (1988).
6. T. M. Mulligan, H. A. Delemarre-Van de Waal, M. L. Johnson, and J. D. Veldhuis, *Am. J. Physiol.* **467,** R202 (1994).
7. J. D. Veldhuis, M. L. Carlson, and M. L. Johnson, *Proc. Natl. Acad. Sci. U.S.A.* **84,** 7686 (1987).
8. R. J. Urban, M. L. Johnson, and J. D. Veldhuis, *Endocrinology (Baltimore)* **124,** 2541 (1989).
9. N. Santoro, M. Filicori, and W. F. Crowley, *Endocr. Rev.* **7,** 11 (1986).
10. J. D. Veldhuis and M. L. Johnson, *in* "Methods in Enzymology" (L. Brand and M. L. Johnson, eds.), Vol. 210, p. 539. Academic Press, San Diego, 1992.
11. C. J. Partsch, S. Abrahams, N. Herholz, M. Peter, J. D. Veldhuis, and W. G. Sippell, *Eur. J. Endocrinol.* **131,** 263 (1994).
12. M. L. Hartman, A. C. S. Faria, M. L. Vance, M. L. Johnson, M. O. Thorner, and J. D. Veldhuis, *Am. J. Physiol.* **260,** E101 (1991).
13. R. J. Urban, W. S. Evans, A. D. Rogol, D. L. Kaiser, M. L. Johnson, and J. D. Veldhuis, *Endocr. Rev.* **9,** 3 (1988).

[4] Methods for Validating Deconvolution Analysis of Pulsatile Hormone Release: Luteinizing Hormone as a Paradigm

Thomas Mulligan, Michael L. Johnson, and
Johannes D. Veldhuis

Importance of Validating an Instrument

Whether making a yardstick in a basement workshop or exploring duplex Doppler ultrasonography in a biomedical engineering laboratory, all assessment instruments must be compared (validated) to an established "gold standard." Otherwise, the yardstick may appear to measure 36 inches while actually extending to 34. Therefore, the yardstick must be validated against a known 36-inch measure, and the duplex Doppler must be validated against known normal and abnormal physiology. However, defining the gold standard in the life sciences is often challenging. For example, in medicine, knowing for certain which individual has an atherosclerotic cavernosal artery and which has a perfectly normal artery can, at times, only be determined by postmortem examination. Nevertheless, every new assessment device must be evaluated against the best available standard(s) to determine how accurately the device approximates the reality that it is purported to measure. Validity, therefore, is the degree to which an assessment corresponds with the true state of the phenomenon (1). Another term for validity is test instrument accuracy (Fig. 1a). Validity should not be confused with precision (reliability), which is related to the ability of a test to perform with limited measurement variability. A highly precise instrument may be making invalid (irrelevant or inappropriate) measurements.

Given that no assessment instrument is perfectly accurate under all conditions or in all applications, sensitivity and specificity are useful descriptors of instrument validity. Sensitivity represents the relative frequency with which the instrument correctly identifies the true event. For example, if a discrete pulse identification instrument correctly identifies 80 of 100 luteinizing hormone (LH) secretory pulses, then the instrument is said to have a sensitivity of 80%. Specificity, on the other hand represents the relative frequency with which an instrument correctly identifies the particular event rather than unrelated events. If a discrete pulse identification instrument identifies 100 interpulse nadirs, and 80 of these are indeed interpulse nadirs,

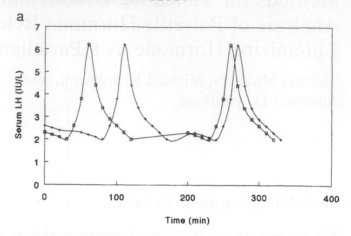

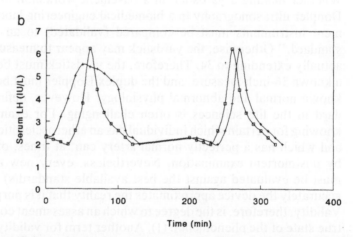

FIG. 1 (a) Schematized comparisons of the positions and shapes of known serum LH concentration peaks (□) versus derived (predicted) LH peaks (+) using deconvolution analysis. Note that the two peaks on the left-hand side fail to coincide exactly in location (low validity of timing estimates), despite similar shapes (high validity of waveform estimates). In contrast, the two peaks on the right-hand side (known and derived) coincide to a reasonable degree in both shape and timing (higher validity of both waveform and timing estimates). (b) Illustrative comparisons of putatively known serum LH concentration peaks (□) versus derived (predicted) LH peaks (+) using deconvolution analysis. Note that the two peaks at left are not congruent in shape; the deconvolution-derived peak is prolonged inordinately compared to the known peak (low reliability of waveform recovery). In contrast, the two peaks at right (known and derived) are similar in both shape and timing (high validity).

then the instrument has a specificity of 80%. For an instrument to be of substantial value, it should have high sensitivity and specificity, as well as known sensitivity and specificity under the various conditions of its expected use. The latter consideration is important, because test sensitivity and specificity may vary significantly when the application context is altered. Of paramount importance, therefore, is the fact that an analytical tool is of unknown value until its validity is established (2).

Reliability, on the other hand, represents the ability of an instrument to consistently arrive at the same assessment (Fig. 1b). For example, if the instrument is applied twice to the same data, it should arrive at the same, or closely similar, result both times (test–retest reliability). Additionally, if an instrument is applied to the same data by two different operators, it should also arrive at the same, or closely similar, results for both applications (interoperator reliability). Reliability, therefore, adds to validity as a helpful descriptor of an assessment instrument. The best instrument should be both valid (accurately measuring the phenomenon) and reliable (consistently measuring the phenomenon). The reliability of deconvolution analysis is discussed in Chapter 3, this volume.

Deconvolution Analysis of Hormone Time Series

Deconvolution analysis as applied in neuroendocrine research represents an important mathematical technique for assessing hormone secretion. In essence, deconvolution techniques in the physical sciences are designed to quantitate the behavior and magnitude of two (or more) component processes contributing to an observed outcome. For example, in seismology, an earthquake felt at a site 200 miles away from the epicenter of the disturbance is a function of the initial disturbance and the parameters of dissipation en route to the distant site. The initial disturbance and the dissipation parameters are said to be convolved. Mathematically disentangling (deconvolving) and estimating the values of these parameters is the basis of deconvolution analysis.

Similar to the seismology example, a serum LH concentration obtained at any given time from a peripheral blood sample is a function of prior gonadotrope secretion of the hormone, as modified by its metabolic clearance up to the time the blood sample is obtained. Importantly, if one continuously monitors serum LH levels *in vivo*, episodic peaks and valleys are observed, suggesting multiple secretory events at a distant site (Fig. 2). In its usual application, deconvolution analysis of endocrine time series attempts to reconstruct the parameters of the secretory events that gave rise to the fluctuations in serum hormone levels, assuming known clearance rates (3).

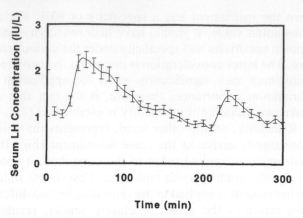

FIG. 2   Serial plasma LH levels derived from a healthy young man. Note that LH levels vary with time and suggest pulsatile LH secretory events at a distant site associated with metabolic clearance.

Some deconvolution methods are designed to calculate hormone secretory rates and metabolic clearance rates simultaneously (4). Here, we illustrate testing validity of multiparameter deconvolution analysis, which is directed at estimating both secretion and removal rates. These principles will apply to other methodologies as well.

Operationally, multiparameter deconvolution analysis relates two or more processes (e.g., hormone secretion and clearance) to the outcome (serum LH concentration) by estimating the individual algebraic descriptors of the processes, which are mathematically related as a dot product. The integral of the dot product of the secretion and clearance functions (convolution integral) gives the net output of the system at any given time. For example, serial serum LH concentrations are the result of prior LH secretory impulses, all antecedent LH clearance, and relevant experimental uncertainty (e.g., preassay and intraassay random variations). The convolution integral relates these contributing processes to the observed LH concentration at any instant in time by way of the following formula:

$$C(t) = \int_0^t S(z)E(t-z)\,dz + \varepsilon.$$

In the equation, $C(t)$ is the hormone concentration at time t, $S(z)$ is the secretion function, $E(t-z)$ is the elimination function, and ε is the experimental uncertainty. Various secretion and elimination functions can be proposed (5). When using a waveform-defined deconvolution technique, such as

Gaussian secretory bursts and exponential elimination kinetics, the above formula can be expanded to a family of convolution integrals, which are solved simultaneously in relation to all of the available $C(t)$ hormone concentrations, so as to provide estimates of the number and locations of statistically significant hormone secretory bursts, the amplitude and mass of each burst, the half-duration (duration of the secretory pulse at half-maximal amplitude), and the half-life of hormone removal. In brief, multiparameter deconvolution analysis is a test instrument intended to estimate secretory burst frequency, amplitude, mass, and duration, as well as hormone half-life according to a particular *a priori* model of hormone secretion and clearance.

Validating Deconvolution Techniques

As mentioned previously, all assessment instruments must be validated, which may constitute a considerable challenge. Methods for detecting peaks in serum hormone concentrations have been subjected to validation (6), but very limited validating data exist for deconvolution methods. Validating deconvolution analysis of endocrine time series requires that the technique be applied to various relevant data scenarios (e.g., data series in which secretory events are independently known to be absent or present, as well as data containing confirmed secretory events of varying amplitude, duration, shape, and frequency). Results of the analysis are then compared to the known data to determine the accuracy of the technique. Importantly, the investigator applying deconvolution analysis must be blinded to all identifying information other than the serum hormone concentrations, so as to avoid bias.

In many experimental contexts, independent *in vivo* assessment of the mere presence or absence of a secretory event can be difficult. Ideally, the gold standard for LH pulses should be direct monitoring of the entire pituitary gonadotrope population to determine the exact time course and quantity of LH release. Of course, this standard is rarely if ever attainable in experimental contexts of interest. Despite the above limitations, for validation purposes we have accepted independent approximation of the timing of the LH release *in vivo* as determined in at least three ways. First, we have applied data from Knobil and co-workers (7), who studied rhesus monkeys by electrophysiological monitoring of mediobasal hypothalamic multiunit neuronal activity, which allows for precise presumptive timing of gonadotropin-releasing hormone (GnRH) release; LH release in peripheral blood occurs with very high reliability a few minutes later (7). Second, we have utilized data from Clarke and Cummins (8), who implanted catheters in the hypothalamo–pituitary–venous plexus of ewes in order to identify the timing of GnRH release, with the finding that LH release into jugular blood follows shortly thereafter with high uniformity (8). Third, multiple investigators have injected GnRH intravenously (i.v.) into subjects with Kallmann's syndrome (isolated hypothalamic GnRH deficiency), which provides a human model that can be virtually devoid of spontaneous pulsatile LH release. Exogenous GnRH stimulation then allows for known and controlled timing of LH secretory bursts (see Fig. 3) (9). Although imperfect, these three *in vivo* paradigms are among the best available biological approximations for independently timing *in vivo* LH secretory events.

Other possible paradigms for creating known pulses of LH with which to test the pulse detection components of deconvolution analysis would include suppressing gonadotropin secretion via GnRH analogs (e.g., leuprolide) and then injecting the subject with recombinant LH at timed intervals. Alternatively, but less rigorously, the investigator could compare the pulses of one hormone with those of a coupled event (e.g., LH and α subunit or testosterone pulses). However, uniform coupling must be expected and demonstrated, and independent confirmation of the occurrence of the companion event must be available (e.g., LH and α subunit must be consistently copulsatile, and α subunit pulses must be independently identified correctly).

In addition to *in vivo* assessment is the use of computer-simulated endocrine time series, which is based on a relevant biophysical model. Computer-simulated data sets can be created by using the previously noted convolution integral (see equation above), which can be integrated numerically via an adaptive, nine-point, Newton–Cotes numerical quadrature integration procedure or can be integrated algebraically in the case of a Gaussian secretory burst model with exponential elimination kinetics (10). With this technique, the investigator can create an unlimited number of simulated data sets each with a defined mean interpulse interval, secretory pulse duration, mass, and amplitude, and hormone half-life. Each of the parameters of interest can be made to vary about a particular mean with an *a priori* variance according to a desired probability distribution, including the normal distribution as defined by a particular mean and stated standard deviation (*SD*), and hence, a corresponding coefficient of variation (11). In some circumstances, other distributions, such as a Poisson probability function in which the mean equals the *SD* in presumed Bernoulli processes, may be appropriate.

Specific Testing of Multiparameter Deconvolution Analysis

To test the sensitivity and specificity of deconvolution analysis when applied to distinct LH secretory events, an investigator should be blinded to the primary indicators of LH secretory bursts (e.g., time of GnRH injection). He/

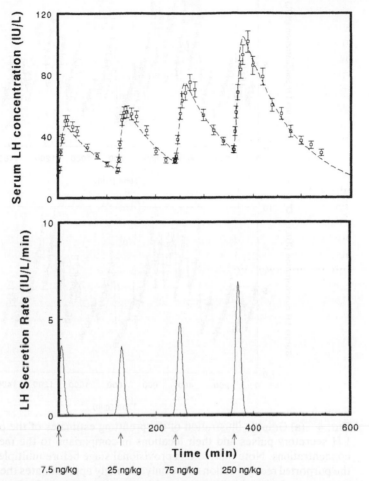

FIG. 3   Graphic example of the results of deconvolution analysis of human serum LH levels obtained serially over 10 hr. Data are serial immunoreactive LH concentrations (W. F. Crowley, Jr., 1990, unpublished) obtained by sampling a patient with GnRH deficiency at frequent intervals before and during i.v. injection of four different doses of GnRH. The top graph demonstrates the sample LH concentrations (mean ± intrasample *SD*) and the deconvolution-derived fitted curve. The bottom graph depicts the location, amplitude, and duration of LH secretory events as assessed by multiparameter deconvolution analysis. Arrows and values denote the times and doses of injected GnRH.

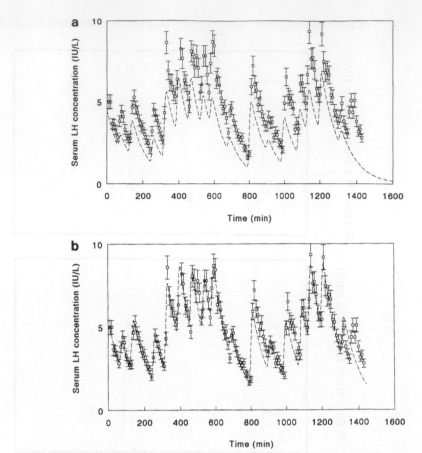

FIG. 4   (a) Graphic illustration of the prefitting estimates of the putative number of LH secretory pulses and their locations in comparison to the measured serum LH concentrations. Note that, at this provisional stage before multiple parameter fitting, the purported reconvolution curve only sparingly approximates the LH concentration time series. (b) Graphic illustration of the putative number of pulses and their locations in comparison to the serum LH concentrations after the first fitting step, fitting only the T_0 concentration and the amplitude of the secretory bursts. Note that the derived secretory peaks now more closely approximate the LH concentration time series, but the position of the peaks is not yet accurate. (c) Graphic illustration of the putative number of pulses and their locations in comparison to the serum LH concentrations after fitting both amplitudes and locations of secretory bursts. Note that the derived secretory peaks now more closely approximate the LH concentration time series, with the exception of the missed probable peak at 1400 min. Indeed, the consecutive positive residuals did require inclusion of a peak here. (d) Graphic illustration of the putative number of pulses and their locations in comparison to the serum LH concentrations after fitting all parameters. Note that the derived secretory peaks closely approximate the LH concentration time series.

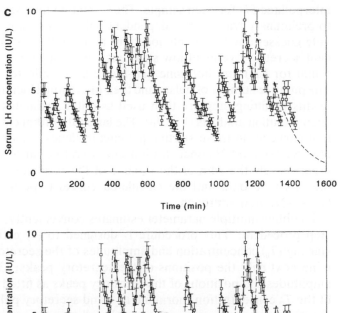

Fig. 4 *(continued)*

she would then apply deconvolution analysis to the serum LH concentration profiles, using a standardized approach for parameter estimation, including the calculation of statistical confidence limits (12). Pertinent objectives would be to determine the presence of individual LH secretory bursts and to estimate the amplitude and duration characteristics of the secretory events. For the multiparameter algorithm (4), we use the following fitting procedure described broadly below.

First, all duplicate or triplicate sample LH concentration measurements obtained at each time of blood withdrawal are used in each time series to estimate dose-dependent intrasample variance as a power function of the sample mean. This procedure provides a mean serum LH value and an interpolated *SD* for each sample. The formatted time series is then submitted to preliminary waveform-independent pulse estimation [PULSE (3)] to provide guesses about possible locations and amplitudes of statistically significant secretory events at low stringency assuming a nominal burst *SD* (e.g., 5 min for LH) and hormone half-life (e.g., 60 min for LH in the human). The putative number of pulses and their provisional locations (times) obtained in this prefitting step are then used as initial estimates for multiparameter deconvolution analysis (Fig. 4a). The latter algorithm calculates a reconvolution curve using these starting parameter values, which is compared to the observed serum LH concentration data. Iterative curve fitting so as to reduce progressively the sum of the squares of the residuals (differences between the observed data and the calculated curve) is performed by a modified Gauss–Newton approach.

To obtain multiple parameter estimates conveniently, we often use a four-step procedure. This procedure is designed to (1) estimate first only the starting (T_0) concentration and amplitudes of the secretory peaks; (2) determine next only the positions of the secretory peaks; (3) evaluate both the amplitudes and positions of the secretory peaks as fitted parameters; and (4) fit the T_0 concentration, hormone $t_{1/2}$, and secretory pulse amplitudes, half-duration, and positions simultaneously (all parameters). Although, in principle, parameters may be fit individually, and then in small and larger groups, the foregoing is a practicable and efficient order for estimating multiple parameters. For example, when deconvolving an LH time series, we first allow only the T_0 concentration and amplitudes of the secretory peaks to vary while holding all other parameters (e.g., peak position, duration, and half-life estimates) fixed (Fig. 4b). We then fit the peak positions only and, thereafter, both the amplitudes and positions together (Fig. 4c). Notably, during the fourth step, all parameters are fit simultaneously (Fig. 4d).

At any time before, and always after the four steps have been completed, or after any indeterminate parameter value is observed, we plot the raw data and the fitted curve. We plot the weighted or unweighted residuals above the fitted curve, or the latter can be simply inspected for consecutive positive residuals. The data analyst inserts a provisional new peak wherever *n* (e.g., 2 for 10-min data) or more consecutive positive residuals (difference between fitted curve and observed data) are observed. This corrected set of parameter estimates is then reevaluated using the same four-step fitting procedure described above. Throughout the fitting procedure, tentative hormone pulses whose amplitudes or locations are mathematically indeterminate [i.e., lower or upper bounds could not be distinguished from -10^{38} or $+10^{38}$ (approaching negative and positive infinity) respectively], are deleted individually and the data refit (steps 1–4 above).

The foregoing tentative "best fit" of the data, constrained only by determinable peak amplitudes and positions and a common half-duration and half-life, is then used in the calculation of 95% confidence limits. Although various levels of statistical stringency can be applied, for 10-min LH data we recommend that the amplitudes of all resolved LH secretory bursts have nonzero lower bounds when joint statistical confidence limits are calculated for the secretory burst half-duration, hormone $t_{1/2}$, and all secretory burst amplitudes simultaneously. Optimal stringency depends not only on the sampling intensity, but also on the data characteristics (e.g., hormone half-life, pulse shape, duration, frequency and amplitude, and noise). Secretory pulses not conforming to a given stringency test are deleted one at a time beginning with the smallest or least determinate value. Parameters are then reestimated by iterative fitting (as outlined above). After completing this strategy, the investigator can describe the location and amplitude of the LH secretory pulses, the duration of the secretory events (and hence the mass), and the $t_{1/2}$ of LH for the individual series under study.

By comparing the deconvolution-derived burst positions with the known (true-positive) burst positions, the investigator can classify derived bursts as either true positives (derived burst position occurs at the same time as the known burst within $\Delta t/2$, where Δt is the width of the sampling interval) or false positives (inferred burst center removed from known position by more than $\Delta t/2$ (Fig. 5). By comparing the deconvolution-derived valleys (the interval between consecutive bursts) to the known valleys, the investigator can classify derived valleys to be either true negatives (an inferred valley containing no true-positive peaks) or false negatives (a derived valley that

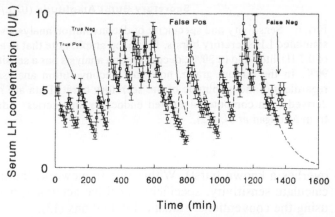

FIG. 5   Illustrative endocrine time series (computer-simulated 24-hr serum LH profile) demonstrating true-positive, true-negative, false-positive, and false-negative peaks and valleys.

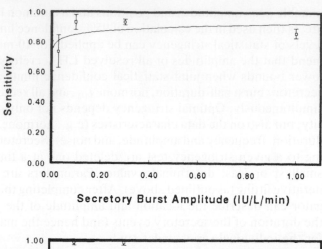

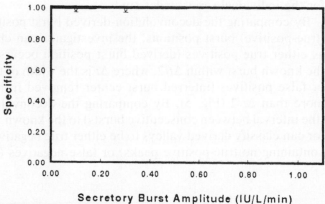

FIG. 6 Sensitivity and specificity of deconvolution analysis versus the amplitude of simulated LH secretory bursts. In the top graph, note that at a mean burst amplitude of 0.1 IU/liter/min (±30%), deconvolution analysis has a sensitivity of approximately 90%. In the bottom graph, note that deconvolution analysis has high specificity regardless of the amplitude of the secretory burst. Data are means ± SEM (*n* = 3) derived from computer-simulated endocrine time series. Adapted with permission from Mulligan *et al.* (12).

overlies a known burst). Within this framework, the investigator can then calculate sensitivity, specificity, positive accuracy, and negative accuracy using the conventional statistical definitions (13).

Sensitivity = true positives/(true positives + false negatives)
Specificity = true negatives/(true negatives + false positives)

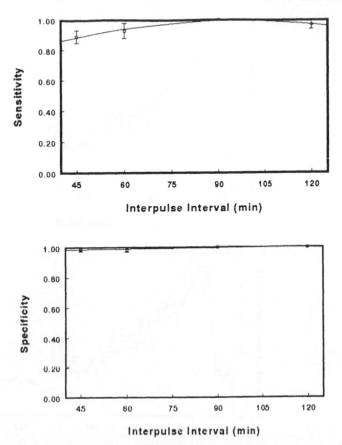

FIG. 7 Sensitivity and specificity of deconvolution analysis versus the interpulse interval of simulated LH secretory bursts. Note in the top graph that at a mean interpulse interval of 60 min (±30%), deconvolution analysis retains a sensitivity approximating 90%. In the bottom graph, note that specificity remains nearly 100% regardless of the interpulse interval. Data are means ± SEM ($n = 3$) derived from computer-simulated endocrine time series. Adapted with permission from Mulligan *et al.* (12).

Positive accuracy = true positives/(true positives + false positives)
Negative accuracy = true negatives/(true negatives + false negatives).

Sensitivity and specificity were defined intuitively earlier in this chapter. In brief, sensitivity denotes the expectation of identifying all true positives, ideally, while limiting the number of false negatives, whereas specificity is

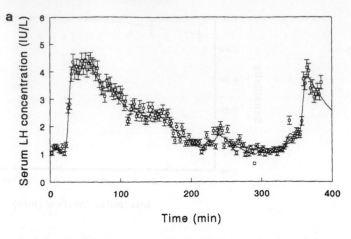

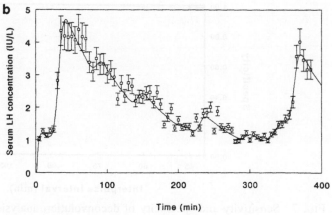

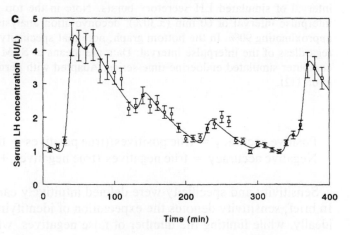

FIG. 7.   Sensitivity and specificity of deconvolution analysis versus the interpulse interval of simulated LH secretory bursts. Data in the top graph that is a mean interpulse interval of not exceeding access to longer relative sensitivity exceeding 90%. In the bottom graph, absolute specificity remains nearly 100% regardless of the interpulse interval. The detected series (*n* = 3) derived from different simulated endogenous time ... analyzed with permission from McMillan [1].

concerned with identifying all true negatives, if possible, while limiting the number of false positives. Positive and negative accuracy, respectively, specify the probabilities that an identified peak and valley are truly peaks and valleys. These last two terms define the predictive accuracy of the analytical method.

Multiple factors influence the sensitivity and specificity of deconvolution analysis. We recommend that for validation testing, (e.g., of human LH series), the investigator explore the influence of at least each of the following factors on test sensitivity and specificity: (1) random experimental variation (e.g., 3–20%), (2) secretory burst amplitude (0.03–1.0 IU/liter/min), (3) burst frequency (interpulse interval 45–120 min), (4) hormone half-life (35–125 min), and (5) sampling interval (every 2.5–30 min). If possible, the influence of varying burst shape and/or duration should also be appraised. For example, low-amplitude secretory events are less likely to be perceived as real secretory episodes (Fig. 6). As is intuitively obvious, the larger the secretory signal, the easier it is for any detector to perceive an event. At the nominal amplitude of 0.3 IU/liter/min, multiparameter deconvolution analysis has a sensitivity approximating 90% for the particular LH half-life, mean burst frequency, and sampling interval (10 min) simulated here.

In addition, a long secretory pulse duration or high frequency of secretory events *ceterus paribus* (other things being equal) impairs sensitivity. This reduced ability to accurately discern secretory events is caused in part by the obligate superimposition of the onset of one secretory peak on the trailing end of its predecessor. For the nominal conditions of analysis chosen, we observed that even with hourly secretory events, deconvolution analysis retains a sensitivity of 90%. The sensitivity of deconvolution analysis for 10-min sampling dips below 90% at an interpulse interval of 45 min (Fig. 7). Higher sensitivity despite higher pulse frequency would require greater

FIG. 8   (a) Illustrative profile of serum LH concentrations and the reconvolution curve obtained from a healthy young man whose blood was sampled every 2.5 min while he slept at night. Note that with intensive blood sampling (every 2.5 min), deconvolution analysis reports 11 secretory bursts (T. Mulligan, A. Iranmanesh, and J. D. Veldhuis, 1992, unpublished data). (b) Separate illustrative analysis of only the samples obtained at 5-min intervals from (a). Note that with less intensive sampling (every 5 min), deconvolution analysis estimates that there are now 9 secretory bursts; 2 bursts were missed (false negatives). (c) Subanalysis of only the samples obtained at 10-min intervals from (a). Note that with less intensive sampling (every 10 min), deconvolution analysis identifies 7 secretory bursts, and 4 bursts were missed (false negatives).

sampling intensity (e.g., every 2.5 or 5 min), a shorter hormone half-life (e.g., α subunit rather than LH), and/or reduced noise in the data series.

Importantly, there are two factors which are under the control of the investigator, and which he/she can use to maximize both sensitivity and specificity, namely, sampling intensity and random experimental variation. For example, consider an experiment in which a subject has blood sampled every 2.5 min, the time series is analyzed using deconvolution analysis, and the LH release profile is found to have 11 secretory bursts (Fig. 8a). However, when the same time series is analyzed using only the samples obtained at the 5- or 10-min intervals, the LH release profile appears to contain only 9 or 7 secretory bursts, respectively (Fig. 8b,c).

Intuitively, the more often serum LH measurements are obtained without diminishing experimental precision (the more data provided of similar precision to the detector), the more frequently a validated and robust (not ill-conditioned) deconvolution analysis will correctly identify a secretory peak. For example, for the nominal LH simulations examined here, obtaining samples every 30 min would result in an inadequate sensitivity of 60%. Even sampling every 20 min may be inadequate (sensitivity of 80%), given a presumptive hormone half-life as tested here of 60 min. Only sampling at intervals less than or equal to 10 min results in 90% sensitivity at this particular half-life and burst frequency, duration, and amplitude (Fig. 9). Accordingly, we recommend among other considerations (see below) careful assessment of the impact of sampling frequency on sensitivity and specificity of the deconvolution analysis for detecting pulsatile release of any particular hormone. As a provisional guideline, we suggest sampling sufficiently often so as to obtain at least five samples for every half-life (e.g., every 10 min for a 50-min half-life).

In addition to sampling intensity, the precision of the assay contributes to sensitivity of the deconvolution. For example, a highly precise assay such as an immunofluorometric technique (Fig. 10) might provide a reduced intraassay coefficient of variation and thereby maximize sensitivity. Importantly, assays with a high intraassay coefficient of variation have a detrimental effect not only on sensitivity but also on specificity, especially in the presence of reduced secretory burst amplitude. For example, a very high (30%) intraassay coefficient of variation can result in a sensitivity of 70% and specificity of less than 90% under the current nominal simulation conditions (Fig. 11).

Finally, for multiparameter deconvolution to be valid, it must not only correctly identify the presence and absence of secretory events but also recover good estimates, when desirable, of both the hormone half-life and production rate. In our experiments, using computer-simulated LH time series with a wide spectrum of hormone half-lives (Fig. 12) and production

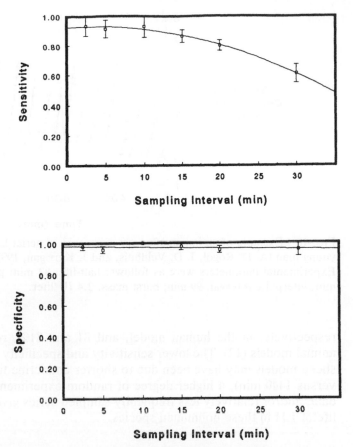

FIG. 9 Sensitivity and specificity of deconvolution analysis versus the sampling interval (e.g., every 10 min). For a hormone with a simulated half-life of 65 min, pulse detection sensitivity (top graph) approximates 60% when simulated samples are obtained every 30 min and the true pulse frequency is 90 min. Importantly, specificity remains high regardless of the sampling interval (bottom graph). Data are means ± SEM (n = 3) derived from computer-simulated endocrine time series. Adapted with permission from Mulligan *et al.* (12).

rates (Fig. 13), multiparameter deconvolution analysis accurately estimates both these parameters (r = +0.994 and r = +0.990, respectively).

Importantly, deconvolution analysis has high sensitivity and specificity not only when applied to computer-simulated endocrine time series, but also when applied to the previously mentioned *in vivo* models. We found deconvolution analysis to have a sensitivity and specificity of 91 and 90%,

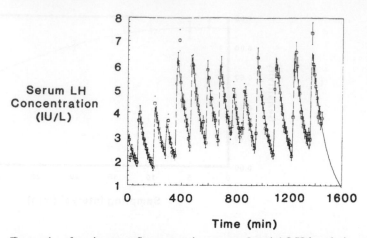

FIG. 10   Example of an immunofluorometric assay of serial LH levels in a healthy young man (A. D. Rogol, J. D. Veldhuis, and J. Kerrigan, 1992, unpublished data.) Experimental parameters were as follows: half-life, 82 min; pulse half-duration, 5 min; interpulse interval, 99 min; burst mass, 3.4 IU/liter.

respectively, in the human model, and 81 and 81%, respectively, in the animal models (12). The lower sensitivity and specificity in the monkey and sheep models may have been due to shorter endocrine time series (300 min versus 1440 min), a higher degree of random experimental variation in the data, and/or relatively less frequently sampled series given the shorter half-life of LH in these nonhuman species.

Summary

Deconvolution analysis is an important analytical tool which can provide valuable information regarding pulsatile hormone secretion. However, as with all analytical tools, it must be properly validated for the application under consideration. To maximize sensitivity and specificity, the investigator should obtain representative samples as frequently and as for as long a duration as practical, use a precise and specific assay, and apply a validated algorithm with statistical confidence limits chosen *a priori*. In addition, the investigator must be aware of the limitations inherent in deconvolution analysis when the amplitude of the secretory pulse is small, the interpulse interval is short, the sampling frequency is low, the amount of experimental uncertainty (noise) is large, and/or the assumptions of the model (3) are violated.

We emphasize that no analytical tool can perform in a valid manner under

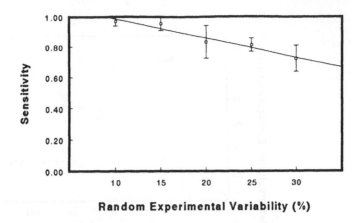

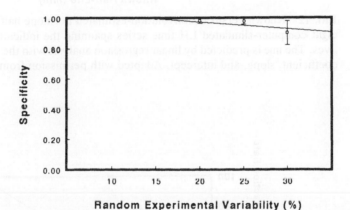

Fig. 11   Sensitivity and specificity of deconvolution analysis versus simulated random experimental variability (noise). Noise here is represented as intrasample coefficients of variations for Gaussian-distributed experimental uncertainty. Data are means ± SEM ($n = 3$). In the top graph (sensitivity), note that sensitivity approaches 80% as noise approximates 20%. In the bottom graph, specificity is hindered as noise approaches 30%. Adapted with permission from Mulligan *et al.* (12).

conditions that violate the assumptions that underlie it. For example, a pure burst model of hormone secretion cannot be expected to estimate pulse amplitude and mass accurately if the data to be analyzed consist of predominantly nonpulsatile hormone release. Thus, the model must be shown to be relevant to the data. The data must be collected optimally (e.g., at adequate sampling frequency) and derived by exemplary measurement techniques (high assay precision, sensitivity, and specificity). Finally, optimal analytical

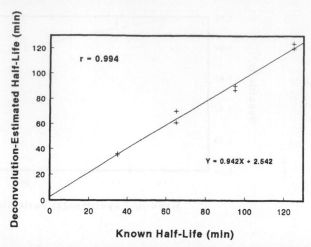

FIG. 12   Known versus deconvolution-estimated hormone half-lives. Data are from eight computer-simulated LH time series spanning the indicated range of LH half-lives. The line is predicted by linear regression analysis with the indicated correlation coefficient, slope, and intercept. Adapted with permission from Mulligan *et al.* (12).

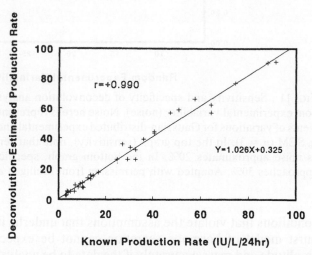

FIG. 13   Known versus deconvolution-estimated daily hormone production rates. Each data value derives from an LH production rate estimate for a single 24-hr computer-simulated LH time series. A total of 35 series were studied. The line is predicted by linear regression analysis, which yielded the indicated correlation coefficient, slope, and intercept.

parameters should be documented as illustrated here to permit high test performance sensitivity and specificity as well as accurate recovery of estimated half-life and production/secretion rates.

Acknowledgments

The authors thank Drs. William F. Crowley for the human GnRH-stimulation series, Iain Clarke for the sheep data, and Ernst Knobil for the monkey data. This work was supported in part by the U.S. Department of Veterans Affairs (T.M.), NIH RCDA 1 KO4 HD00634 from NICHHD (J.D.V.), the National Science Foundation Science and Technology Center for Biological Timing (J.D.V., M.L.J.), and the Diabetes and Endocrine Research Center NIDDK D.K.-38942 (J.D.V., M.L.J.).

References

1. R. H. Fletcher, S. W. Fletcher, and E. H. Wagner, "Clinical Epidemiology" Williams & Wilkins, Baltimore, Maryland, 1988.
2. S. B. Hulley and S. R. Cummings, Planning the measurements: Precision and accuracy. in "Designing Clinical Research." Williams & Wilkins, Baltimore, Maryland, 1988.
3. J. D. Veldhuis and M. L. Johnson, in "Methods in Enzymology" (L. Brand and M. L. Johnson, eds.), Vol. 210, p. 539. Academic Press, San Diego, 1992.
4. J. D. Veldhuis, M. L. Carlson, and M. L. Johnson, *Proc. Natl. Acad. Sci. U.S.A.* **84,** 7686 (1987).
5. J. D. Veldhuis, W. S. Evans, J. P. Butler, and M. L. Johnson, *Methods Neurosci.* **10,** 241 (1992).
6. R. J. Urban, W. S. Evans, A. D. Rogol, D. L. Kaiser, M. L. Johnson, and J. D. Veldhuis, *Endocr. Rev.* **9,** 3 (1988).
7. R. C. Wilson, J. S. Kesner, K. M. Kaufman, T. Uemura, T. Akema, and E. Knobil, *Neuroendocrinology* **39,** 256 (1984).
8. I. J. Clarke and J. T. Cummins, *Endocrinology (Baltimore)* **111,** 1737 (1982).
9. N. Santoro, M. Filicori, and W. F. Crowley, *Endocr. Rev.* **7,** 11 (1986).
10. J. D. Veldhuis, A. E. Lassiter, and M. L. Johnson, *Am. J. Physiol.* **259,** E351 (1990).
11. J. D. Veldhuis and M. L. Johnson, *Am. J. Physiol.* **255,** E749 (1988).
12. T. Mulligan, H. A. Delemarre-Van de Waal, M. L. Johnson, and J. D. Veldhuis, *Am. J. Physiol.* **267,** R202 (1994).
13. B. M. Bennett, *Biometrics* **28,** 793 (1972).

[5] Complicating Effects of Highly Correlated Model Variables on Nonlinear Least-Squares Estimates of Unique Parameter Values and Their Statistical Confidence Intervals: Estimating Basal Secretion and Neurohormone Half-Life by Deconvolution Analysis

Johannes D. Veldhuis, William S. Evans, and Michael L. Johnson

Introduction

Neurohormone axes typify complex feedback systems in which two or more model variables (e.g., secretion and clearance) are required to account for the nonlinear behavior of the system output (neurohormone concentrations over time). For example, episodic fluctuations in blood concentrations of a neurohormone are controlled jointly by neurohormone secretory event frequency, amplitude, duration, and waveform, as well as by neurohormone disappearance rates from the blood (1, 2). In addition, a variable admixture of basal and pulsatile neurohormone release may further determine circulating effector concentrations (3–5). A major complicating factor is the extent to which model parameters (e.g., secretory burst frequency, basal secretion rate, neurohormone half-life, and secretory event amplitude, mass, and/or duration) are highly correlated, as such parameter correlations make it difficult to determine unique model values to characterize any particular set of observed data. For example, a quantitative model of admixed pulsatile and basal neurohormone secretion and removal may require simultaneous estimation of both basal and pulsatile neurohormone secretory rates with or without concomitant estimates of neurohormone half-life (see Chapter 2, this volume). However, in relation to any particular data set, any given estimate of the basal secretory rate has a strong statistical dependency on the half-life estimate, and vice versa. Such parameter correlations are expected to challenge nonlinear least-squares methods of parameter estimation in at least two respects: (a) the determination of unique solutions to multiparameter model estimates and (b) the valid estimation of statistical confidence intervals

Methods in Neurosciences, Volume 28

that define the precision of parameter estimates whether considered alone or jointly (6–9).

Here, we consider briefly the implications of high parameter correlations by evaluating episodic neurohormone release for two typical luteinizing hormone (LH) time series, in which both basal and pulsatile hormone release are assumed to coexist and the hormone half-life is unknown. We show by systematic parameter grid searches consisting of approximately 1250 combinations of half-life and basal secretion rates that any given estimated value of the basal secretion rate is associated with a substantial range of plausible LH half-lives, and vice versa. Moreover, by estimating statistical confidence intervals for the parameter values rigorously, compared to a conventional asymptotic standard error estimation method, we show that asymptotic standard errors are inappropriate descriptors of the asymmetric and multidimensional confidence interval contours generated by this nonlinear-estimation problem.

Statement of the Problem

As a paradigm of multiparameter nonlinear least-squares estimation, we can evaluate a model of putatively combined basal and pulsatile LH release in men and women (3–5). Suppose that we implement a model of randomly dispersed Gaussian (or skewed) neurohormone secretory bursts (1) superimposed on basal time-invariant hormone secretion (2; see also Chapter 2, this volume) associated with monoexponential LH disappearance kinetics. The neurohormone concentration in plasma is then determined at any given instant by three principal components in addition to experimental uncertainty: (i) basal secretion rate, (ii) pulsatile hormone profile, and (iii) the hormone half-life. The pulsatile hormone secretory profile in turn is determined by pulse event number, waveform, duration, and amplitude (the last three model-dependent variables control the final pulse mass, or integral of the pulsatile secretory event over time). Assume further that the pulse shape (e.g., Gaussian), duration [e.g., as defined by a standard deviation (SD) or half-duration], and number are all fixed by independent or prior knowledge, so that the set of interdependent/highly correlated variables is reduced for illustrative purposes to secretory pulse positions (timing), secretory pulse amplitudes (maximal secretory rate attained within an event), basal secretion rate (a time-invariant value, i.e., a constant), and hormone half-life (minutes, or ln 2/disappearance rate constant).

The simplified problem then is to determine the LH half-life and basal LH secretory rate, and their statistical confidence intervals from typical serum

LH time series; see Figs. 1 and 2, which show 24-hr profiles of serum LH concentrations obtained by 10-min blood sampling in a normal young woman and man (10, 11). In these LH pulse profiles, we find that the correlation coefficients between the basal LH secretory rate and LH half-life are −0.983 (Fig. 1, midluteal phase LH release) and −0.995 (Fig. 2, male pattern of pulsatile LH release).

Correlation is not a biological or biochemical consequence. It is a consequence of the mathematical form of the models and the experimental data. Important aspects of the experimented data include the elimination half-life in relation to the pulse frequency and the level of experimental uncertainty in relation to the burst mass.

Analysis

Given that LH secretory burst duration, frequency, and waveform are fixed, our task is to obtain parameter estimates for basal LH secretion and LH half-life. To this end, we evaluated a large range of combinations of basal

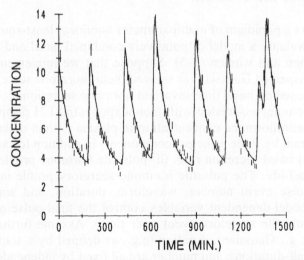

FIG. 1 Multiparameter deconvolution fit of a 24-hr serum LH profile obtained by sampling a young woman at 10-min intervals during the midluteal phase of a normal menstrual cycle. Data are mean sample serum immunoreactive LH concentrations with dose-dependent intrasample standard deviations estimated by a power function fit of the intrasample variance (square of duplicate *SD*) versus sample LH mean. The continuous curve represents a nonlinear least-squares fit of the observed data using a convolution model (DECONV) comprising admixed basal and pulsatile LH release (2).

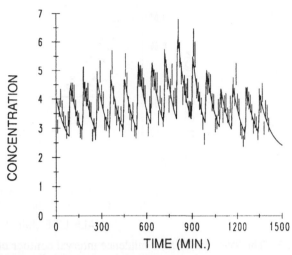

FIG. 2   Profile of fitted serum LH concentrations as in Fig. 1 except that data were derived from a normal man.

secretion and half-life estimates and generated 95% confidence interval contours defining the paired parameter values. Specifically, the basal secretion versus half-life confidence contours (Figs. 3 and 4) were created by a grid search method applied in conjunction with multiparameter deconvolution analysis, namely, the DECONV algorithm (1, 2; see also Veldhuis and Johnson, Chapter 2, this volume). Our construction of the contours required that all pairs of half-life and basal secretion estimates within the contour fall at or within the 95% confidence level. To produce the contours, the half-life axis and the basal secretion axis were each divided into grids of approximately 35 equally spaced intervals. For each of the approximately 1250 combinations of half-life and basal secretion, the DECONV algorithm was used iteratively to evaluate the remaining secretion pattern parameters (e.g., secretory pulse positions and amplitudes), and the variance-of-fit was recorded. A cubic-spline interpolation of this grid was subsequently used to delineate the points contained within the 95% (±2 SEM) statistical confidence interval. For comparison, conventionally used asymptotic standard errors were also calculated.

Many authors utilize asymptotic standard errors to provide an estimate of the precision of parameters determined by nonlinear least-squares procedures. The application of asymptotic standard errors makes a number of implicit assumptions (6–9), to wit that (1) the fitting function is linear, (2) there is a nearly infinite number of data points, and (3) there is no correlation between the parameters being estimated. If these assumptions are met, then the confidence contours shown in Figs. 3 and 4 should be elliptical in shape

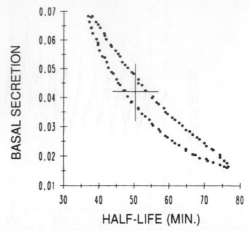

FIG. 3   The 95% (±2 SEM) confidence interval contour of estimates of LH half-life and basal secretion measures for the single pulsatile LH data set shown in Fig. 1 as estimated by deconvolution analysis using the DECONV algorithm. For this analysis, the grid search embodied 31 half-life intervals and 33 basal-secretion intervals. At each point in the grid search, the amplitudes and positions of seven putative secretion events were estimated by the DECONV algorithm. The secretion event standard deviation was fixed at 4.6 min (which corresponds to a half-duration of 11 min), and the number of secretory events was held constant at seven. The + symbol corresponds to the basal secretion and half-life values, where the size of the + symbol corresponds to the 95% confidence region predicted by the more conventional asymptotic standard error method for the evaluation of the precision of parameters estimated by linear least-squares techniques. The correlation coefficient between the basal secretion and the half-life parameters as determined by the least-squares algorithm was −0.983.

and align with the axes. Clearly for the present examples, none of these assumptions is valid or realistic. Indeed, the poor correspondence between directly estimated parameter confidence interval contours and the asymptotic standard errors shown in Figs. 3 and 4 is a clear demonstration of how inappropriate the use of asymptotic standard errors is for this type of nonlinear problem.

Other Complicating Factors

As alluded to in the introduction, when multiple parameters jointly determine model output, their interdependence/correlation will influence both the uniqueness of the nonlinear fit of a data set and the statistical confidence

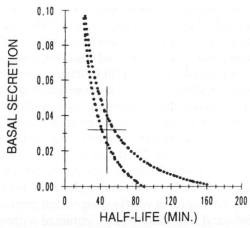

FIG. 4 The 95% (±2 SEM) confidence interval contour of estimates of the half-life and basal secretion rate for the data shown in Fig. 2. In this analysis, the grid search utilized 40 half-life intervals and 40 basal-secretion intervals. At each point in the grid search, the amplitudes of 15 secretion events were estimated by the DECONV algorithm. The number of secretory bursts was held constant at 15, and the secretion event SD was fixed at 1.5 min. The latter corresponds to a secretory burst half-duration of 2.4 min. The + symbol corresponds to the basal secretion and half-life as estimated by the DECONV algorithm, where the size of the + symbol corresponds to the 95% confidence region predicted by the more conventional asymptotic standard error method for the evaluation of the precision of parameters estimated by linear least-squares techniques. The correlation coefficient between the basal secretion and half-life, as determined by the least-squares algorithm, was −0.995.

intervals [properly determined (see Refs. 8 and 9)] of the parameter estimates. These challenges are illustrated for two serum LH concentration time series in Figs. 1 and 2, when a model is demanded to define both half-life and basal secretion. Note that these two-parameter grid searches represent a simplified presentation of our thesis. Indeed, only a single pair of parameters is considered here, whereas a fully realistic representation would require an *n*-dimensional grid space defined by basal secretion rate, hormone half-life, pulse number (and their individual positions in time), and secretory burst mass (determined by amplitude, duration, and waveform). Determining the number of secretory bursts *a priori* or independently for any given data set simplifies this nonlinear fitting problem considerably.

If the assumption of known secretory burst number is combined with an assumed pulse shape (secretory burst waveform) and duration, any given paired half-life and basal secretion-rate estimates still require least-squares fitting estimations of burst amplitudes and positions. Thus, an important methodological consideration wherever possible is the need to obtain independently estimates of the following: hormone half-life; secretory burst number, shape, and/or duration; and basal hormone secretion rates. In addition, whenever possible, it is helpful to have neurohormone data collected under conditions yielding relatively homogeneous (and independently determined) effector half-lives; data containing regions with evidently only basal release, for example, daytime serum GH concentrations (12), which then allow estimates of a basal secretory component and independently estimated hormone pulse number [e.g., by Cluster analysis (13)]; and/or a reasonable assumed secretory burst waveform (and duration), for example, as based on venous sampling proximal to the secretory gland (2). Furthermore, computer simulations may be helpful in verifying algorithm performance (5, 14, 15). Independent biological information (e.g., obtained without nonlinear curve fitting of the experimental data) will help reduce the complexity of the n-dimensional parameter space and obviate the impact of high interparameter correlations. Even so, the influences of multiparameter correlations include broadening and/or distorting the confidence interval contours for final joint parameter estimates, with the result that conventional asymptotic standard-error estimates may be quite inappropriate.

Figure 5 shows the implication of failure to recognize coexistent basal and pulsatile hormone release accurately when fitting half-life parameters. As predicted intuitively, for example, the inappropriate application of excess basal hormone release in a model of combined pulsatile and basal secretion leads to an underestimate of the true half-life value. This illustrates the importance of the foregoing considerations.

Summary

We discuss the methodology for evaluating the impact of highly correlated model variables on nonlinear least-squares parameter estimates and statistical confidence intervals. We illustrate that the joint parameter contours for hormone half-life and basal secretion rates (determined for 24-hr serum LH concentration profiles) do not necessarily conform to predictions of asymptotic standard errors. Consequently, nonlinear curve fitting of neuroendocrine data should employ rigorous methods for error propagation (e.g., Monte Carlo perturbations) that do not depend on the assumptions of linearity of the fitting function and the statistical independence of fitted parameters as required for asymptotic error theory. Moreover, whenever possible, experimentally independent estimates of model parameters should be obtained

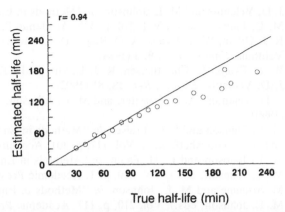

True half-life (min)

FIG. 5   Influence of excessive estimates of basal LH secretion on half-life calculations in a model of combined basal and pulsatile secretion. Data were simulated as described previously (5, 15), except that very low fixed basal secretion was added to the convolution integral so as to achieve a blood LH concentration of 0.25 IU/liter and allow admixed pulsatile and basal LH release. A range of LH half-lives was imposed in the simulations (x-axis values, true half-life). The solid line shows the line of identity. Circles mark the calculated half-lives when basal secretion estimates are artificially (erroneously) set to twice the true values. Thus, half-life under-estimation occurs when an incorrect secretion model of excess basal release is applied (J. D. Veldhuis and F. Schaefer, 1994, unpublished).

prior to nonlinear least-squares curve fitting of data arising from highly correlated model variables.

Acknowledgments

We thank Patsy Craig for skillful preparation of the manuscript and Paula P. Azimi for the artwork. This work was supported in part by National Institutes of Health Grant RR 00847 to the Clinical Research Center of the University of Virginia, NICHD RCDA 1 KO4 HD00634 (J.D.V.), NIH Grants GM-35154 and RR-08119 (M.L.J.), AG05977 (W.S.E.), the Baxter Healthcare Corporation, Roudlake, IL (J.D.V.), the Diabetes and Endocrinology Research Center Grant NIH DK-38942, the NIH-supported Clinfo Data Reduction Systems, the Pratt Foundation, the University of Virginia Academic Enhancement Fund, and the National Science Foundation Center for Biological Timing (NSF Grant DIR89-20162).

References

1. J. D. Veldhuis, M. L. Carlson, and M. L. Johnson, *Proc. Natl. Acad. Sci. U.S.A.* **84,** 7686 (1987).

2. J. D. Veldhuis and M. L. Johnson, *in* "Methods in Enzymology" (L. Brand and M. L. Johnson, eds.), Vol. 210, p. 539. Academic Press, San Diego.

3. R. J. Urban, W. S. Evans, A. D. Rogol, D. L. Kaiser, M. L. Johnson, and J. D. Veldhuis, *Endocr. Rev.* **9**, 3 (1988).

4. W. S. Evans, E. Christiansen, R. J. Urban, A. D. Rogol, M. L. Johnson, and J. D. Veldhuis, *Endocr. Rev.* **13**, 81 (1992).

5. J. D. Veldhuis, A. B. Lassiter, and M. L. Johnson, *Am. J. Physiol.* **259**, E351 (1990).

6. M. L. Johnson and S. G. Fraiser, *in* "Methods in Enzymology" (C. H. W. Hirs and S. N. Timasheff, eds.), Vol. 117, p. 301. Academic Press, San Diego, 1985.

7. M. L. Johnson and L. M. Faunt, *in* "Methods in Enzymology" (L. Brand and M. L. Johnson, eds.), Vol. 210, p. 1. Academic Press, San Diego, 1992.

8. M. Straume and M. L. Johnson, *in* "Methods in Enzymology" (L. Brand and M. L. Johnson, eds.), Vol. 210, p. 117. Academic Press, San Diego, 1992.

9. M. L. Johnson, *in* "Methods in Enzymology" (M. L. Johnson and L. Brand, eds.), Vol. 240, p. 1. Academic Press, San Diego, 1994.

10. M. L. Sollenberger, E. C. Carlson, M. L. Johnson, J. D. Veldhuis, and W. S. Evans, *J. Neuroendocrinol.* **2**, 845 (1990).

11. J. D. Veldhuis, M. L. Johnson, and M. L. Dufau, *Am. J. Physiol.* **256**, E199 (1989).

12. A. Iranmanesh, B. Grisso, and J. D. Veldhuis, *J. Clin. Endocrinol. Metab.* **78**, 526 (1994).

13. J. D. Veldhuis and M. L. Johnson, *Am. J. Physiol.* **250**, E486 (1986).

14. J. D. Veldhuis, J. Moorman, and M. L. Johnson, *Methods Neurosci.* **20**, 279 (1994).

15. T. Mulligan, H. A. Delemarre-van de Waal, M. L. Johnson, and J. D. Veldhuis, *Am. J. Physiol.* **267**, R202 (1994).

[6] Techniques for Assessing Ultradian Rhythms in Hormonal Secretion

Jeppe Sturis and Eve Van Cauter

Introduction

The fact that virtually all hormones are secreted in a pulsatile or oscillatory fashion calls for the use of methods to enable quantification of these temporal variations. Hormonal secretion involves temporal variations in different period ranges, including seasonal, circadian, and ultradian variations, as well as changes that are of the order of a few minutes or less. This chapter focuses on the analysis of ultradian variations in hormonal levels. In its original definition, the term ultradian designated the entire range of periods shorter than circadian (i.e., circadian being 20–28 hr). Currently, the term ultradian is primarily used to refer to periods of fractions of hours to several hours. The term circhoral has been used to designate hormonal variations with periods of about 1 hr.

A careful analysis of ultradian oscillations in peripheral hormonal levels may reveal important characteristics of the physiological mechanisms underlying their generation and help formulate hypotheses regarding their functional significance. For hormones under direct hypothalamic control, such as luteinizing hormone (LH) and adrenocorticotropic hormone (ACTH), secretory oscillations in the ultradian range generally reflect phasic activity of neuronal pulse generators. For other hormones, such as insulin and glucagon, ultradian oscillations may represent an optimal functional status of the local regulatory network. Among important questions that mathematical analysis can help address are the accuracy of period control (i.e., is the generating mechanism a strictly periodic oscillator or a sloppy pacemaker?) and the relative importance of intermittent versus continuous release.

The analysis of ultradian hormonal variations may be considered at two levels. Hormonal measurements obtained in the peripheral circulation can be analyzed directly, thereby defining and characterizing significant variations in blood levels on the basis of estimations of the size of the measurement error, the major source of which is usually assay error. However, when possible, it is often desirable to derive mathematically secretory rates from the peripheral concentrations, a procedure commonly referred to as deconvolution. Whether peripheral concentrations or secretory rates are examined, there are two major approaches to analyzing the episodic fluctuations. The first,

Methods in Neurosciences, Volume 28

and most commonly used, is time domain analysis, where the data are plotted against time and pulses are detected and identified. The second is frequency domain analysis, where amplitude is plotted against frequency or period. These two approaches differ fundamentally both in the mathematical treatment of the data and in the questions they may help to resolve and should therefore be viewed as complementary. In this chapter, we give an overview of a number of methods that in our experience have been and continue to be useful in the quantification of oscillations of *in vivo* hormonal secretion with particular emphasis on rhythms in the ultradian range.

Analyzing Peripheral Concentrations versus Secretory Rates

When the clearance kinetics of the hormone under study are known with good accuracy, secretory rates can be derived from peripheral levels by deconvolution using a mathematical model to remove the effects of hormonal distribution and degradation. This procedure may be based on a one-compartment model (i.e., single exponential decay), or on a two-compartment model (i.e., double exponential decay). Deconvolution will often reveal additional pulses of secretion which were not apparent in the profile of peripheral concentrations. It will also more accurately define the temporal limits of each pulse (i.e., the onset and the offset of secretion) and therefore provide more meaningful comparisons with the dynamics of other simultaneous physiological processes, such as the secretion of another hormone and the transitions between wake and sleep stages.

An example is shown in Fig. 1, which compares the profile of plasma growth hormone (GH) (top) to that of the GH secretory rates (middle) calculated based on a single compartment model for GH clearance and the profile of transitions between wake and sleep stages (bottom). In this study (1), pulsatile GH secretion at sleep onset was stimulated by an injection of growth hormone-releasing hormone (GHRH). The subject experienced a spontaneous awakening within 60 min after sleep onset. The analysis of the profile of plasma concentrations clearly suggests that the awakening was associated with a reduction in GH levels, followed by a new surge when sleep was again initiated. The analysis of the profile of secretory rates, however, reveals that the awakening resulted in an immediate and total inhibition of GH secretion, which resumed only when the subject fell asleep again. Thus, in this case, deconvolution permitted definition of the effects of sleep–wake transitions on GH secretion much more accurately than analysis of the plasma levels.

Several deconvolution procedures have been proposed for hormonal data. The earliest and simplest approach involves calculating the amount of hor-

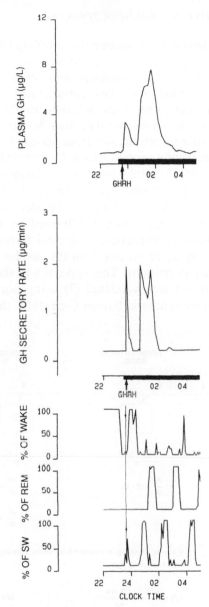

FIG. 1 Profiles of plasma GH (top), GH secretory rates (middle), and sleep stages [wake, rapid eye movement (REM), and slow wave (SW); bottom] in a normal young man who received an injection of 0.3 µg/kg of growth hormone-releasing hormone (GHRH, timing of injection shown by the arrow) after 60 sec of slow-wave sleep and experienced a spontaneous awakening subsequently. Note that the deconvolution of the plasma concentrations demonstrates that a complete interruption of the secretory process followed the awakening and that secretion resumed occurred following the reinitiation of sleep. The dark bar represents the scheduled bedtimes. Adapted from (Ref. 1) E. Van Cauter *et al*. Sleep, awakenings and insulin-like growth factor I modulate the growth hormone secretory response to growth hormone-releasing hormone. *J. Clin. Endocrinol. Metab.* **74,** 1451–1459, 1992. © The Endocrine Society.

mone secreted between successive sampling times by solving the diffferential equations using numerical approximations. Figure 2 gives the model representations, differential equations, and continuous solutions for both a one-compartment model and a two-compartment model.

The analytical form of the solution for the two-compartment model for deconvolution of endocrine data was first proposed by Eaton et al. (2) to estimate insulin secretion rates from plasma C-peptide levels. The advantage in estimating insulin secretion from C-peptide levels rather than from plasma insulin is 2-fold: (1) insulin undergoes a large and variable hepatic extraction and therefore the plasma levels are an unreliable index of the amount of hormone secretion; in contrast, C-peptide, which is cosecreted with insulin on an equimolar basis, is not extracted by the liver; (2) in the range of concentrations representative of normal fasting levels, the C-peptide assay is considerably more precise than the insulin assay, even when performed under optimal conditions. This approach to the estimation of *in vivo* human insulin secretion was validated (3) using bolus injections and continuous infusions of biosynthetic human C-peptide, thereby allowing the C-peptide

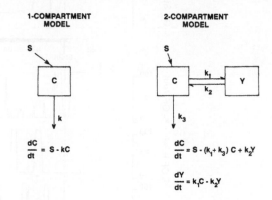

FIG. 2   Model, differential equations, and analytical solutions for deriving hormonal secretory rates (S) from peripheral concentrations (C) based on a one-compartment model (left) or a two-compartment model (right). It is assumed that the kinetic parameters (K or K_1, K_2, K_3) are known. Adapted from Ref. 6.

kinetics to be accurately defined on an individual basis and the rate of infusion to be recovered by deconvolution analysis. It also allowed the demonstration that postprandial insulin secretion occurs in a series of pulses (4). We have subsequently shown average kinetic parameters that take into account the sex, age, and degree of obesity can be successfully used without significant loss of accuracy (5). We have utilized this method using both individually estimated and average kinetic parameters in our studies of ultradian insulin secretory oscillations.

Although this straightforward approach to deconvolution is attractive because of its simplicity, it may also provide absurd results, such as negative secretory rates, and in certain cases may seriously overestimate the frequency of secretory pulses. Negative secretory rates will be found if the kinetic model is inadequate or if there was a drop of concentration due to dilution rather than decreasing or interrupted secretion. Even if none of these problems exist, negative secretory rates may still occur because deconvolution involves amplification of measurement error, particularly when secretory rates are low. However, if one assumes that the measurement errors on the concentrations are normally distributed, statistical errors on the calculated secretory rates can be estimated and secretory rates which are significantly negative can be identified (6).

The other problem with straightforward deconvolution is the fact that errors on the secretory rates increase when the sampling interval decreases. This paradoxical phenomenon reflects the fact that the secretory rate is estimated separately in each sampling interval. To identify significant pulses of secretory rates, pulse analysis may be repeated using a modified version of the pulse detection algorithm where the thresholds for pulse detection are based on the calculated errors on the instantaneous secretory rates (6). In all cases, it is essential that the clearance kinetics of the hormone under consideration be well defined in order to obtain reliable conclusions.

A number of complex algorithms which attempt to circumvent some of the above-mentioned difficulties have been proposed (7–10), and some of them are presented and discussed in other sections of this volume. Some approaches to hormonal deconvolution have been model-based or waveform-specific (7, 8). In such model-based methods, the waveshapes of underlying pulses are postulated and the number of pulses, their location and amplitudes, and the half-life of the hormone are estimated by multiparameter optimization techniques.

Pretreatment of Data

In a number of instances, it is desirable or necessary to pretreat the data before using one of the procedures for quantification of the pulsatile variation.

The objective of such pretreatment procedures is to minimize the impact of measurement error and other sources of variability on the characterization of the process under study. The most common of these procedures are smoothing and detrending.

Smoothing

The smoothing procedure will dampen fluctuations that are more rapid than those under study as well as minimize the impact of measurement error. For example, if the objective of the study is the characterization of oscillations in insulin levels in the ultradian range of 100–140 min and the sampling frequency was 10 min, variability due to the ultrafast insulin oscillations in the range of 10–15 min will be apparent and will need to be minimized to optimize the estimation of the frequency and amplitude of ultradian oscillations.

Moving averages are standard smoothing procedures that have been used in many studies of hormonal oscillations. Each data point is replaced by the arithmetic mean of the data points included in a n-point window centered on the data point under consideration. For example, in a three-point moving average, each data point is replaced by the mean of its own value and the value of the two neighboring data points [i.e., $X_{i\,\text{smooth}} = (X_{i-1} + X_i + X_{i+1})/3$]. An n-point moving average will strongly dampen all fluctuations shorter than n sampling intervals and improve the estimation of slower variations, albeit at the expense of a modest reduction in their amplitude. It will also decrease measurement error at each time point by a factor of $n^{1/2}$. Thus, this procedure can be very useful before performing deconvolution on the data, particularly if the sampling interval is much shorter than the longer half-life of the hormone under consideration (e.g., less than one-half of the longer half-life for a two-point moving average, less than one-third of the longer half-life for a three-point moving average).

Detrending

When analyzing data for ultradian oscillations, a significant slow trend onto which the pulses of interest are overlaid is often apparent. Such long-term variations may reflect diurnal variations, modulation by the sleep–wake cycle, effects of prolonged fasting conditions, or other conditions. These long-term trends can have masking effects on more rapid variations, such as ultradian oscillations, and may give rise to artifacts in the analysis. Thus, removal of a slow trend is often necessary in order to obtain interpretable results and is essential if an analysis in the frequency domain is contemplated. Sometimes the trend may be a simple linear increase or decrease, but usually it is more complicated. In such cases, a best-fit curve can be calculated, for example, using an algorithm proposed by Cleveland (11), or repeated periodogram calculations (12).

To remove the slow trend, the raw data can then be divided by the best-fit curve, or the best-fit curve can be subtracted from the raw data and the mean level added to obtain a constant, nonzero, overall level. An example is shown in Fig. 3, which illustrates the profile of insulin secretion rates over 53 hr (obtained by deconvoluting plasma C-peptide levels with a two-compartment model) of a single obese subject who was studied during constant glucose infusion for a prolonged period including 8 hr of nocturnal sleep, 28 hr of continuous wakefulness, and 8 hr of daytime recovery sleep. Long-term trends reflecting effects of sleep and circadian rhythmicity are apparent and were quantified by a regression curve obtained using the Cleveland procedure (11) with a window of 6 hr. The middle profile in Fig. 3 shows the series detrended by subtraction of the regression curve and subjected to pulse analysis as described in the next section of this chapter.

Digital filters can also be used to eliminate temporal variations in the low-frequency range; however, they generally involve a substantial reduction in the number of data available for subsequent analyses and thus are of limited use for hormonal series, which are generally relatively short due to limitations on the amount of blood loss. An exception is the simple first-difference filter, where each data point is replaced by the difference between itself and the previous data point. We have used this method to remove slow trends before performing spectral analysis of glucose and insulin profiles.

Time Domain Analysis of Pulses of Plasma Concentrations

Pulse-by-Pulse Analysis

Since the pioneering approach proposed by Santen and Bardin (13) in their classic article describing pulsatile LH release across the menstrual cycle, a number of computer algorithms for identification of pulses of hormonal concentration have been proposed. Several review articles (14–16) have provided comparisons of performance of several pulse detection algorithms, including ULTRA (17), PULSAR (18), CYCLE DETECTOR (19), CLUSTER (20), and DETECT (21). Comparative studies of the performance of various pulse detection algorithms on experimental series conducted by Urban et al. (22) indicated that ULTRA, CLUSTER, and DETECT perform similarly when used with appropriate choices of parameters. Objective as-required and is essential. The variation in frequency domain is conceptualized. Sometimes the trend may be a linear, either increase or decrease, but usually it is more complicated. In such cases, a smooth curve can be calculated, for example, using a nonparametric procedure by Cleveland (11), or repeated psychological...

To remove the trend, the raw data can either be divided by the trend curve, or the latter may instead be subtracted from the raw data and the mean level added to obtain a zero or positive overall level. An example is shown in Fig. 3, which illustrates the profile of insulin secretion rates over 53 hr obtained by monitoring plasma insulin levels with a two-compartment model... analof sampling... who was studied during constant glucose infusion... at intervals including a 1-hr of nocturnal sleep. Secretory rates... one... at the time of every sleep. Long-term variations... and circadian hysteresis are apparent and were quantized by a regression curve obtained using the Cleveland procedure (11) with a window of 6 hr. The middle profile in Fig. 3 shows the secretory series after subtraction of the regression curve and subjected to pulse analysis as described in the next section of this chapter.

Digital filters can also be used to eliminate temporal variations in the low frequency ranges; however, they generally involve a substantial reduction in the number of degrees... and thus are of limited use for... relatively short due to fluctuations. In the simplest first difference filter, where each data point is replaced by the difference between itself and the previous data point. We have used this method to remove slow trends before performing spectral analysis of glucose and insulin profiles.

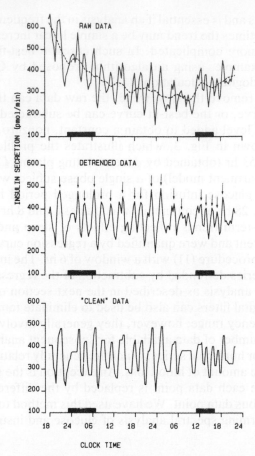

FIG. 3   Performance of ULTRA on a profile of insulin secretory rate observed in an obese subject during a 53-hr period of constant glucose infusion (E. Van Cauter, unpublished data). The black bars represent the sleep periods. The top graph presents the raw data with the long-term trend quantified by a regression curve shown as a dashed line. The middle graph shows the secretory profile detrended after subtraction of the regression curve. The arrows indicate the significant secretory pulses identified by analysis with the ULTRA algorithm for pulse identification. The bottom graph shows the "clean" series, in which all nonsignificant variations have been eliminated.

... in their ... concentration have been proposed. Several review articles (14–16) have provided comparisons of performance of several pulse detection algorithms including ULTRA (17), PULSAR (18), CYCLE DETECTION (19), (17–18)...

sessment of the performance of a pulse detection program may be by testing computer-generated profiles including both pulses and noise and examining false-positive and false-negative errors. This approach has been used by the authors of ULTRA, CLUSTER, and DETECT (17, 23–25).

We have mainly used ULTRA as our tool for the quantification of pulsatile hormonal secretion. Briefly, it operates under the following principles: (1) it takes into account both the increasing limb and the declining limb in determining the significance of a pulse; (2) it allows for variable precision of the assay in various concentration ranges; (3) the significance of a pulse is evaluated independently of its width or waveshape; and (4) its performance is not affected by the existence of a fluctuating baseline, as may be caused by circadian variation. In summary, a pulse is considered significant if both its increment and its decline exceed, in relative terms, a threshold expressed in multiples of the intraassay coefficient of variation *(CV)* in the relevant range of concentration. All changes in concentration which do not meet the criteria for significance are eliminated by an iterative process, providing a clean profile as an output. The middle and lower graphs of Fig. 3 illustrate the performance of ULTRA on the detrended profile of insulin secretion. The coefficient of variation of the C-peptide assay averaged 6% throughout the relevant range of concentrations. A threshold of $3CV$, instead of the more standard $2CV$ threshold, was used to account for amplification of measurement error by the deconvolution procedure.

To test ULTRA for false-positive and false-negative errors (17), the program was applied to a large set of computer-generated series including both signal (i.e., pulses) and noise (i.e., measurement error). These simulation studies indicated that for series with a medium to high signal-to-noise ratio (i.e., large and frequent pulses), a threshold of 2 times the local intraassay *CV* minimizes both the false-positive and false-negative errors. For series with low signal-to-noise ratios (i.e., low pulse amplitude and/or frequency), a threshold of $3CV$ is preferable.

The usefulness of detrending along with subsequent pulse analysis is further demonstrated in an analysis of insulin secretory and serum insulin data from a group of nondiabetic subjects shown in Fig. 4 (26). Two experimental protocols were used, one to investigate the ultradian oscillations in insulin secretion and one to investigate the coexisting rapid pulses. Subjects were studied both during fasting and during glucose infusion. In the 8-hr studies investigating the ultradian oscillations, no slow trend was apparent, so pulse analysis was performed directly on the insulin secretory profiles. It was found that the relative amplitude of the insulin secretory rate (ISR) pulses increased progressively as the rate of glucose infusion was increased. In the 2-hr studies investigating the rapid pulses, a slow trend representing the ultradian oscillations was apparent in the insulin profiles, as illustrated in the top graphs of Fig. 4. A best-fit curve was therefore calculated for the raw data using the Cleveland procedure, and detrending was performed by dividing by the regression curve. The results of the subsequent pulse analysis on the detrended profiles (depicted in the middle graphs of Fig. 4) show that

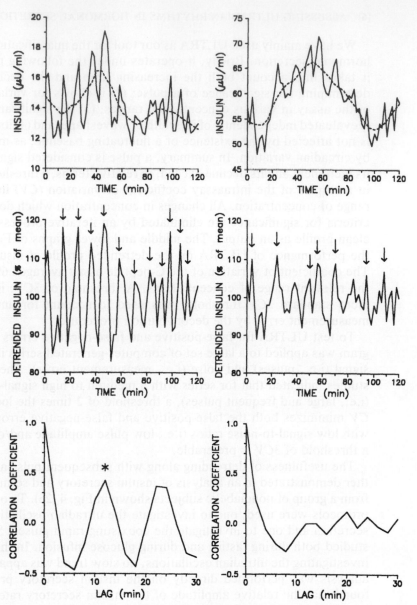

FIG. 4 Serum insulin data and results of pulse and autocorrelation analyses during fasting (left) and during constant glucose infusion (right) in one normal subject studied on two separate occasions. The top graphs show the raw data with the best-fit curves obtained using the Cleveland algorithm as a dashed line. The middle graphs show the detrended data, with significant pulses marked by arrows. The lower graphs illustrate the estimations of the autocorrelation function. The asterisk denotes statistical significance at $p < 0.05$. Data from Ref. 26.

the relative amplitude of the insulin pulses was smaller during glucose infusion (right) than during fasting (left). Thus, this analysis permitted the demonstration that intravenous glucose infusion has differential effects on rapid pulses and ultradian oscillations in insulin secretion.

Methods of pulse-by-pulse analysis have the advantage that each pulse is individually evaluated with respect to significance and amplitude. Furthermore, measures such as duration of a pulse and interpulse interval can be obtained. However, because these methods try to gain as much insight as possible into each individual pulse, they are often not very sensitive to the presence or absence of a significant overall periodic component in the time series. For this purpose, autocorrelation analysis can be a very useful tool.

Autocorrelation Analysis

Autocorrelation analysis can give information about the presence of a significant periodic component in a data set. The procedure consists of calculating the correlation coefficient between the time series and a copy of itself at lags of 0, ± 1, ± 2, ± 3, etc., sampling intervals (27). In cases where the time series has a periodic component of X min, the autocorrelation function (defined as the calculated correlation coefficients as a function of the lag) will exhibit a negative correlation coefficient at a lag of approximately $X/2$ min and a positive coefficient at a lag of X min. The significance of the correlation coefficient at a lag of X min can be evaluated with Fisher's Z-values (28). For this particular type of analysis, the presence of trends slower than the periodic pulsatile behavior of interest can distort the analysis, and removal of long-term trends is necessary.

Because they reveal different types of information, the combined use of autocorrelation analysis and pulse-by-pulse analysis is often advisable. Returning to Fig. 4, whereas analysis with ULTRA showed that 11 and 7 significant pulses were present during fasting (left) and glucose infusion (right), respectively, autocorrelation analysis of the detrended data revealed that during fasting a 12-min periodic component was present and that no significant component was detectable in the data obtained during glucose infusion. This result is illustrated in the bottom graphs of Fig. 4. Thus, although autocorrelation analysis did not reveal any significant periodic component in the data during glucose infusion, this does not mean that the series is not pulsatile, as clearly demonstrated by pulse analysis. This example demonstrates the usefulness of applying both types of analytical tools to the same pulsatile hormonal profile.

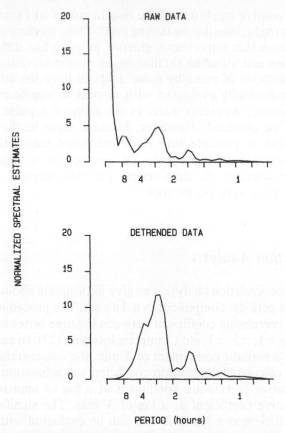

Fig. 5 Spectral analysis of the profile of insulin secretory rates shown in Fig. 3. The top graph shows the spectral power estimations for the raw data, and the bottom graph shows the spectral power estimations for the detrended data. Note that the detrending resulted in the elimination of low-frequency components and facilitated the detection of the ultradian periodicity.

Frequency Domain Analysis

Analysis in the frequency domain may, like autocorrelation analysis, indicate whether the underlying process is periodic, but in addition spectral analysis can indicate whether the process is quasi-periodic or random (27, 29). To examine spectral power in the ultradian range, detrending to eliminate low-frequency variations is essential. This is illustrated in Fig. 5, which compares spectral estimations for the original series (Fig. 3, top) and for the detrended series (Fig. 3, middle). It is evident that the spectral peak at the ultradian frequency, with a maximum at 148 min, is largely masked by the variability in the low-frequency range in the raw data and becomes much more apparent in the detrended data.

We have found spectral analysis to be particularly useful in the analysis of insulin secretory data using an experimental protocol in which glucose is infused as a sine wave with a period of 144 min (30). The method we employed was the so-called window closing procedure using a Tukey window (27). The data were detrended with the first difference filter before being subjected to the spectral analysis. The purpose of that study was to see if the ultradian insulin secretory oscillations that occur during constant glucose infusion can be entrained by an exogenous oscillatory glucose infusion in patients with non-insulin-dependent diabetes (NIDDM), in subjects with impaired glucose tolerance (IGT), and in weight-matched control subjects with normal glucose tolerance.

By submitting the insulin secretory data to spectral analysis, we were able to show that, in subjects with IGT and NIDDM, there is an abnormality in the degree to which entrainment is possible. Figure 6 thus shows representative individual insulin secretory and glucose profiles from each of the three groups. Figure 7 illustrates the mean power spectra of the insulin secretory and glucose data from each of the groups. The comparison of the spectra for the three groups shows that, although the glucose profiles are clearly entrained to the exogenous glucose infusion pattern in all subjects, the normalized spectral power at 144 min for the insulin secretory profiles is significantly less in the IGT and NIDDM groups compared to the control group. This failure of insulin secretion to entrain to oscillatory glucose changes is thought to be one of the earliest defects of glucose regulation that can be detected prior to the onset of diabetes.

Conclusions

In summary, we have presented methods of analysis for assessment of ultradian rhythms in hormonal secretion which we have successfully used and are continuously using in our research. The choice of the method of analysis should be primarily motivated by the major question of interest. Time domain analysis is particularly suitable for studies aiming at comparing number and amplitude of secretory pulses or oscillations in various subject populations or under various experimental conditions. If rapid variations occur in addition to ultradian oscillations, smoothing prior to pulse analysis is essential to obtain unbiased estimations of pulse frequency and amplitude. Long-term trends will affect estimations of frequency but generally not estimations of

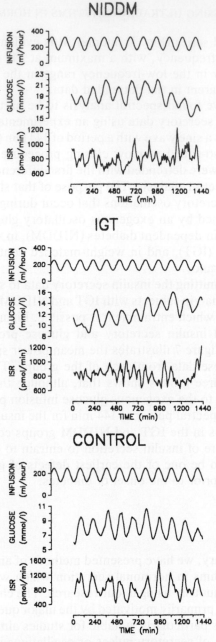

FIG. 6 Profiles of glucose infusion rates, plasma glucose levels, and insulin secretory rates (ISR) from a patient with non-insulin-dependent diabetes (NIDDM; top), a subject with impaired glucose tolerance (IGT; center), and a weight-matched control subject (bottom). Glucose was infused in an oscillatory fashion with an ultradian period of 144 min. Insulin secretion rates entrained to the glucose periodicity in the control subjects but not in the IGT and NIDDM patients. Adapted from Ref. 30.

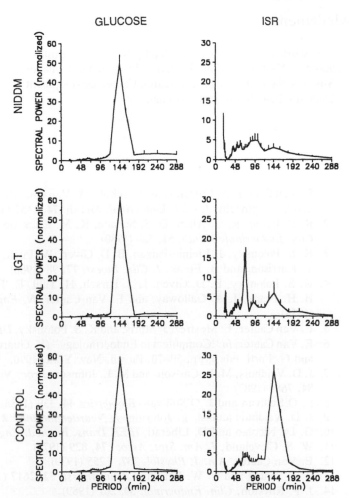

FIG. 7   Spectral analysis of the profiles of plasma glucose and insulin secretion from the three groups of subjects exemplified in Fig. 6. The data shown are the mean spectra (+SEM) for each group. Adapted from Ref. 30.

amplitude. Detrending is also important in examining estimations of autocorrelation. Frequency domain analysis is the method of choice when the question of regularity and periodicity is central to the objectives of the study. In this case, detrending to eliminate slower variations is crucial, while smoothing is usually of lesser benefit.

Acknowledgments

This work was partially supported by National Institutes of Health Grant DK-41814 and by a NATO Scientific Exchange Programme–Collaborative Research Grant. Dr. Sturis is the recipient of a Research Career Development Award from the Juvenile Diabetes Foundation International.

References

1. E. Van Cauter, A. Caufriez, M. Kerkhofs, A. Van Onderbergen, M. O. Thorner, and G. Copinschi, *J. Clin. Endocrinol. Metab.* **74,** 1451 (1992).
2. R. P. Eaton, R. C. Allen, D. S. Schade, K. M. Erickson, and J. Standefer, *J. Clin. Endocrinol. Metab.* **51,** 520 (1980).
3. K. S. Polonsky, J. Licinio-Paixao, B. D. Given, W. Pugh, P. Rue, J. Galloway, T. Karrison, and B. Frank, *J. Clin. Invest.* **77,** 98 (1986).
4. K. S. Polonsky, B. D. Given, L. J. Hirsch, H. Tillil, E. T. Shapiro, C. Beebe, B. H. Frank, J. A. Galloway, and E. Van Cauter, *N. Engl. J. Med.* **318,** 1231 (1988).
5. E. Van Cauter, F. Mestrez, J. Sturis, and K. S. Polonsky, *Diabetes* **41,** 368 (1992).
6. E. Van Cauter, *in* "Computers in Endocrinology" (V. Guardabasso, D. Rodbard, and G. Forti, eds.), pp. 59–70. Raven, New York, 1990.
7. J. D. Veldhuis, M. L. Carlson, and M. L. Johnson, *Proc. Natl. Acad. Sci. U.S.A.* **84,** 7686 (1987).
8. F. O'Sullivan and J. O'Sullivan, *Biometrics* **44,** 339 (1988).
9. J. D. Veldhuis and M. L. Johnson, *J. Neuroendocrinol.* **2,** 755 (1990).
10. G. De Nicolao and D. Liberati, *IEEE Trans. Biomed. Eng.* **40,** 440 (1993).
11. W. S. Cleveland, *J. Am. Stat. Assoc.* **74,** 829 (1979).
12. E. Van Cauter, *Am. J. Physiol.* **237,** E255 (1979).
13. R. J. Santen and C. W. Bardin, *J. Clin. Invest.* **52,** 2617 (1973).
14. J. P. Royston, *Clin. Endocrinol.* **30,** 201 (1989).
15. R. J. Urban, W. S. Evans, A. D. Rogol, D. L. Kaiser, M. L. Johnson, and J. D. Veldhuis, *Endocr. Rev.* **9,** 3 (1988).
16. W. S. Evans, M. J. Sollenberg, R. A. J. Booth, A. D. Rogol, R. J. Urban, E. C. Carlsen, M. L. Johnson, and J. D. Veldhuis, *Endocr. Rev.* **13,** 81 (1992).
17. E. Van Cauter, *Am. J. Physiol.* **254,** E786 (1988).
18. G. R. Merriam and K. W. Wachter, *Am. J. Physiol.* **243,** E310 (1982).
19. D. K. Clifton and R. A. Steiner, *Endocrinology (Baltimore)* **112,** 1057 (1983).
20. J. D. Veldhuis and M. L. Johnson, *Am. J. Physiol.* **250,** E486 (1986).
21. K. E. Oerter, V. Guardabasso, and D. Rodbard, *Comput. Biomed. Res.* **19,** 170 (1986).
22. R. J. Urban, D. L. Kaiser, E. Van Cauter, M. L. Johnson, and J. D. Veldhuis, *Am. J. Physiol.* **254,** E113 (1988).
23. A. D. Genazzani and D. Rodbard, *Acta Endocrinol.* **124,** 295 (1991).
24. R. J. Urban, M. L. Johnson, and J. D. Veldhuis, *Am. J. Physiol.* **257,** E88 (1989).

25. V. Guardabasso, G. De Nicolao, M. Rochetti, and D. Rodbard, *Am. J. Physiol.* **255,** E775 (1988).
26. J. Sturis, N. M. O'Meara, E. T. Shapiro, J. D. Blackman, H. Tillil, K. S. Polonsky, and E. Van Cauter, *J. Clin. Endocrinol. Metab.* **76,** 895 (1993).
27. G. M. Jenkins and D. G. Watts, "Spectral Analysis and Its Applications." Holden Day, San Francisco, 1968.
28. R. A. Fisher, "Statistical Methods for Research Workers." Oliver and Boyd, Edinburgh, 1934.
29. D. R. Matthews, *Acta Paediatr. Scand Suppl.* **347,** 55 (1988).
30. N. M. O'Meara, J. Sturis, E. Van Cauter, and K. S. Polonsky, *J. Clin. Invest.* **92,** 262 (1993).

[7] Frequency Domain Analysis of High-Frequency Ultradian Plasma Adrenocorticotropic Hormone and Glucocorticoid Fluctuations

Molly Carnes and Brian Goodman

Introduction

Measurement of plasma hormone concentrations at time scales of several minutes reveals complex patterns of fluctuations (1). The field of neuroendocrinology has experienced a conceptual shift from homeostasis toward homeodynamics (2) which has influenced both the measurement and interpretation of temporal variations in plasma hormone concentrations. Traditionally, mean plasma hormone concentrations or the area under the concentration versus time curve were used to assess the activity of a neuroendocrine system. Refinements in assay methodology enabled repeated measures of plasma hormone concentrations to be performed within individual subjects. Sampling paradigms using 10- to 20-min intervals subsequently revealed fairly large amplitude spontaneous fluctuations in plasma concentrations of many hormones (1). Because these fluctuations occur at time scales less than 24 hr, they were termed ultradian rhythms. The presence of these fluctuations, usually referred to as pulses (3), implied that investigating the pattern of hormone fluctuations would reveal information regarding the physiology of neuroendocrine systems unobtainable through static measurements.

A number of peak detection algorithms based predominantly on excursions of concentrations beyond some multiple of assay variability were developed to identify pulses within a concentration time series (4). These algorithms dealt well with the relatively high-amplitude pulses resolved with sampling intervals of 10–20 min but could not capture the complex dynamics that emerged in the concentration time series of virtually all neuroendocrine hormones when sampling intervals were reduced to no longer than 5 min (5, 6).

With the current understanding of neuroendocrinology, it would be premature to assume that low-amplitude, high-frequency fluctuations, which are often at or near assay detection limits, are unimportant in carrying biochemical information. For this reason, it may be useful to view hormone concentration time series as complex signals with periodic, aperiodic, and noise compo-

Methods in Neurosciences, Volume 28

nents and to search for ways to evaluate not just the relatively high-amplitude pulses but all the information contained in a concentration time series. One method that has been used extensively in the physical sciences for evaluating signals is frequency domain or spectral analysis (2, 7). In general, peak detection algorithms which operate in the time domain are more appropriate for gauging incremental changes within a continuous process, whereas the frequency domain is more appropriate for gauging structural changes within a continuous process.

This chapter presents the application of spectral analysis to concentration time series of adrenocorticotropic hormone (ACTH) and glucocorticoids. Although spectral analysis has been applied to circadian rhythms of these hormones (8) completing cycles in approximately 24 hr, we discuss its application only to higher frequency fluctuations at time scales of several minutes to several hours observable with 1- to 2-min measurements over 3–4 hr. We present the following: background on the concepts of spectral analysis with its strengths and weaknesses related to analysis of hormonal time series, a step-by-step process for analyzing the frequency structure of a time series, examples of previous applications of spectral analysis from our laboratory, methods of comparing individual spectra, and the use of mathematical modeling to assist with interpretation of spectral observations.

The sequence of presentation parallels efforts from our own laboratory, first to analyze the data we observed experimentally and then to decipher the meaning of our observations. The presentation focuses on ACTH because our work has been predominantly with this hormone. Glucocorticoid time series will be addressed mainly in discussion of cross-spectra. The spectral methodology presented, however, can be applied to any hormonal time series.

Background

Defining a Spectrum

The mathematician Baron Jean Baptiste Joseph Fourier invented Fourier transform functions in the early 1800s to provide a means for determining the periodicity of cyclic changes Fourier observed in the water level of Lake Geneva while fishing from the dock of his summer home (9). His pioneering developments laid the groundwork for what is now commonly referred to as spectral analysis.

The result of a spectral analysis is a spectrum. A spectrum is the distribution of a characteristic of a physical system or phenomenon as a function of scale (wavelength) or inverse scale (frequency). White light refracted through a prism produces a spectrum of its composite colors distributed as a function of the wavelengths. The longer wavelength (lower frequency) red colors are sorted out from the shorter wavelength (higher frequency) blue colors. The contribution of each color to the overall brightness of the white light is represented by its intensity. Comparably, spectral analysis of a time series produces a spectrum of the amount of variance contributed to the total time series variance by the different individual time scales contained within the time series. The contribution of each time scale to the overall variance of the time series is represented by its power (i.e., intensity). Available methods for performing spectral analysis on time series are variations on those introduced by Schuster (10), Walker (11), and Fisher (12) and are discussed in detail by any textbook written on the subject [e.g., Anderson (13)].

Fourier Transforms

A physical process can be described either in the time domain by a series of values $h(t)$ or in the frequency domain by a series of complex numbers $H(f)$. The real parts of these complex numbers contain amplitude information, whereas the imaginary parts contain phase information as a function of frequency. Spectral analysis of a time series represents a linear operation in which the observed variability as a function of time is mapped from the time domain into the frequency domain, where the same information is expressed as a function of frequency. One goes back and forth between these two representations through the Fourier transform equations:

$$H(f) = \int_{-\infty}^{\infty} h(t)\, e^{2\pi i f t}\, dt,$$

$$h(t) = \int_{-\infty}^{\infty} H(f)\, e^{-2\pi i f t}\, df.$$

If t is measured in time units (e.g., seconds), then f is measured in inverse time units (e.g., cycles per second). The result of computing the Fourier transforms for a range of frequencies represents the spectrum of a time series. The total power or variance of a time series is the same whether it is computed in the time or frequency domain. This result is known as Parseval's theorem:

$$\text{Total power} = \int_{-\infty}^{\infty} |h(t)|^2\, dt = \int_{-\infty}^{\infty} |H(f)|^2\, df.$$

In the frequency domain, the total power equals the amplitude squared of the Fourier transform $H(f)$ integrated over all frequencies. Often of interest is the amount of power within a finite frequency interval between f and $f + df$. This quantity is referred to as the power spectral density. Both the periodic and aperiodic signal components of $h(t)$ will additively contribute to the total power, and they can be analytically derived from their appropriate Fourier transform functions (14).

Digital Fourier Transform

Whereas the fluctuation of hormone concentrations in plasma is a continuous process, experimental data are collected at discrete times. In practice, spectral analysis is not applied to a continuous function $h(t)$, but rather to a sequence of repeated measures of that continuous process. The Fourier transform of a discretely sampled time series is referred to as the discrete or digital Fourier transform (DFT). The DFT approximates the integral of the continuous Fourier transforms as a discrete sum which maps N complex numbers in the time domain (h_k) into N complex numbers (H_n) in the frequency domain, where the real part of the complex number contains the amplitude information and the imaginary part contains the phase information:

$$H_n = \sum_{k=0}^{N-1} h_k\, e^{2\pi i k(n/N)},$$

$$h_k = (1/N) \sum_{k=0}^{N-1} H_n\, e^{-2\pi i k(n/N)}.$$

With N number of points, $N/2$ is the maximum number of independent amplitudes and phases which can be estimated. The DFT and the continuous Fourier transform integral have similar symmetry properties. The DFT also satisfies a discrete form of Parseval's theorem.

The DFT algorithm we use derives amplitude and phase spectral estimates for a given frequency by applying multiple regression techniques to a time series. The regression coefficients in this algorithm represent the amplitudes of a sine–cosine pair. The amplitudes of the sine and cosine terms can be algebraically manipulated to construct the desired amplitude and phase spectral estimates for a cosine function:

$$h(t_k) = H_0 + \sum_{n=1}^{N/2} A_n \sin(2\pi f_n t_k) + \sum_{n=1}^{N/2} B_n \cos(2\pi f_n t_k),$$

$$h(t_k) = H_0 + \sum_{n=1}^{N/2} C_n \cos(2\pi(f_n t_k + \phi_n)),$$

$$C_n = (A_n^2 + B_n^2)^{1/2},$$

$$\phi_n = \tan^{-1}(A_n/B_n),$$

$$f_n = (n/N)/P,$$

where H_0 is the time series mean, A_n, B_n, C_n are the spectral amplitude estimates for the nth frequency f_n, ϕ_n are the corresponding spectral phase estimates, and P is the time period covered by the data length being analyzed. The amplitude squared, C_n^2, is analogous to the $|H(f)|^2$ term in the continuous formulation and represents the digital equivalent of the spectral power density, although it is not integrated over any incremental range of frequency but in fact represents a discrete power estimate for the nth frequency. Small measurement errors in the time domain will create large errors in the computation of spectral estimates in the frequency domain. The use of multiple regression techniques allows for the direct computation of standard errors and probabilities for each pair of regression coefficients and of goodness-of-fit statistics (e.g., correlations, significance probabilities, standard error of estimates) for each frequency fit.

Fast Fourier Transform

A special case of the more general class of DFT is the fast Fourier transform (FFT). The FFT algorithm is computationally more efficient than the DFT algorithm. For example, to analyze a million data points, the DFT algorithm will take roughly 40,000 times longer to compute than the FFT algorithm. The tremendous computational efficiency of the FFT, however, sacrifices flexibility. The FFT is limited to deriving spectral estimates only at harmonic frequencies for data sets with 2^N data points, where N is an integer.

For large data sets this is not a problem. If a data set does not contain exactly 2^N data points, it can either be padded with zeroes up to the next power of two or the number of data points reduced down to the next nearest power of two, and still retain an acceptable spectral resolution from only the harmonic frequencies. However, for small data sets ($N < 1000$) the above limitations can impair the utility of analyzing data in the frequency domain, because the smaller values of N will result in lower spectral resolution. In this situation it is better to use the more general DFT algorithm. The DFT is also more flexible than the FFT in handling missing data points. On smaller data sets and with the speed of modern computers, the longer computation times are not an obstacle.

Spectral Analysis Output

Because Fourier transforms (continuous or discrete) are made up of sine–cosine pairs, they naturally lend themselves to the detection of periodic sinusoidal waveforms. However, Fourier analysis can be used to represent the spectral composition of any signal measured over a finite interval of time, not just a signal composed of periodic sinusoidal waveforms. Spectral estimates computed from a finite data set will completely represent the temporal pattern of fluctuations observed in the time series.

The results of a spectral analysis are typicaly presented as a spectral power plot where the ordinate is either an absolute (amplitude squared, C_n^2) or relative (amplitude squared normalized by the total variance, C_n^2/σ^2) measure of the amount of total variance explained by a sinusoidal waveform at the nth frequency and where the abscissa is frequency or its inverse, period. Often the corresponding amplitude (C_n) and phase (ϕ_n) spectral estimates are also plotted versus frequency or period. Different information can be gleaned from examination of each of these spectra.

Periodic fluctuations in a time series will cause peaks or lines of spectral power at their corresponding frequencies. Aperiodic fluctuations in a time series comprise the background continuum. The spectral power plots will show spectral lines superimposed on a smoothly varying background continuum. A spectral power (or amplitude) plot tends to accentuate the spectral line pattern more clearly than it does the spectral background continuum. Therefore, it is most often used when investigating the periodic structure of a time series. When investigating the aperiodic structure of a time series, it is the spectral background continuum which is of interest. The spectral background continuum can be accentuated by plotting the base 10 logarithmic transforms of the spectral power estimates ($\log C_n^2$) and their corresponding frequencies ($\log f_n$) on the ordinate and abscissa axes, respectively. This style of power plot is referred to as a log–log spectral plot (15).

Hormonal clearance integrates a neuroendocrine secretory signal because plasma clearance of a hormone occurs at time scales longer than the time scales of secretion. The spectrum of any signal involving an integrating process will have a background continuum which follows a $1/f^x$ relationship (i.e., frequency and power are inversely related), resulting in a negative slope if a line is fit to the log–log plot (15). The slope of the log–log plot may provide a measure of the fractal dimension of a signal as well as an indirect assessment of hormone clearance rate.

The identified spectral estimates must be interpreted with care. Collectively, they form a frequency domain signature which completely describes the average spectral characteristics of a physiological process during the experimental period. It is not always clear, however, whether an individual spectral line can be ascribed to a true periodic sinusoidal signal or whether it represents a harmonic of some other fundamental frequency. This mathematical artifact results from the application of a Fourier transform to a periodic nonsinusoidal signal. For example, a spectral analysis of a sawtooth waveform which repeats every 48 min will display a spectral peak not only at 48 min, but also at its subsequent harmonic frequencies (24, 16, 12, ..., 2 min) with logarithmically decreasing amplitudes. In this situation, the fundamental frequency and its harmonics will all be present in the Fourier transform. Even when the input into a system is a process containing true periodic sinusoidal waveforms, the spectra of the observed output time series can easily become complicated by nonlinear and nonstationary response characteristics, which could result in a variety of amplitude, frequency, and phase modulations.

These limitations make it difficult to work backward from a spectrum to derive physiological insight without some *a priori* knowledge of the biological system being studied. Nevertheless, spectral analysis can be used to gain insight into and test hypotheses about the physiology of a biological system, because any hypothesis generated must be able to explain the observed spectral properties. The computation of spectral estimates is fairly straightforward. The interpretation of the computed spectral estimates, however, is not an exact science, and it still should be viewed in some part as an art form.

Design of Spectral Analyses

Sampling Rate and Aliasing

The faster a signal varies, the faster the sampling rate must be in order to characterize the signal accurately. For any sampling interval Δ, there exists a special frequency f_c, called the Nyquist critical frequency, defined as $f_c = 1/2\Delta$. The Nyquist frequency is the highest frequency that can be resolved with a given sampling interval (16). Expressed otherwise, critical sampling of a sine or cosine wave is two sample points per cycle. For example, to resolve a signal completing 1 cycle in 48 min, the experimental design to identify this signal will require a minimum sampling rate of 24 min, once at the peak and once at the nadir of the cycle. If a signal is less regular or contains noise (as is the case for all hormone concentration time series), then it is recommended that the experimental design be modified to increase the sampling rate up to 6 times the highest frequency to be resolved. For example, resolving a quasi-regular or noisy 48-min cycle would require a sampling rate of 8 min. Figure 1 illustrates from our data how the impression of the dynamics of ACTH pulsatility is greatly altered by the length of the sampling interval.

When a continuous signal is discretely sampled, the potential exists for observing a spurious rhythm (13). This occurs because all of the power spectral density which lies at frequencies larger than f_c (i.e., signals occurring at time scales less than twice the sampling interval) are spuriously folded back onto frequencies less than f_c. This phenomenon is known as aliasing (Fig. 2). There is little that can be done to remove aliased power after a continuous process has been discretely sampled. A high sampling density relative to the signal frequency and the use of continuous blood collection rather than an instantaneous bolus sample collection are two strategies to minimize the possibility of an aliasing error in a time series analysis (2). Continuous blood collection integrates the hormone concentration between measurements and averages together the unresolvable power arising from all signals occurring on time scales less than the sampling interval.

Sampling Duration and Number of Data Points

In addition to the density of sampling, the duration of sampling is also a crucial factor in analyzing a time series. In searching for truly cyclic processes, a minimum of at least two successive cycles must be collected in order to identify its existence statistically (2). As such, the duration of sampling defines the lowest frequency which can be resolved within a time series. For example, if the total length of a data set covers a 4-hr period, then only cycles with periods less than 120 min can be statistically resolved. When designing an experiment it will be necessary to consider the time scales of the phenomenon to be observed to ensure that both the sampling rate and duration are sufficient to resolve them. Specifically with respect to hormonal plasma concentrations, other ancillary factors that must also be considered are (1) the volume of blood necessary to perform an assay, (2) the rate at which blood can be drawn from a subject without altering the physiological signal to be studied, (3) the number of assay replicates, and (4) the cost of each assay. The results of balancing all these factors in an experimental design yields the number of data points which will make up the time series to be analyzed.

According to information theory, it is the number of data points which

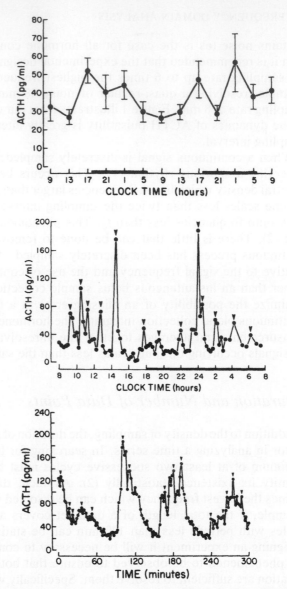

FIG. 1 The top data set represents the means ± SE of plasma ACTH concentrations collected as bolus samples every 4 hr from 10 rats over 48 hr. The middle plot shows plasma ACTH concentrations from one rat collected as bolus samples every 15 min with replacement of red blood cells resuspended in plasma substitute after each sample. The bottom data set shows 2-min plasma ACTH concentrations from one rat sampled continuously with simultaneous plasma volume replacement. These concentration time series illustrate the importance of sampling density in resolving the pulsatile event of interest and how the sampling interval affects the preception of the investigator of the dynamics of a neuroendocrine system. From Carnes, *et al.*, *Neuroendocrinology* **50,** 19–20 (1989).

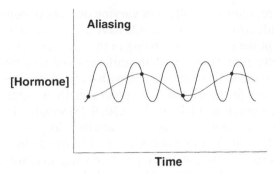

FIG. 2 Undersampling a signal can result in the detection of an aliased rhythm.

determines the degrees of freedom of a time series, and therefore the information content of a spectrum (17). The information content of a spectrum influences the width of the spectral lines. The greater the number of data points to describe a continuous process containing a fixed periodic cycle, the more clearly that periodic cycle will be able to be defined. In general, the width of a spectral line is inversely proportional to the square root of the number of data points.

Experimental and Computational Error

The constant presence of random measurement errors when discretely sampling a continuous process can also limit the resolution of physiologically significant low-amplitude, high-frequency fluctuations. If the signal-to-noise ratio is too small, then it becomes difficult to establish the statistical validity of these spectral features. For plasma hormone concentration time series, these random measurement errors usually enter through the different tasks associated with the sample collection and assay procedures. A small amount of computational error due to truncation and round-off errors will also enter the spectral analysis operations, but these are of secondary importance compared to the other experimental errors.

Assay error can be dealt with in several ways. If multiple assay replicates are available, then simply averaging them will reduce the magnitude of the associated error component by the inverse of the square root of the number of replicates being averaged. An alternative methodology to reduce the error component in single or multiple replicates is to apply a low-pass filter specifically designed to remove $2\Delta t$ fluctuations. Random errors will appear in a time series as an alternating up-and-down component (i.e., $2\Delta t$ signal). Many different low-pass filters of varying degrees of complexity are available (18). The ideal low-pass filter would have a constant unit response at all frequencies except the one corresponding to the $2\Delta t$ signal, where its response would be exactly zero. In practice, the different low-pass filter spectral characteristics approximate this ideal profile with varying degrees of success. The simplest low-pass filters are the three-point equally weighted moving average and the three-point binomially (0.25, 0.50, 0.25) weighted moving average. Although somewhat more complicated, another low-pass filter we have had good results with is the lowess filter (19) which has better spectral response properties than the three-point moving average filters. It also generates N filtered values from N original data points, as opposed to the $N - 4$ filtered values generated by the three-point moving average filters. Although useful in removing noise, filtering procedures will also reduce the smallest resolvable time scales.

Another option for handling noise is to include it in the spectral analysis. The spectrum of random error, or what is more often referred to as white noise, has a characteristically flat spectral pattern, such that in the frequency domain it can be modeled and removed. This option is most effective when the assay coefficient of variation and the noise-to-signal ratio are small. When the magnitudes of these factors increase, the spectral estimates become significantly biased, and it is difficult to separate the true signal spectrum from the white noise spectrum. We have determined through numerical simulation experiments that noise levels greater than or equal to 20% of the signal amplitude will obscure the spectral signature of that signal.

Several different methods have been proposed for modeling the error component in a spectrum. A permutation random test (PRT) method has been proposed by Odell *et al.* (20) for empirically modeling the random error spectral contribution. The technique is based on a bootstrap approach where the sequence of the original data points is randomized and a spectral analysis is performed on the randomized sequence. This procedure is typically repeated a minimum of 100 times. The distribution of the repeated spectral estimates can be used to determine confidence intervals for each of the frequencies analyzed due to random errors.

The location, magnitude (amplitude or power), and phase of individual spectral peaks, which satisfy some significance criteria, represent a set of spectral parameters which quantitatively characterize the periodic component of a time series. The significance criteria are classically defined as greater than the 95% white noise significance level. For power spectra the white noise level is by definition equal to $(2/N)$, where N is the number of data points. The 95 and 99% white noise confidence levels would be 1.96 and 2.575 times this amount, respectively. The goodness-of-fit statistics which can be calculated directly from the DFT regression results are also very useful in quickly determining statistical significance. The PRT method provides a more elegant and empirically more rigorous derivation for statistical significance criteria, and we recommend this method as the most appropriate one when specifically searching for physically real periodic components in a time series.

Nonstationarities and Detrending

Spectral analysis requires a time series to be stationary; that is, the statistical properties of the time series cannot vary over the sampling period. Singular nonstationarities or bad data points typically represent experimental error, whereas systematically varying nonstationarities or trends may result from environmental, mechanical, or physiological processes which operate on time scales longer than the data sampling period. These nonstationarities will bias the spectral amplitude and phase estimates for the periodic component of a time series, because they are computed using the entire data record, and as such, represent time-averaged quantities. For example, spectral analysis of a time series containing a periodic cycle which has zero phase during the first half of the time series and which during the last half switches to 180° out-of-phase will show no spectral power at that periodicity, because the two halves will cancel one another out.

We identify nonstationarities within a time series by applying a moving data window and statistically comparing the distribution of the resultant means, variances, minima, and maxima against the respective expected normal distributions. Typically the window size is selected as 10% of the total number of data points. Pincus and Keefe (21) have formalized this approach with specifically defined probabilistic measures of regularity and stationarity.

Singular nonstationarities can simply be flagged as missing values and removed from any further analysis. One common method of detrending a time series is to use a linear regression model for the trend. This method should be used with caution, however, as the linear regression coefficients are extremely sensitive to outliers. Linear regression trend estimates are static; ideally, the trend estimates should be dynamic. A more dynamic approach is to detrend by the first-order differences of the time series (22). If the trend can be attributed to a known physiological process, then a third detrending alternative is to explicitly reconstruct and subtract its contribution to the variance of the time series.

Comparing Spectra

It is always useful to first qualitatively compare spectra through a visual inspection of the spectral plots. Although this approach is subjective, it provides the investigator with some general insight into the nature of the various processes which may be involved in producing the observed spectral signatures. This insight can prove invaluable when interpreting the statistical and physical significance of the various derived spectral parameters.

In general, investigators will want to compare the pattern of fluctuations of either the same hormone between different experimental groups or of different hormones, such as ACTH and glucocorticoid, within the same subjects. In the latter case, the dynamic relationship between the hormones should also be measured.

Independent Spectra

If spectral analysis has been applied to each time series for the same sequence of frequencies, the different experimental groups or different hormones can be compared with respect to their spectral parameters. Comparisons that can be made include the distribution of the location of spectral lines, the amplitudes and phases of individual spectral lines, and the slope of the log–log plots. If linear regression is used to fit lines to the behavior of the background continuum on the log–log plots, then, in addition to the negative sloped integrating line, a flat zero-sloped line at the higher frequencies and a break point between them are often identified (23). These parameters appear to be related to the magnitude of the plasma clearance rate and the stochastic part of the signal. In this case the location of the break points and the amount of power on either side of them can be compared.

A second approach for comparing different experimental groups or hormones is to generate composite spectra. The best technique is to average the individual sine and cosine amplitudes and recompute spectral amplitude and phase estimates for each frequency. This avoids any problems which may have arisen when averaging the phase estimates because of the nonlinearities associated with how phase is defined. For composites of log–lot plots, we have found it best to first logarithmically transform the individual spectral estimates, and then average the transformed quantities for each frequency. Analogously to the individual spectra, the composite spectra can have the periodic and aperiodic components characterized by the various spectral parameters.

A correlation coefficient for two individual or composite spectra can be computed to measure how well the two spectral signatures agree. One technique we have used is borrowed from geology (24) and involves calculating the cross-correlation function for the log–log representations of two spectra. The lag at which the maximum positive correlation coefficient occurs represents a time stretching factor (or time compression factor, depending on the sign of the lag). This factor is related to the log–log cross-correlation by the relation $S = 10^{-L(\Delta \log f)}$, where S is the time stretching factor, L is the lag at which the maximum positive correlation occurs, and $\Delta \log f$ is the logarithmic frequency interval.

Cross Spectra

When analyzing coincident concentration time series of different hormones, such as ACTH and glucocorticoids, which are suspected to be causally related as a stimulus–response (input–output) system, it is desirable to have an index of the extent to which the response is due to the stimulus (i.e., is coherent with the stimulus). The response $y(t)$ of a linear system to a stimulus $x(t)$ is given by the convolution integral $y(t) = \int h(\tau) x(t - \tau) d\tau$, where $h(\tau)$ represents the transfer or filter function of the system. This relationship holds whether the stimulus is periodic, transient, or random. Taking the Fourier transform of the convolution integral results in the multiplicative equation $Y(f) = H(f)X(f)$, where Y, H, and X represent the complex power spectra for the response, filter, and stimulus power spectra, respectively, of a system. If the system consists of a series of coupled subsystems, then it can be shown (25) that in the frequency domain the overall system filter response is simply equal to the complex product of their individual filter response functions $H(f) = H_1(f)H_2(f) \ldots H_n(f)$. The reverse procedure (deconvolution) can be used to derive the spectral characteristics of any of the product terms given knowledge of the spectral characteristics of the other terms. This information can be acquired either empirically from spectral analysis of observed time series or theoretically from *a priori* knowledge of the functional form of the spectral representation of the processes involved.

If the system input and output spectral characteristics are known, then the system filter spectral characteristics can be derived by dividing the complex output spectra by the complex input spectra. Such would be the case when deriving the transfer function describing glucocorticoid secretion in response to fluctuating plasma concentrations of ACTH. The complex division results are typically presented as gain spectra, $G(f)$, phase spectra, $\Phi(f)$, and squared coherency spectra, $\kappa^2(f)$:

$$G(f) = |Y(f)|/|X(f)|,$$
$$\Phi(f) = \phi_Y(f) - \phi_X(f),$$
$$\kappa^2(f) = \cos(\phi_Y(f) - \phi_X(f))^2,$$

where $|Y(f)|$, $\phi_Y(f)$ and $|X(f)|$, $\phi_X(f)$ represent the output and input spectral amplitude and phase estimates, respectively. The gain spectra represents the in-phase power component, whereas the phase spectra represents the out-of-phase power component. The squared coherency spectra represents the fraction of power in the response signal due to the stimulus signal at each frequency. For a linear system without noise the squared coherency will equal unity. When it is less than unity, either the system is nonlinear, the signals are contaminated by noise (random error), or the two signals do not have a causal relationship.

Because small errors in the time domain can produce large errors in the frequency domain, some smoothing of the individual spectra is suggested before computing cross-spectra (26). Smoothing also avoids problems in computing the gain spectra due to division by amplitude values close to zero. Typically, a three-point binomially (0.25, 0.50, 0.25) weighted moving average filter is used. However, any type of low-pass filter will suffice. Individual cross-spectra can be characterized for comparison purposes in an analogous manner to the ones defined for the individual power spectra.

Spectral Modeling

As mentioned above, glandular secretion rates can be deconvolved from plasma hormone concentration time series if the functional form of the clearance process is known or can be assumed. Several deconvolution methodologies are available in the time domain (27, 28) for both exponential or biexponential clearance processes. The convolution integral of an exponential clearance function has been shown (29) to be represented by a first-order autoregressive (AR) process.

The spectral signature of an AR process approximates the spectral signature of a first-order Markov process when its lag − 1 autocorrelation coefficient is defined by the AR coefficient:

$$|S(f)|^2 \approx |\bar{s}^2|\{(1 - r_1^2)/[1 + r_1^2 - 2r_1 \cos(\pi f/f_c)]\}^2,$$
$$r_1 \equiv \exp(-1/t_d).$$

Here, $|S(f)|^2$ is the power estimate appropriate to an exponential clearance function, $|\bar{s}^2|$ is the average power estimate from the spectra of the observed plasma concentrations series, r_1 is the lag − 1 autocorrelation coefficient, f is the frequency, f_c is the Nyquist critical frequency, and t_d is the exponential decay time ($t_d = t_{1/2}/0.693$). The corresponding

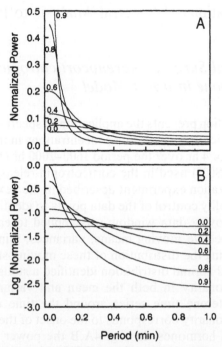

FIG. 3   Power spectra (A) and log–log power spectra (B) of random time series generated by a simple autoregressive (AR) process for a range of lag −1 autocorrelation coefficients (0.0, 0.2, 0.4, 0.6, 0.8, 0.9). All curves are normalized such that total variance is one.

spectral signature is displayed in a power plot format (Fig. 3A) and a log–log format (Fig. 3B).

The spectral signature of a biexponential clearance function can be derived from the weighted sum of the spectral signatures corresponding to the two separate exponential decay times. The weighting factors represent the fraction of the input signal shunted through each of the clearance compartments. In the frequency domain, noise corruption of the observed plasma concentration signal contributes to the spectral power additively, and it can therefore be separately estimated by incorporating it as an additional term in our fitting algorithm, $|S(f)|^2 + |N(f)|^2$. We use a cross-validation scheme in our fitting procedure similar to the one described in Goodman et al. (28). However, a nonlinear regression scheme as outlined in Johnson and Frazier (30) could also be applied.

Specific Applications of Spectral Analysis to Plasma Hormone Time Series

Onset of Diurnal Surge of Adrenocorticotropic Hormone and Corticosterone in a Rat Model

This section presents the application of spectral analysis to coincident ACTH and corticosterone plasma concentrations in integrated 2-min blood samples drawn for 4 hr over the period 1600–2000 hr (120 data points) from a control rat (NRS007) used in the corticotropin-releasing hormone (CRH) immuno-neutralization experiment described in the next section. The analysis begins with quality control of the data points to identify nonstationarities. An eight-point moving data window was applied to each hormonal time series and the respective moving means, variances, minima, and maxima calculated. Comparing the distribution of these moving statistical measures against the expected normal distribution identified a statistically significant step change nonstationarity in both the mean and variance levels of the ACTH and corticosterone time series around the time of lights on at 1800 hr. This nonstationarity corresponds to the onset of the diurnal surge known to occur for both hormones. In Fig. 4A,B the power spectra corresponding to the first and second halves of the ACTH and corticosterone plasma concentration time series are plotted, respectively. Figure 4 illustrates the similarities and differences in the spectral signatures of the hormonal signals before and after the onset of the diurnal surge. As our experiment was designed to characterize the spectral signature of the diurnal surge inclusive over this 4-hr period, the spectral analysis was performed on the entire data set.

The results of applying our DFT spectral analysis algorithm to the NRS007 ACTH and corticosterone plasma concentration time series are displayed in Fig. 5A–C and 5D–F, respectively. In Fig. 5A,D, the individual contributions from the four most significant spectral signature periodic components are displayed, and their combined contributions are overlayed on the time series of the original data values. For ACTH, these spectral signature components were found at periods of 240, 80, 56, and 26 min and explained 90% of the observed variance (the 240-min component explained 72% by itself). For corticosterone the four periods were found at 240, 113, 85, and 21 min and explained 65% of the observed variance (the 240-min component explained 53% by itself). In Fig. 5B,E, the normalized power spectra from which the four significant periodic spectral components were selected are displayed. The 95% white noise normalized power confidence limit appropriate for 120 data points is 0.0327. As expected, because of the integrating nature of the clearance rates convolving the secretion rates, the plasma hormone power

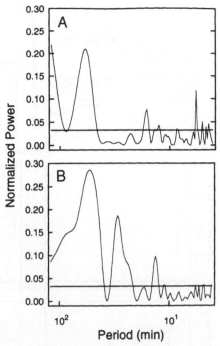

FIG. 4 Power spectra corresponding to the 2 hr before (A) and 2 hr after (B) the onset of the diurnal surge observed in the integrated 2-min sampled plasma ACTH concentrations of rat NRS007. The power spectra are normalized such that the total variance is one.

spectra demonstrate amplified values at the lower frequencies and suppressed values at the higher frequencies.

This can be seen more clearly in Fig. 5C,F, where the normalized power spectra for ACTH and corticosterone, respectively, are displayed on log–log plots along with their bilinear background continuum model fits. Bilinear fits to the ACTH and corticosterone log–log spectral background continua yielded, respectively, negative slopes of −1.937 and −2.379 and break point periods of 15.13 and 38.7 min. The steeper slope and lower frequency break point for the corticosterone background continuum are related to its slower plasma clearance rate relative to ACTH. The higher normalized power for the flat section of the corticosterone background continuum could be related to higher noise-to-signal ratios in that particular assay and/or to greater contributions by the high-frequency components to the secretion rate.

The cross-spectral analysis results for the squared coherency, gain, and phase spectra are displayed in Fig. 6. The individual amplitude and phase

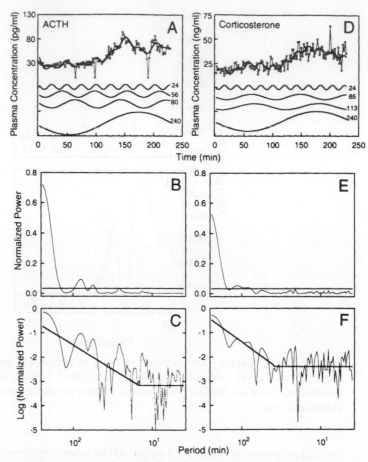

FIG. 5   Integrated 2-min sampled plasma ACTH (A) and corticosterone (D) concentration time series are plotted along with the reconstructed combined and individual temporal contributions from their respective four most significant periodic spectral components. The corresponding ACTH (B) and corticosterone (E) DFT derived power spectra are plotted along with the appropriate 95% white noise confidence level (horizontal line). The ACTH (C) and corticosterone (F) log–log power spectra are plotted along with the respective bilinear model fits to each background continuum. The power spectra are normalized such that the total variance is one.

ACTH and corticosterone spectra were smoothed with a Hanning filter before computing the cross-spectral estimates. Departures of the squared coherency values are indicative of either nonlinearities, high noise-to-signal ratios, or lack of a causal relationship. Squared coherency values close to one are

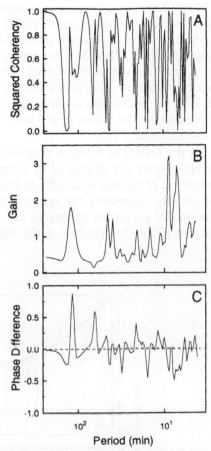

FIG. 6   Squared coherency (A), gain (B), and phase (C) spectra are plotted for the data displayed in Fig. 5. Gain is defined as the ratio of the corticosterone amplitude over the ACTH amplitude at each frequency. Phase difference is defined as the corticosterone phase minus the ACTH phase at each frequency. Positive differences imply that the corticosterone signal leads, whereas negative differences imply that the ACTH signal leads.

indicated for several frequency components which correspond to spectral peaks in the individual power spectra. This implies that a considerable amount of the underlying corticosterone signal structure is coherent with the ACTH signal structure. The majority of the periods with high coherency values appear to correspond to zero phase differences and 50% gain responses. Most likely these components represent the spectral signature of the positive feedforward process of ACTH on corticosterone. The phase spectrum shows two large positive phase differences (corticosterone leads ACTH) near periods of 120 and 20 min with gain responses of 150 and 100%, respectively, and high coherency values. These features can be interpreted as the spectral signature of negative feedback processes of corticosterone on ACTH. High gain values are not always synchronized with high coherency values. For example, the peaks near 7 and 9 min in the gain spectra correspond to low coherency values.

In Fig. 7A the results of fitting a biexponential clearance function to the ACTH power spectra are displayed in a log–log plot. This technique will supersede our use of bilinear model fits to the background continua, because it more accurately represents the spectral signature of the clearance processes. The model parameter estimates are 1 and 30 min for the clearance decay times, 0.37 for the fraction of the input signal shunted through the

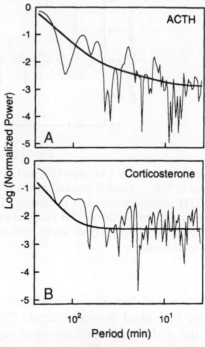

FIG. 7   The log–log power spectra derived from the 2-min plasma ACTH (A) and corticosterone (B) time series for rat NRS007 are plotted along with the respective biexponential and single exponential clearance function spectral models.

fast clearance, and a noise level which corresponds to 0.12% of the total variance contribution to each frequency. In Fig. 7B the results of fitting a single exponential clearance function to the corticosterone power spectra are displayed as a log–log plot. The model parameters are 80 min for the clearance decay time and an error level which corresponds to 0.34% of the total variance contribution to each frequency.

Effects of Corticotropin-Releasing Hormone Immunoneutralization on Plasma Adrenocorticotropic Hormone Signal Structure

To determine the dependence of ultradian ACTH rhythms on corticotropin-releasing hormone (CRH), six rats were passively immunized against CRH (0.5 ml intravenously 3 hr before blood sampling), and six rats were similarly treated with normal rabbit serum (NRS) (31). Two-minute samples were collected over 4 hr by continuous withdrawal with a peristaltic pump with simultaneous infusion of charcoal-stripped, filtered, heparinized rat plasma warmed to 37°C. The ACTH was measured by radioimmunoassay. Resultant hormone concentration time series were subjected to DFT analysis.

A series of amplitude, phase, normalized power expressed as an explained fraction of total variance (R^2), and F ratio values were generated with the DFT algorithm for frequencies corresponding to equally spaced periods between 4 and 240 min. R^2 values greater than twice the white noise level and F ratio values greater than the 95% confidence interval for the appropriate number of degrees of freedom were used to evaluate the statistical significance of the derived spectral estimates (12). Composite power and amplitude spectra for the two treatment groups were constructed for the individual rats (Fig. 8A,B, respectively). We also investigated the possibility of a systematic frequency modulation in the pattern of the signal response (Fig. 8A) with CRH immunoneutralization using the technique of Kwon and Rudman (24).

We tested for two types of mathematical artifacts in the identified spectral coefficients. The first dealt with the possibility that the identified spectral peaks represented aliasing of unresolvable high frequencies. An eight-point (16-min) moving average was calculated for each 2-min time series, and the resultant filtered time series were reanalyzed with DFT at the same frequencies used to analyze the raw data. If the spectral peaks identified in analysis of the raw data were due to aliasing, they would not be present. Second, to address the possibility of nonstationarity of an oscillator at approximately 13 min, a moving regression analysis was performed with a 39-min window on each 2-min time series.

Significant periodic patterns were identified within the individual and composite time series in both treatment groups. Whereas most of the explained

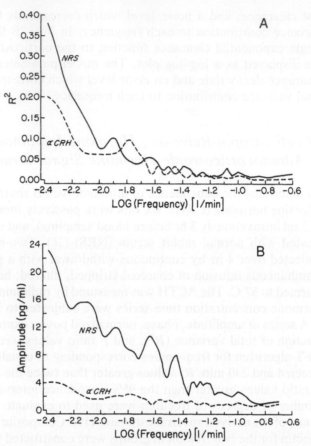

FIG. 8 Composite power (A) and amplitude (B) spectra derived from the individual amplitude and power spectra are plotted for two groups of male rats immunoneutralized against CRH (αCRH; $n = 6$) and normal rabbit serum (NRS; $n = 6$). Note the compression of the time scale by approximately 23% in the αCRH group. From ref. 23, M. Carnes, B. M. Goodman, and S. J. Lent, *Endocrinology* (*Baltimore*) **128**, 902 (1991). © The Endocrine Society.

variance in the spectrum of each rat appeared at the lower frequencies, significant high-frequency structure of pulsatile ACTH activity was detected between 4 and 140 min. The most prominent spectral bands were 4–15, 30–50, 50–80, 110–140, 160–180, and 200–220 min, as well as a longer period that could not be resolved. Individual spectra revealed greater intersubject variation than the treatment group difference. Using the significant spectral peaks identified to reconstruct the time series for individual rats, we found that the DFT results predicted between 29 and 94% of the total observed variance and between 85 and 98% of the observed varaince when the stochastic contribution to the variance was removed.

Immunoneutralization against CRH primarily reduced the amplitude of the waveforms identified in individual rats and removed the low-frequency trend caused by the diurnal surge of ACTH. Comparison of composite spectra revealed a consistent reduction of $81.6 \pm 5.3\%$ ($\bar{x} \pm SD$) in the amplitude response with immunoneutralization for waveforms above periods of approximately 15 min (Fig. 9). Amplitudes of waveforms with periods less than 15 min appeared to be unaffected by immunoneutralization. In addition to the amplitude reduction in waveforms in response to immunoneutralization, comparison of composite spectra suggested frequency modulation as well. The frequency structure itself did not appear to be altered, but the time scale was systematically condensed so that periods were shorter ($22.6 \pm 9.1\%$) in the immunoneutralized rats. This frequency shift was seen only for waveforms with periods above approximately 15 min.

A systematic break in the slope of the linear fit to the plots of background continua occurred between 10 and 15 min in the individual time series, and at approximately 15 min in the composite 2-min time series. This suggests that the background continuum has two components. Oscillations with periods shorter than 15 min appeared to be superimposed on a flat (zero-slope) white noise continuum suggestive of random secretory firing of corticotropes in addition to a coherent rhythmic behavior. Longer periods appeared to be superimposed on a negative-sloped integrating line suggestive of an inte-

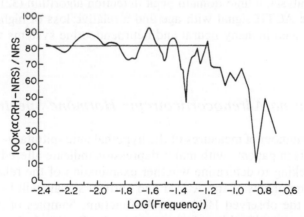

FIG. 9 The frequency-dependent amplitude responses of the immunoneutralization treatment relative to the control treatment are plotted as percent departures, $100(\alpha CRH - NRS)/NRS$. From ref. 23, M. Carnes, B. M. Goodman, and S. J. Lent, *Endocrinology* (*Baltimore*) **128,** 902 (1991). © The Endocrine Society.

grative lagged secretory firing response. This break was unaffected by CRH immunoneutralization.

Digital Fourier transform analysis of data filtered with a 16-min moving average revealed preservation of waveforms with periods above 13 min, indicating that these peaks were not a result of aliasing of a higher frequency oscillator in the initial analysis. Moving regression analysis revealed a locking of the phases for a 13-min oscillator. This finding supports the assumption of stationarity within the time series, a necessary prerequisite for the performance of spectral analysis. A heteroskedastic amplitude modulation with respect to the mean concentration of ACTH was also revealed throughout the time series. Higher ACTH levels corresponded to larger amplitudes, which could not be accounted for by differences in assay variability.

Comparison of Plasma Adrenocorticotropic Hormone Signal Structure in Young and Old Rats

Using a similar experimental paradigm, ACTH time series from 10 young (3–5 months of age) and 14 old rats (20–26 months of age) were compared (5). Despite no difference in mean levels, spectral analysis of the old rats revealed a reduction in amplitudes of spectral peaks, less power at the higher frequency end of the spectrum, and a stretching of the composite spectra by approximately 20%. These findings, which were undetectable with cluster analysis, a time domain peak detection algorithm (32), suggest a slowing of the ACTH signal with age and a relative loss of high-frequency variability as seen in many neural and neuroendocrine systems with age (33).

Human Plasma Adrenocorticotropic Hormone Spectra

A number of measures of the hypothalamic–pituitary–adrenocortical (HPA) axis in patients with major depression indicate altered function (34). We are seeking to determine whether examination of the relatively high-frequency components of the ACTH and cortisol signals will be illuminating in terms of the observed HPA axis dysfunction. Samples of ACTH concentrations from a male subject during an episode of major depression and during remission were collected continuously at 1-min intervals over 3 hr from the antecubital vein. The time series were filtered with a 3-min lowess filter to reduce noise due to assay variability and then subjected to spectral DFT analysis

(Fig. 10). The spectrum of the ACTH time series during depression revealed a peak of spectral power near 10 min accounting for approximately 15% of the overall variability in the time series. This 10-min feature was absent from the ACTH signal measured during the nondepressed state in the same subject.

It would be premature to assign any physical attributes to the identified 10-min cycle, but its presence in the depressed state and absence in the

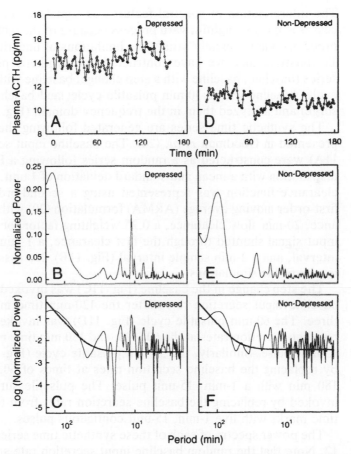

FIG. 10   Time series of integrated 1-min plasma ACTH concentrations collected over a 3-hr period from an elderly male in both a depressed (A) and nondepressed (D) state are plotted after smoothing with a 3-min lowess filter. Note that the spectral peak near 10 min in the power spectrum of the time series collected during depression (B) is absent during remission (E). The depressed (C) and nondepressed (F) log–log power spectra are also plotted along with the respective biexponential clearance function spectral models and indicate that clearance is similar.

nondepressed state is intriguing and certainly underscores the ability of spectral analysis to bring to light characteristics of a hormonal time series otherwise unapparent to the investigator.

Using Spectral Models to Assist with Interpretation of Spectral Features

The interpretation of spectral features as signatures of physiological processes is not a straightforward process because of the fact that physiological process do not typically satisfy the limitations of linearity and stationarity. To illustrate this, we have contructed five synthetic plasma ACTH time series (baseline, baseline with a step change, baseline with a 60-min pulsatile cycle, baseline with a 30-min pulsatile cycle, and baseline with a pulsatile surge) and analyzed them in the frequency domain (Fig. 11).

The synthetic time series are generated following the simulation model presented in Goodman *et al.* (28). The baseline input secretion rates (Fig. 11A) were constructed as a random series following a Poisson probability distribution with a mean and standard deviation of 1 unit. The biexponential clearance function was represented using a second-order autoregressive, first-order moving average (ARMA) formulation (29) with a 1-min fast clearance, 20-min slow clearance, a 0.25 weighting factor for the fraction of the input signal shunted through the fast clearance, a 1-min computation time interval, and a 2-min sample interval (Fig. 11B). No assay error was added to any of the simulations.

The step change in the baseline (Fig. 11C) was invoked by multiplying the baseline input secretion rates after the 120-min time mark by a factor of three. The 60-min pulsatile cycle (Fig. 11D) was invoked by replacing the baseline secretion rates at the 60, 120, and 180 min time marks with a 1-min, 15-unit pulse. Similarly, the 30-min pulsatile cycle (Fig. 11E) was invoked by replacing the baseline secretion rates at times of 60, 90, 120, 150, and 180 min with a 1-min, 15-unit pulse. The pulsatile surge (Fig. 11F) was invoked by replacing the baseline secretion rates from the 120- to 125-min time marks with five 1-min, 15-unit contiguous pulses.

The power spectra of each of these synthetic time series are shown in Fig. 12. Note that the random baseline input secretion rate spectrum (Fig. 12A) has most of its power spread out fairly evenly over the entire spectral range, as expected. Filtering this random series through an integrating clearance function amplifies the lower frequencies and suppresses the higher frequencies. Also note the significant spectral line components between periods of 20–100 min. These are the result of a finite data record and the filter amplification. They do not represent true periodicities, but rather that this particular

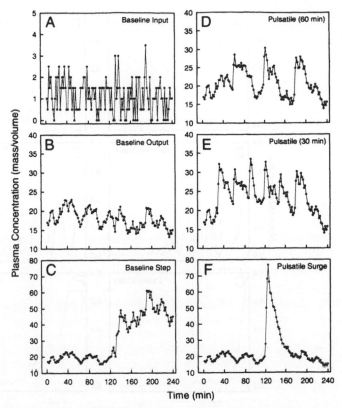

FIG. 11 Plasma concentration time series are plotted for a random baseline secretion rate (A) and the corresponding baseline plasma hormone concentrations (B) for a biexponential clearance function with a 1-min fast clearance, 20-min slow clearance, and 25% of the input signal shunted through the fast clearance. Plots C–F illustrate the plasma concentration response to different pulsatile secretion patterns (see text).

finite series of random numbers demonstrates a significant drift component at those time scales. Taking a different series of random numbers would result in different placement of these "spectral peaks." When a step change nonstationarity is included, the resulting spectrum (Fig. 12C) shows that most of the spectral power is concentrated in the lowest frequencies, which are the components most closely aligned with the shape of the step change.

The 60-min pulsatile cycle spectrum (Fig. 12D) does in fact generate a spectral peak at a period of 60 min, and it is the largest peak. Because of the nonsinusoidal nature of this waveform, the harmonics and subharmonics of the 60-min cycle are artificially enhanced by the Fourier transformation

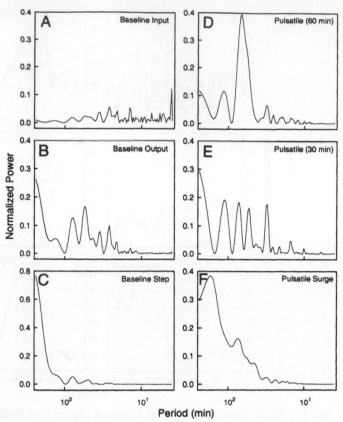

FIG. 12   Corresponding power spectra are plotted for the six different synthetic time series displayed in Fig. 11.

and the spectral filter response of the clearance function. The same is true for the 30-min pulsatile cycle (Fig. 12E), except now the convolution due to the slow clearance is more severe because of the shorter cycle time relative to the slow clearance time. A spectral peak does occur at a period of 30 min, and its harmonics and subharmonics are also artificially enhanced; however, now the largest amount of spectral power is concentrated at the lowest frequencies. The pulsatile surge spectrum (Fig. 12F) is a good example of the difficulty the frequency domain has in representing large-amplitude transient features. The resulting spectrum concentrates most of the power in the lowest frequencies, with a peak at a period of 180 min and logarithmically decreasing with increasing frequency. The peak at a period of 180 min is four times the 45-min duration of the surge feature in the plasma concentration time series.

The corresponding log–log spectral plots are displayed in Fig. 13 with two spectral clearance models overlayed. The dashed line is the biexponential clearance model appropriate for the defined clearance parameters with no additional noise component. The solid line is the same model with a noise component. Including a noise component improves the fit for all the synthetic series. The noise component comes from the computational error inherent in the analysis, as no assay error was specifically added. The relative importance of the computational error is still secondary to the assay error, but it

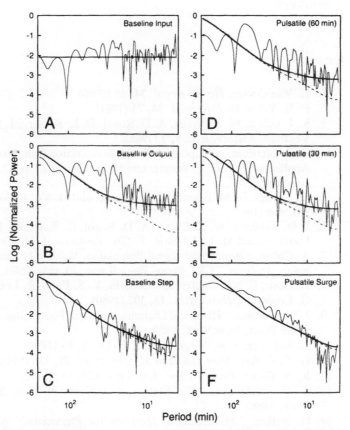

FIG. 13 Corresponding log–log power spectra are plotted along with the spectral signature of the defined biexponential clearance model without a noise component (dashed line) and including a noise component (thick line).

is not insignificant. The spectral signature of the biexponential clearance function can be clearly seen in each of the different synthetic time series.

Summary

The secretion of neuroendocrine hormones and their fluctuations in plasma represent a rich language which offers the possibility of transmitting subtle nuances in meaning to target cells. Investigators studying the complexities of this language must accurately intercept and analyze the constituent hormonal signals. We have presented frequency domain analysis as one method to approach complex hormonal time series using ACTH and glucocorticoid as prototypes.

References

1. E. Van Cauter, *Horm. Regul.* **34,** 45 (1990).
2. F. E. Yates, *Biol. Reprod.* **24,** 73 (1981).
3. R. J. Urban, W. S. Evans, A. D. Rogol, D. L. Kaiser, M. L. Johnson, and J. D. Veldhuis, *Endocr. Rev.* **9,** 3 (1988).
4. G. R. Merriam, *in* "Episodic Hormone Secretion from Basic Science to Clinical Application" (T. O. F. Wagner and M. Filicori, eds.), p. 37. TM-Verlag, Hameln, Germany, 1987.
5. M. Carnes, B. M. Goodman, S. J. Lent, and H. Vo, *Endocrinology* (*Baltimore*) **133,** 608 (1993).
6. J. D. Veldhuis, W. S. Evans, A. D. Rogol, C. R. Drake, M. O. Thorner, G. R. Merriam, and M. L. Johnson, *J. Clin. Endocrinol. Metab.* **59,** 96 (1984).
7. A. Cohen, "Biomedical Signal Processing: Volume I, Time and Frequency Domains Analysis." CRC Press, Boca Raton, Florida, 1986.
8. D. Desir, E. Van Cauter, J. Goldstein, V. S. Fang, R. Leclerq, S. Refetoff, and G. Copinschi, *Horm. Res.* **13,** 302 (1980).
9. J. W. Cortada, "Historical Dictionary of Data Processing: Biographies." Greenwood Press, New York, 1987.
10. A. Schuster, *Terr. Magn. Atmos. Electr.* **3,** 13 (1898).
11. G. T. Walker, *Mem. Indiana Meteor. Dept.* **21,** 13 (1914).
12. R. A. Fisher, *Proc. R. Soc. London A* **125,** 54 (1929).
13. T. W. Anderson, "The Statistical Analysis of Time Series." Wiley, New York, 1988.
14. G. Arfken, "Mathematical Methods for Physicists." Academic Press, New York, 1970.
15. R. G. Athay, "Radiation Transport in Spectral Lines." Reidel, Dordrecht, The Netherlands, 1972.

16. R. A. Meyers (ed.), "Encyclopedia of Physical Science and Technology," Vol. 12, p. 22. Academic Press, Orlando, Florida, 1983.
17. M. Bellanger, "Digital Processing of Signals: Theory and Practice." Wiley, New York, 1984.
18. R. K. Otnes and L. Enochson, "Applied Time Series Analysis: Volume 1, Basic Techniques." Wiley, New York, 1978.
19. P. R. Bevington, "Data Reduction and Error Analysis for the Physical Sciences." McGraw-Hill, New York, 1969.
20. R. H. Odell, S. W. Smith, and F. E. Yates, *Ann. Biomed. Eng.* **3,** 160 (1975).
21. S. M. Pincus and D. L. Keefe, *Am. J. Physiol.* **262,** E741 (1992).
22. R. McCleary, R. A. Hay, Jr., E. E. Meidinger, and D. McDowell, "Applied Time Series Analysis for the Social Sciences." Sage Publ., Beverly Hills, California, 1980.
23. M. Carnes, B. M. Goodman, and S. J. Lent, *Endocrinology (Baltimore)* **128,** 902 (1991).
24. B.-D. Kwon and A. J. Rudman, *Math. Geol.* **11,** 373 (1979).
25. P. Z. Marmarelis and V. Z. Marmarelis, "Analysis of Physiological Systems: The White-Noise Approach." Plenum, New York, 1978.
26. M. Rosenblatt, "Time Series." Wiley, New York, 1963.
27. J. D. Veldhuis and M. L. Johnson, *Bio/Technology* **8,** 634 (1990).
28. B. M. Goodman, M. Carnes, and S. J. Lent, *Life Sci.* **54,** 1659 (1994).
29. V. Guardabasso, G. DeNicolao, M. Rocchetti, and D. Robard, *Am. J. Physiol.* **255,** E775 (1988).
30. M. L. Johnson and S. G. Frazier, *in* "Methods in Enzymology" (C. H. W. Hirs and S. N. Timasheff, eds.), Vol. 117, p. 301. Academic Press, San Diego, 1985.
31. M. Carnes, S. J. Lent, B. M. Goodman, C. Mueller, J. Saydoff, and S. Erisman, *Endocrinology (Baltimore)* **126,** 1904 (1990).
32. J. D. Veldhuis and M. L. Johnson, *Am. J. Physiol.* **250,** E486 (1986).
33. L. A. Lipsitz and A. L. Goldberger, *JAMA, J. Am. Med. Assoc.* **267,** 1806 (1992).
34. P. W. Gold, F. K. Goodwin, and G. P. Chrousos, *N. Engl. J. Med.* **319,** 413 (1988).

[8]   Monitoring Dynamic Responses of Perifused Neuroendocrine Tissues to Stimuli in Real Time

A. R. Midgley, R. M. Brand, P. A. Favreau, B. G. Boving,
M. N. Ghazzi, V. Padmanabhan, E. Y. Young, and
H. C. Cantor

Introduction

Although neuroendocrine chemical signaling is widely viewed as time-dependent changes in concentration, over a decade ago the importance of stepping back and viewing these sequential changes from a broader perspective, as patterns, began to receive emphasis. Indeed, in 1981, Eugene Yates encouraged the neuroendocrine community to study the "pharmacolinguistics" of cellular communication (1). Yates posed that the episodic fluctuations and pulse trains so common to endocrine signaling likely contain meanings that are hidden from us; these time-dependent, concentration patterns contain functional commands that modify cellular behavior. We have made much progress in understanding signal-generated, receptor-coupled intracellular mechanisms since then, but we still cannot extract the informational content in trains of episodic signals. Difficulties with testing hypotheses include the need to isolate cellular systems, keep them functioning under conditions closely mimicking those found *in vivo,* apply test signals in definably controlled and complex patterns for sufficient periods, record the response of the cells to applied inputs over long periods, and devise new methods to monitor high-frequency signals. Although not precluding such studies, logistical problems of controlling the environment, delivering definable patterns of signals, and monitoring responses by collecting and assaying discrete samples have thwarted the efforts of would-be endocrine linguists.

We started developing our perifusion system with the objective of finding solutions to these problems. Although we cannot claim to have accomplished our goal, by building on the efforts of many predecessors (including Refs. 2–9), we have developed a miniature perifusion system that can maintain cells in a functionally responsive state for many days, minimize dispersion of delivered and response signals, deliver defined and complex signal patterns automatically in an unattended mode, and acquire data reflective of changing functional states in near real time. We begin a description of the system with

Methods in Neurosciences, Volume 28

a few comments on the effects of a moving fluid on the integrity of a signal and on problems associated with sampling and the acquisition of necessary data.

Fluid Flow

The distribution of capillary networks ensures that, with a few special exceptions (e.g., granulosa cells of the ovarian follicle, chondrocytes, and cells of the lens of the eye), no cell is more than about 100 μm from a source of oxygen and other nutrients and concomitantly a site for removal of waste products (10). Transfer of gases from pulmonary alveoli to blood ensures that concentrations of oxygen are never toxic, yet circulating hemoglobin ensures the transfer of enormous quantities of oxygen and carbon dioxide. The average rate of blood flow in tissue [roughly 1 mm/sec through capillaries with mean lengths of 0.5–1 mm (11)] ensures that local concentration of waste products like ammonia and lactate do not build up. Contrast these conditions to those in static culture where cells sit in their waste products in the absence of dynamic changes in paracrine and other factors. Further, if the medium exceeds 1 mm in depth, oxygen is exhausted (12).

Soluble molecules move into and within interstitial fluid by several means including convection, electrophoresis (13), and, most importantly, diffusion. While convection across capillary endothelium accounts for 60 μl/100 g tissue/min, diffusion, the result of random molecular motion, accounts for 5000 times more fluid movement [300,000 μl/100 g tissue/min (11)]. Diffusion rates for any molecule can be accurately modeled according to Fick's law or described in terms of mass transfer coefficients (14).

Perifusion (or superfusion) systems, like perfusion (flow through vascular components of a tissue), have the potential to mimic conditions *in vivo* far more closely than can static culture systems. By means of a moving stream, perifusion systems replace nutrients and remove cellular waste products continuously and have the potential to equal or exceed rates observed *in vivo*. This may be one reason why aggregates of interstitial cells from rat testes produced six times more steroid in perifusion culture than in static culture (15) and why the dose of growth hormone-releasing hormone antagonists causing 50% inhibition of growth hormone release from pituitary cells was 11 times greater in static than in perifusion cultures (16).

In all moving streams, including the fluid in a perifusion system, molecules at the leading and trailing fronts of a signal are spread forward and backward creating two parabolic boundaries. This spreading, axial dispersion, results from the combined effects of diffusion and laminar flow: the center of a stream moves faster than its periphery. As a consequence, some molecules in a signal reach a given point prior to, and other molecules later than, the time predicted solely on the basis of convective flow. The leading and trailing boundaries created by axial dispersion are modified further by radial diffusion of molecules between the boundary and flow layers (17). In a perifusion system, these processes distort a signal as molecules progress through the tubing. Taylor showed that, under appropriate conditions, this distortion can be modeled accurately as a function of molecular diffusivity, flow rate, tubing radius, and system volume (17).

While not encompassed by the Taylor model, other problems can arise from turbulence and from solute interacting with the tubing wall, either by adsorption or absorption. The point at which turbulence occurs can be calculated. If the Reynolds number (Re) determined as

$$Re = \frac{\text{flow rate} \times \text{density} \times \text{diameter}}{\text{cross-sectional area} \times \text{viscosity}}$$

is less than 2000 at all points in the system, then turbulence is not a problem. These various considerations dictate that optimization of perifusion systems requires attempts to (1) minimize distortion (reduce tubing diameter, eliminate dead zones, avoid changes in tubing diameter to reduce eddy currents, and keep distances short); (2) adjust flow rate to be compatible with detecting output signals, removing waste, avoiding end-product inhibition, and providing adequate nutrients—all while avoiding shear forces and turbulence; (3) minimize adsorption and absorption through careful selection of tubing and connector composition; and (4) reduce nonspecific binding.

To address these concerns, we let principles of fluid flow derived from chemical engineering guide our selection of tubing diameters and distances, cell compartment dimensions, and flow rates. To minimize dispersion we kept diameters small and distances short while restricting the number of changes in diameter (each of which can cause local eddy currents and more dispersion). Flow rates were selected to ensure that the mean velocity of medium in the cell compartment would provide fluid exchange that exceeded known diffusional rates calculated for interstitial fluid *in vivo* (3 μl/mg tissue/min) and yet not exceed the known mean velocity of fluid in capillaries (1 mm/sec) to avoid shear and turbulence. Upward flow past cells was selected to minimize obstruction at the chamber outflow filter. The cells were supported by nonporous, hollow, inert glass beads with near neutral buoyancy (1.02 g/cm^2) to create a packed bed fluidized reactor that reduces dispersion and permits modeling with principles of chemical engineering. Also addressed were concerns for maintaining a stable environment free of perturbing influences, a means for controlling flow rates and signal introduction, and means for sampling at high frequency including use of on-line sensors, special fraction collectors, and computer-control systems.

Data Acquisition

Defining the patterns of endocrine signaling requires describing a rich variety of episodic pulse shapes, amplitudes, and periodicities. To define the frequency components of a pulsatile signal, the Nyquist theorem (18) specifies that the frequency of sampling must be at least twice that of the component with the highest frequency. Thus, if cellular responses were characterized by secretory bursts every 20 min, to assess the profile and avoid aliasing, sampling would need to occur at least once every 10 min. Because pulsatile signals are not sinusoidal, the highest frequency components can be much faster than estimated from interpulse intervals. For example, if the components of interest are the slope of the rising phase of a pulse or the amplitude of a peak and one desires to reconstruct an accurate view of these components free of distortion, then the needed frequency of sampling is at least twice that of a sinusoidal wave fitting this slope. Yates recommends sampling at 6 times the highest frequency of interest for most endocrine studies (1) and Zuch recommends at least 10 times the frequency of the fastest component (19). This places severe constraints on sampling, sample processing, assay, and analysis.

We are implementing two approaches to facilitate higher frequency sampling. The first utilizes immunoassays that require little manual effort and provide results in an electronically readable form. The second utilizes automated methods to monitor related parameters of cellular activity such as changes in metabolism and active secretion. Metabolic activity can be assessed by monitoring changes in dissolved gases like oxygen, carbon dioxide, and ammonia and ions like hydrogen, calcium, potassium, and lactate. Miniature commercial electrodes are available for a number of these ions. Because many molecules oxidize at characteristic potentials determined by their chemical structure, easily constructed electrodes can be used to monitor the secretion of a number of these agents or cosecreted molecules by amperometry, an approach used for monitoring the release of catecholamines from adrenal medullary cells in perifusion (2, 3). By applying a cyclical potential over time, and measuring resultant oxidatively derived current flow as a function of applied potential, these various molecular species can be quantified and partially identified. We have taken advantage of this approach to measure electroactive (oxidizable/reducible) molecules in the perifusate immediately before and after the cell chamber. The combination of high frequency monitoring and specific immunoassays of analytes in perifusate collected in discrete fractions has the potential to reveal new insights into the dynamics of intercellular signaling.

The rest of this chapter is organized into three major sections: the first provides a detailed description of our current system, the second describes how the system is operated, and the last illustrates the capabilities of the system, including its physical characteristics and ability to reveal insights into the behavior of living cells.

The System

Overview and Design

Our miniature perifusion system was designed to (1) provide constant temperature with minimum electromagnetic field noise; (2) control medium and gas tension around the cells; (3) provide optimum medium flow rate with minimal dispersion; (4) deliver defined signal patterns independently of flow rate; and (5) acquire data from sensors on-line, at high frequency, and in real time. These objectives were met mainly by custom development of hardware and software.

The diagrammatic illustration of the system provided as Fig. 1A serves as an overview prior to detailed description. In this diagram the ends of tubing, generally with attached fittings, are numbered. (Throughout the following description, all numerals enclosed in braces will refer to the tubing ends in Fig. 1 or to the component adjacent to the tubing terminus; for example, {1} refers to the connector attached to a syringe containing control medium.) As illustrated here, control and test media, preequilibrated with 5% (v/v) CO_2 in air at 37°C and sealed in syringes, are placed in pumps in a refrigerator. The pumps deliver the media through separate, chemically inert tubes {1,2} into an incubator that is relatively free of electromagnetic fields. The control and test media are joined via a pneumatically activated injection valve just prior to entering the cell chamber (Fig. 1A). The latter is loaded with freshly dispersed cells just prior to the initiation of perifusion. Outflow from the chamber is directed to a fraction collector. Optionally, ion-selective or amperometric sensors are included immediately prior to and after the cells. In some cases, a separate, continuously operating syringe pump is used as illustrated in Fig. 1B. In this case, fluid leaving the two pumps (each with two syringes) {1,2} joins at a Y connector {11} and then passes into the chamber assembly without an intervening valve. Each of these and alternative components will now be described.

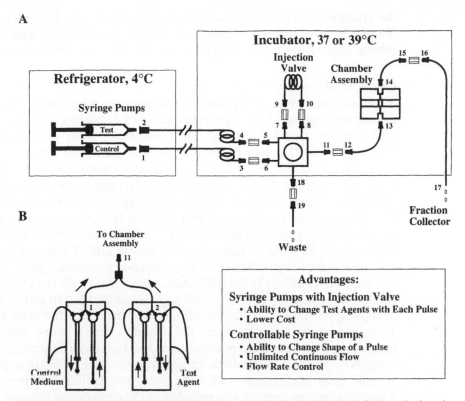

FIG. 1   Exploded, diagrammatic illustration of the overall medium flow path showing fittings and connectors. Two alternative systems are used for controlling the delivery of perifusate with intermixed pulses of test agents and controls, each of which can be run automatically with a computer (see text). (A) Conventional syringe pumps with injection valve. This approach uses conventional syringe pumps with a commercially available low-pressure, high-performance injection valve to deliver rectangular pulses. This arrangement provides the opportunity to change test agents at any time but cannot be used to modify pulse shape and requires periodic refilling of the syringes. (B) Controllable, continuous duty syringe pump. This approach can be operated indefinitely since, as one syringe of a pair empties, the other can be filled. It can deliver pulses of different shapes and flow rates but, in addition to being more expensive, does not permit easy changing of agents.

Chamber Assembly without Sensors

Definitions

For clarity in the following description, the word chamber refers to the space in which the cells are confined, chamber module to the module that contains the chamber, and chamber assembly to the chamber module enclosed with filters and bolted between apposing modules with connecting inlet and outlet tubing. Optionally, intervening electrode modules may be included.

Chamber Modules General

As illustrated in Fig. 2A,B, the cell chamber modules are constructed by drilling and tapping blocks of acrylic plastic cast sheeting (type GP holds up well with repeated autoclaving). Inlet and outlet holes were drilled $\frac{3}{16}$ in. deep with a No. 21 drill and a flat tip drill of the same size. They were tapped with a No. 10-32 flat tip tap to receive miniature polypropylene fittings (No. FITM156, Neptune Research, Maplewood, NJ) that compress the flared end of Teflon tubing running through the fitting against the bottom of the hole. Chamber modules can be made with different thicknesses and diameters of the chamber hole to provide chambers of different capacities. The one most often used is 32 μl (2.69 mm diameter, No. 36 drill, 5.563 mm high). This volume is reduced by including sufficient hollow, nonporous, glass, microcarrier beads (No. 102-915; SoloHill, Ann Arbor, MI) to fill the chamber. The beads, with a density of 1.02 gm/ml and measuring 90–150 μm in diameter, occupy approximately 17 μl of volume, leaving 15 μl for cells and fluid. Their nonporous nature reduces the chamber dead volume and provides better flow dynamics than porous beads like Cytodex (Pharmacia, Piscataway, NJ). Watanabe and Orth earlier reported this effect in a perifusion system: ovine corticotropin-releasing factor was excluded by porous beads while arginine vasopressin was retarded (9). The beads and cells are confined within the chamber by Nucleopore No. 111116 PC membranes with 12-μm pores. A drilled and tapped side port, sealed with a 4-40 Allen head stainless steel screw and a Tygon tubing collar or gasket, provides loading access to the chamber module. This modular design not only provides flexibility in chamber sizes but also opens the possibility of studying sequential cellular interactions by inserting additional chamber modules.

Chamber Assembly with Amperometric Sensors

Amperometric Electrode Modules

When amperometric (redox) sensing is desired, we include electrode modules on both sides of the cell chamber (Fig. 2C). Each module contains a lumen that is of the same inner diameter (ID) as the chamber and is spanned with two 26-gauge platinum electrodes (serving as working and counter electrodes) and one 26-gauge silver reference electrode electrolytically coated with silver chloride. Data are collected from both ends of the chamber at intervals from

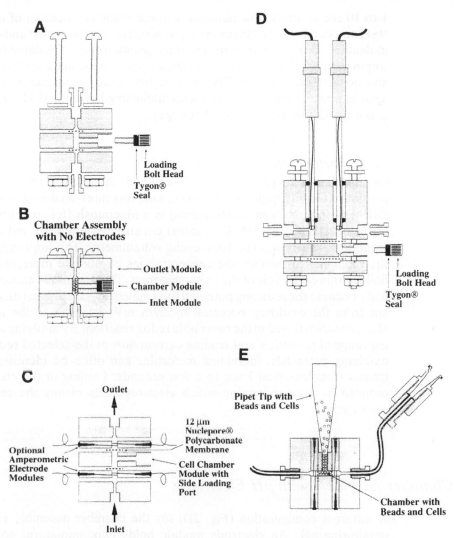

FIG. 2   Five side views, not to scale, of chamber assemblies with optional electrode modules. (A) Exploded view of the assembly without electrodes. (B) Chamber assembly without electrodes clamped together and with the loading port sealed by a bolt. (C) Exploded view of a chamber assembly with an amperometric electrode module on either side of the chamber module. The clamping bolts are not shown. (D) Chamber assembly with dual pH electrodes. Medium delivered to the first electrode passes down through two membranes, across a channel, up through the chamber with cells and beads enclosed by the membranes, past the second electrode, and then out to a fraction collector. (E) Chamber assembly with amperometric electrodes being loaded on its side with a suspension of beads and cells.

1 to 10 sec to study flow patterns, cellular viability, location of introduced test compounds, and changes in output of one or more secreted and oxidizable molecules. In the case of ovine pituitary gonadotropes stimulated by gonadotropin-releasing hormone (GnRH), these appear to include the chromogranins and/or secretogranins (20–22). These highly acidic glycoproteins are copackaged in granules and cosecreted with luteinizing hormone (LH), which itself is not oxidizable at the applied voltages.

Amperometric Data Acquisition

On-line data are collected via a 16-channel, direct memory access, high-resolution (16 bit), high-speed (55 kHz sampling rate), analog-to-digital board (NBMIO-16X; National Instruments) in a Macintosh IIci computer. To improve specificity, we (H. C. Cantor) constructed circuitry and developed computer algorithms to perform cyclic voltammetry. This approach provides the capability to monitor the concentrations of multiple molecules sequentially within each cycle of the voltammogram. This technique enhances specificity because the reducing potential for a molecule is characteristically different from the oxidizing potential by $59/n$ mV, where n is the number of electrons transferred in the reversible redox reaction. By applying an oscillating range of potentials and reading current flow at the selected reducing and oxidizing potentials, individual molecules can often be identified in time frames from less than 1 sec to a few seconds. Fouling of the electrodes is reduced by voltage cycling which electrolytically cleans the sensor with each cycle.

Chamber Assembly with pH Electrodes

A different configuration (Fig. 2D) for the chamber assembly is used for monitoring pH. An electrode module holds two, miniature, combination electrodes (Model MI-710, Microelectrodes Inc., Londonderry, NH) that are each passed through two pairs of O rings (No. ORB-006, Cole Parmer, Chicago, IL) in two holes drilled into a $1 \times 1.5 \times 1$ in. tall block of acrylic plastic. The middle $\frac{7}{32}$ in. thick chamber module with a side loading port contains the cell chamber. The bottom, connector module consists of a $\frac{7}{32}$ in. thick block of acrylic plastic with an inscribed, horizontal channel (~0.3 by 13 mm) lining up with the two vertical channels in the chamber module. Holes are drilled into the corners of all three blocks, permitting them to be bolted together.

Flow Path and Residence Time Distribution

Medium flows into the lower portion of the first electrode chamber, passes by the first electrode, down through a small hole (1 mm ID) in the chamber module, and across the channel in the top of the connector module. Next, it passes up through the bottom membrane, into the cell chamber and through the upper membrane before it reaches the second electrode. The fluid then exits the electrode module via a second side port and passes into a tube leading to the fraction collector or waste. This arrangement positions the miniature combination electrodes vertically to ensure continued slow flow of contained electrolyte through the junctions. Volume contained by tubing and connectors from the inlet of the injection valve or the joining of two tubes at a Y connector to the first electrode approximates 1 μl, each electrode chamber adds about 4 μl, and the channel between the first electrode and the chamber adds approximately another 1 μl.

pH Data Acquisition

Electrometer units, designed and constructed by KMS Fusion, Inc. (Ann Arbor, MI) in accordance with Mosca *et al.* (23), are used to amplify potentials. Data are collected at 1-sec intervals via a 32-channel, analog-to-digital converter (Opto 22, Optomux, Huntington Beach, CA) connected to a Macintosh computer. All data and events are recorded electronically by means of a custom-developed software program and written into a spreadsheet-compatible document. To reduce electronic noise and maintain stable signals, the fluid stream is grounded and a grounded metal screen surrounding the incubator is used to simulate a Faraday cage.

Sensor Data Collection and Reduction

Data Acquisition Software

Data acquisition software developed in-house provides control over and flexibility in setting collection rates, data display modes, printing, and storage. Further, the software permits real-time analysis of the experiment and monitoring of viability and treatment efficacy. Currently, the data acquisition software is run on a Macintosh IIci computer separately from the computer that is controlling the pumps and valves. To facilitate data storage and ease data analysis, the data storage routines were written with algorithms that compress and split data into subsets of convenient size. Data can be displayed continuously as plots or printed and stored to disk in a format compatible with commercial graphing packages, all while an experiment is in progress. Users can interrupt the graphical display of results to insert annotations.

Once an experimental protocol is developed, the perifusion experiment can be run automatically according to the protocol. The software and hardware allow multiple chambers to be monitored simultaneously while collecting up to a total of 16 channels of data. Results from the experiments can be displayed simultaneously in multiple, adjacent, user-sizable windows or in user-selectable single, full-screen windows. The user has full control over which channels of data are to be displayed. Full control of off-line printing/displaying is also provided, thereby permitting examination of different subsets of data.

Pumps and Agent Introduction

Criteria

The ideal pumping system would operate indefinitely, be responsive to external computer-derived signals, provide any desired time-concentration profile and flow rate, introduce little flow fluctuation, preserve concentrations of dissolved gases, not introduce contaminants, and permit introduction of different agents independently with any desired timing and duration. We have compared and developed a number of pumps and have not found any pump or system that meets all of these criteria (24).

Peristaltic Pumps

Our initial focus on peristaltic pumps positioned upstream of the cells was abandoned when we recognized that these pumps require substantial lengths of Silastic, PVC, or similar elastic tubing (which can release potentially harmful nonpolymerized monomers, absorb lipophilic substances, and dissipate dissolved gases) and also introduce pulsatile flow, create intraline variability, and cause significant electronic noise with ion-selective electrodes. The noise results from streaming potentials introduced as the squeezed tubing begins to open or close (25). Some of these problems are diminished when the pumps are positioned downstream of the cells, but this introduces substantial additional dispersion of the response signal as it traverses from the cell chamber to the fraction collector.

Syringe Pumps

Although conventional syringe pumps (e.g., Harvard, Model 975) require periodic refilling, as long as work schedules and experimental demands can accommodate this constraint, they can work well for constant, nonvariable delivery. Additionally, by using coupled, dual syringe pumps it is possible to deliver medium continuously. All syringes are filled with medium which, to reduce bubble formation, is preequilibrated with the desired gas mixture at the temperature to be used for the perifusion. The syringes are then capped until use. This approach ensures that the proper concentration of dissolved gases will be delivered.

Syringe Pumps with Injection Valves

As illustrated in Fig. 1A, we use conventional syringe pumps in combination with low-pressure, pneumatically actuated valves (No. 5020, Rheodyne, Cotati, CA) to inject rectangular pulses of different concentrations and types of agents at defined times. Air-actuated rather than electronically actuated solenoid-switched valves were selected to avoid generating electromagnetic fields within the incubator. Changes in air pressure are controlled by remote solenoids (Rheodyne model 7163) operated in response to computer-controlled, X-10 modules (Radio Shack). The selected injection valves have no dead volume and can be positioned close to the cell chamber. Tubing supplied with the valves was replaced with the shortest feasible length of small diameter (0.3 mm ID) tubing. This brings the volume of the region between the valve and the cell chamber {13} to less than 1 μl.

As illustrated in Figs. 1A and 3, loops of tubing of different lengths can be attached to the valves {9,10} to provide virtually any desired injection volume. These loops can be filled independently while delivery of control medium continues (Fig. 3A). On switching the valve (Fig. 3B), medium enters the loop and the displaced test agent is delivered into the chamber module. Throughout, flow continues at the same rate with almost no detectable interruption. By reversing the valve substantially before all the test agent is displaced from the loop by control medium, the desired agent can be introduced as rectangular pulses. This process can be repeated indefinitely by

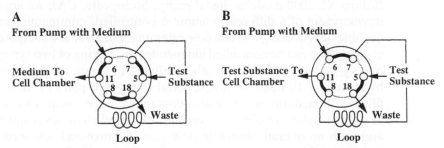

FIG. 3 Diagrammatic illustration of the two alternative flow paths in the injection valve. (A) The loop of tubing can be filled while the cell chamber continues to receive medium. (B) Switching the valve diverts the delivery stream into the loop forcing the contents of the loop into the cell chamber with essentially no interruption in flow.

recharging the loop with fresh agent during a period of control medium flow. To minimize loss of biological activity, syringe pumps containing agents waiting for delivery to the loop are stored in a small refrigerator adjacent to the incubator. Under computer control, the loop is automatically filled by turning on an X-10 module-controlled, refrigerated syringe pump sufficiently in advance of the time of injection to permit the contents of the loop to reach the temperature of the incubator.

Continuous Flow Syringe Pumps

To achieve precise computer control of delivery we initially developed two continuous flow microsyringe pumps (24). Each pump contains up to 12 pairs of 100-μl Hamilton gas-tight glass syringes with Teflon-coated plungers. The 12 pairs of syringes are arranged horizontally, facing in opposite directions, with the barrels fixed and the plungers attached to a slider plate driven by a computer-controlled stepping motor. The syringes are connected to all-Teflon, three-way miniature solenoid valves (Neptune Research, Maplewood, NJ), and configured such that as one member of a syringe pair fills, the other member of the pair empties. The inlet port of each solenoid valve is connected to a reservoir; one outlet port leads to the cell chamber, and the other to a syringe. By moving the solenoid valves just as the syringes reverse direction, fluid is delivered continuously at rates controlled by computer. By controlling the two pumps, one with a test agent and the other with medium, any desired signal pattern can be delivered with mixing immediately prior to the cell chamber. Although this setup is successful in providing a controlled delivery system, bubbles that are difficult to remove tend to appear and become trapped in the syringe tips.

An alternative approach is illustrated in Fig. 1B. Using commercially available, digitally controlled syringe pumps designed for pipetting stations (Cavro XL 3000 modular digital pump; Sunnyvale, CA), we are completing development of a different, computer-controlled, continuous syringe pump capable of delivering variable flow rates under control of a computer. Because each syringe can be controlled independently, pairing of two syringes permits one syringe to fill while the other delivers at a rate that can be controlled by computer through an RS232 serial port. Two pairs of these pumps, all filled with medium, one pair also containing a test agent, can be controlled to deliver pulses of different concentration and shape over indefinite times and with no overall change in flow rate. Control and test media from the two pairs of pumps are joined using flared tubing fittings and tapped plastic T's just prior to entering the chamber of electrode holder. Two 55-cm lengths of Teflon tubing with medium, one from each pump, are coiled in the incubator to achieve temperature equilibration before joining at the cell chamber.

Because each syringe is operated in a vertical position with tips up, bubbles are readily cleared.

Software for Control of the Pumps

Software algorithms to control the various pumps and valves have been developed in BASIC and are being ported to LabView 3. Fluid flow rates can be determined and entered into a table in the computer before an experiment begins, thus allowing unattended operation. The preprogrammed schedule can be changed at any time during an experiment.

Relative Strengths of the Two Pumping Approaches

The two systems of pumping offer different advantages and disadvantages. Each can be used to deliver trains of pulses with different interpulse intervals. Conventional syringe pumps with injection valves provide the opportunity to change test agents as often as desired, solely through rinsing and filling the loop before each pulse. However, this system does not provide a means to control pulse shape (e.g., vary the rate of increase or decrease) and cannot be operated for long periods without interruption since the syringes must be refilled. In contrast, the controllable, continuous duty paired syringe pumps provide the advantage of being able to change dynamically the shape (concentration changes over time) of each pulse and more generally to deliver essentially any desired concentration profile over time, all in accordance with an algorithm or stored table describing an experimental protocol. The disadvantage is that this pumping system is largely restricted to use of only a single agent during an experiment. However, use of further pairs of pumps permits introduction of additional agents, all of which can be superimposed in essentially any desired profiles (e.g., superimposition of pulses of one agent on a gradient of a second agent). By using small syringes, typically one milliliter capacity, the continuously operating pumps also provide more precise fluid delivery than is possible with conventional syringe pumps which, to avoid too frequent refilling, are used with large syringes, typically 60 ml capacity.

Assessment of Pump Performance and Calibration

To test the performance of the pumps, we deliver water into a tared container on an automatic balance with digital RS 232 output (Model SM480, Mettler-Toledo, Inc., Worthington, OH). By recording data in a computer file and later plotting the change in accumulated weight of delivered water over time, an accurate assessment of pump performance (flow rate and variability over time) can be obtained (the slope of the measurement, mass of water in mg divided by time in minutes gives approximate flow rate in microliters/minute). Plots of these accumulated weights have proven to be indistinguishable from a straight line.

Tubing and Connectors

Tubing Selection

The selection of tubing, and design of tubing connections, were aimed at minimizing dispersion, gas diffusivity, adsorption, absorption, and turbulence. Dispersion was minimized by reducing tubing diameter to an extent that did not cause problems with obstruction, avoiding changes in tubing diameter that could create pockets and places for eddy currents to form, and eliminating dead spaces. Gas diffusivity is a function of tubing composition. Teflon and silica have very low diffusivity for oxygen and carbon dioxide. These same materials generally show low adsorption and absorption. Accordingly, all tubing is Teflon [1.59 mm outer diameter (OD) with 0.3 mm ID for tubes connecting the valve to the chamber and the chamber to the fraction collector and 0.5-mm ID from the medium pump to the valve and the valve to waste; Anspec, Inc., Ann Arbor, MI] except for the final tip entering the fraction collector which consists of an inserted section of a 31-gauge needle or 0.18-mm ID, deactivated, fused silica tubing (J&W Scientific, Folsom, CA).

Tubing Connections

Connections between segments of tubing were made by attaching color-coded, flanged-tube fittings designed for low-pressure, high-performance, liquid-chromatography systems (Anspec) and flanging the ends of the tubing with a 1.59-mm OD heated flanging tool (with tip for 0.3-mm ID tubing; Anspec). The flanged ends with attached, externally threaded 1/4-28 chemically inert fittings (Fig. 2E; Anspec) can be clamped end to end with an internally threaded polypropylene coupling (e.g., between {3} and {6}) or screwed into suitably tapped holes in the cell chamber assembly (10–32 fittings {13,14}, part FITM149; Neptune Research).

Temperature Regulation

Incubator

The incubator was constructed as a box (in multiples of 12 in. × 12 in. × 12 in., generally of $\frac{1}{4}$ in. acrylic plastic) with a front, air-tight, access door ($\frac{3}{8}$ in. acrylic) and multiple tubing and cabling ports sealed by stoppers or foam plugs. Acrylic plastic was selected for its high transparency, low cost and low heat loss. Heating is by recirculated thermoregulated air as described later.

Avoidance of Inductive Fields

The inclusion of ion-selective or amperometric sensors requires avoidance of high electromagnetic fields, a need compounded by concerns regarding the possible effect of inductive fields on cellular release of hormones. Our initial attempts to use conventional water baths or commercial tissue culture incubators were abandoned when disruptive field effects were observed as the heaters cycled. Eventually we chose to control the temperature by a rapidly recirculating stream of air. By placing the heater and other high current devices including solenoids at some distance from the cells, placing the cells in a separate incubator box, and connecting the box to the heater by long, large diameter tubing, we achieve a high degree of isolation. Figure 4 illustrates this approach.

Circulating Warm Air

For the heater, we removed and discarded the fan from a hair drier. The heater elements and thermal safety devices were then attached to a 3030 rpm squirrel cage type blower (Model 30-E, Electronic Cooling Devices, Inc., Van Nuys, CA) and placed in a sealed styrofoam box (picnic chest).

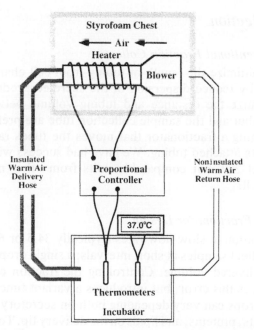

FIG. 4 Diagram of the system used to maintain a closely controlled temperature in an incubator for perifusing cells remote from electromagnetic fields.

Heat-resistant hose (5.5 feet, 1.75 in. ID; vacuum cleaner or comparable hose works well) connects the styrofoam box with the incubator. The hose conducting air from the heater to the incubator was insulated with outer foam tubing to reduce heat loss, whereas the return hose was left uninsulated to provide heat loss for smoother temperature control. Air in the styrofoam chest (~0.9 ft^3) and the incubator (~3 ft^3) provide some degree of thermal mass for temperature buffering. Blower operation is assessed by a small string dangling in front of the warm air supply orifice.

Temperature Control

A Dyna-Sense proportional temperature controller (Cole Parmer, Model L-02156-00) and a thermistor-based air temperature probe (YSI, Yellow Springs, OH, Model 405) provide the control system for the heater. The thermistor probe was mounted near the center of the incubator together with a second temperature sensor serving an external, digital display (Radio Shack). The system is operated at either 37° or 39°C (the latter being closer to the core body temperature of sheep). Typically the system operates with no greater than ±0.2°C deviation over 24 hr.

Fraction Collection

Conventional Fraction Collector

To minimize the length of tubing from cell chamber to collection tube and thereby reduce dispersion of the secreted product, fraction collectors that minimize the distance and tubing volume between the outlet of the cell chamber and the sample collection tube are preferable. This usually means selecting a fractionator that moves the tubes rather than the delivery arm and its attached tubing. We have had success with the Buchler Fractomette model 200, but comparable units from other manufacturers should work as well.

New Fractionator Design

Operation at slow flow rates (typically 34.5 or 69 μl/min) makes it difficult to collect samples at short intervals: a single drop can make a large difference in delivered volume. Controlling the fraction collector by a drop counter reduces this error, but introduces a variant time series. Further, the size of the drops can vary depending both on secretory activity and on the buildup of salts, proteins, and lipids on the delivery tip. To address these concerns and provide other advantages, with the assistance of Allen C. Ward, University of Michigan, we have designed and are now completing a fraction collector based on the use of 96-well microplates. The tubing leaving the incubator attaches to an arm that moves the delivery tip in two directions: to the bottom of each well and to succeeding wells in a single row of eight. When delivery time is complete, the tip rises through the fluid meniscus, thereby wiping the tip, avoiding the inaccuracies introduced by drops and creating an invariant time series for later analysis. The raised tip then moves laterally to the bottom of the next well in the row. On completing delivery into the eighth well of a row, the conveyer belt moves the plate forward by one row and the delivery proceeds in serpentine fashion over the plate. At the completion of each plate, the fractionator drops plates onto a conveyer belt that proceeds to move the plates one row of eight wells at a time. The plates are covered by plastic lids with small holes over each well to reduce evaporation.

Using 96-well microplates for collecting samples has several advantages beyond those mentioned. They reduce labeling needs, handling errors, overall effort, time to completion of an assay, and storage requirements in freezers. They can accommodate sample volumes ranging from a few microliters to 3 ml (with special macroplates), and they can take advantage of the many devices available for working with 96-well microplates.

Operation

Dissociation of Cells

A number of different cell types have been studied successfully in this system including various cell lines, rat adrenal fasciculata cells, and rat and ovine pituitary cells. Somewhat different procedures were used to dissociate the various tissues, but we describe in detail only the procedure with which we have had the greatest experience.

Ovine Pituitary Dissociation with Collagenase

Pituitary glands collected at a local abattoir are placed in 50 ml of filter-sterilized Dulbecco's Minimal Essential Medium (DMEM; Life Technologies, Inc., Grand Island, NY) to which gentamicin (20 μl of 50 mg/ml; Life Technologies, Inc.) and amphotericin B (Fungizone, 50 μl of 250 μg/ml; Life Technologies, Inc.) are added. After transport on ice, the anterior portions of the glands are rinsed several times in antibiotic-containing medium, trimmed, sectioned into approximately 0.5-mm thick slices with a Stadie–Riggs tissue slicer (A. H. Thomas Scientific, Swedesboro, NJ), and dispersed largely as described elsewhere (26). The Stadie–Riggs apparatus is sterilized by dipping the component pieces into 70% ethanol and then drying them under a UV light. Briefly, the tissue is incubated in Ca^{2+}- and Mg^{2+}-free Hank's balanced salt solution (HBSS; Life Technologies, Inc.) with 0.3% collagenase (Type 1, CLS1; Worthington Biochemical Corp., Freehold, NJ) for 1 hr. Cells are centrifuged for 5 min (~400 g), washed in fresh medium, and resuspended in Ca^{2+}- and Mg^{2+}-free HBSS containing 0.25% pancreatin 4X (Life Technologies, Inc.) for 15–20 min. Disaggregated cells are washed four times with complete growth medium (DMEM containing gentamicin, Fungizone, and L-glutamine) and counted with a hemocytometer. In an attempt to promote attachment and reduce the possibility of clogging of the top filter, the cells were previously cultured on the beads in static culture (in growth medium with 10% fetal calf serum) for up to 4 days prior to loading. While this approach worked well, we have found it is no longer necessary with the loading procedure to be described. Cells are now directly coloaded with glass beads into the cell chamber immediately after dissociation and perifusion is started without delay. This provides an opportunity to test possible differences resulting from treatments *in vivo*, an approach not possible after several days of prior static culture.

Preparing and Loading the Chamber Assemblies

Washing, Dispensing, and Sterilizing SoloHill Beads

Hollow, nonporous, glass microcarrier beads (No. 102-915, SoloHill) are washed three times with a few milliliters of double distilled water. The beads are allowed to air-dry overnight. A scoop, made with the same material as the chamber module being loaded and drilled with the same size drill, is used to measure the beads. The scoop is tapped gently to allow the beads to settle and, after removing excess beads, the measured quantity is transferred into each of several small tubes. One milliliter of DMEM or similar medium is pipetted into each tube and the tubes are autoclaved. The sterile tubes with beads and medium are capped and stored at room temperature.

Putting the Chamber Assemblies Together

The amperometric, pH, and nonsensing chamber units are each assembled a little differently. The following discussion applies to a chamber assembly with no electrodes (Fig. 2B); it must be modified slightly when sensors are included. To reduce contamination, wipe the inside surfaces of all three cell chamber pieces with laboratory tissue paper soaked in 70% ethanol. While holding the inlet module with the attached tubing and fitting hanging down {12,13}, place a piece of polycarbonate filter (Nuclepore No. 111116 PC membranes with 12-μm pores) over the hole. Place the aligned chamber module on top and cover its hole with a second filter. Cover with the outlet module and clamp in place with two or four machine screws. Place washers (smooth side to plastic) and nuts on the screws and tighten just sufficiently to eliminate leaking. Check for leaks by attaching a syringe to one fitting {12} and a plug to the other {15}, insert the filling port screw, and pressurize the chamber with distilled water. Optionally, clean the assembled chamber by attaching to a pumping system and flushing with distilled water for an hour or more at 50–70 μl/min.

Loading the Chamber

In the hood, attach 1-ml syringes to both tubing ends {12,15}, one syringe filled with medium. Position the chamber with the filling port up and syringes at each side. Screw a 1-ml, autoclaved, air-displacement pipette tip bearing a sleeve made from a section of Tygon tubing into the filling port (chamber fill tube). The threads will catch and tighten after only a little turning and the Tygon tubing sleeve will help to prevent leakage. Push medium from the filled syringe into the chamber until it flows up the filling port and into the fill tube. Pull some of this medium out into the empty syringe, checking closely to see that no air bubbles remain on either side of or within the chamber. Repeat with both syringes until all air is removed and the chamber fill tube is one-half full of medium. Using a 1000-μl air-displacement pipette, aspirate all beads (usually 200 μl is sufficient). Transfer the beads to the chamber fill tube. Then add to the chamber fill tube the desired number of cells (e.g., 200 μl of 5×10^6 cells/ml to introduce 1×10^6 cells per chamber). Allow the beads and cells to settle in the chamber fill tube and into the chamber. Tapping the side of the chamber assembly a few times helps to spread the cells and beads evenly within the chamber. When the cells and beads have settled and the filling port is mostly clear, use the syringes to empty the chamber fill tube of medium and any remaining cells, leaving the medium level even with the edge of the chamber. Avoid introducing air into the chamber. Remove the chamber fill tube, and insert the screw into the loading port. Spritz with 70% ethanol to wet the connectors and loading surface of the chamber assembly and wipe dry. Transport the chamber with attached syringes to the incubation chamber, remove the syringes, and attach the chamber to the appropriate tubes (fittings {11,12} and {15,16}), again avoiding introduction of air. Again, spritz the surfaces with ethanol and wipe dry.

Running the System

The night before an experiment, fill containers with desired media, cover loosely, and place in a humidified tissue culture incubator with 5% carbon dioxide in air to equilibrate under warm conditions. Optionally, fill the desired number of tubes or wells with 100 μl of phosphate-buffered saline (PBS) containing an inert protein to reduce nonspecific binding (e.g., 0.1% gelatin or 1% bovine serum albumin) and any agents needed to promote stability of the analytes to be measured (e.g., bacitracin, glycerol, EDTA, soybean trypsin inhibitor); cover tightly (e.g., with Parafilm) and refrigerate.

Turn on the heater for the incubator sufficiently in advance to allow the unit to come to the desired temperature. If syringe pumps are to be used for delivery of the main medium, place large syringes (e.g., 60 ml) full of equilibrated medium (no bubbles) in the syringe pump and set desired pumping rate. Connect the tubes without the chamber assembly in place {11 to 16}. Flush the system while loading the chamber, and then connect the loaded chamber assembly to the inlet and outlet tubing as described above {11 to 12} and {15 to 16}. If electrode sensors are included, perifuse the chamber assembly with a standardized pulse or step of medium for calibrating the sensors (e.g., medium containing 10 mM ascorbate or phenol red, 10 mg/ml, for amperometric sensors or medium with a known, small pH shift for hydrogen ion-selective sensors). Meanwhile, fill syringes with medium containing the desired test agent(s) and place them in the syringe pump(s) in the adjacent refrigerated unit. Place sample tubes (or plates) in the fraction collector rack, set the collector to desired collection rate, and begin the experiment by starting the computer and software with preprogrammed instructions.

Termination and Clean-up

Cleaning the Tubes and Valves

At the end of a run, transfer the free end of the tube leading to the fraction collector {17} to a waste container, and turn off the pump(s), heater, and blower. Disconnect the two couplers on either side of the chamber {11 from 12, 15 from 16}, remove the chamber assembly with its attached inlet and outlet tubes, and join the freed tubes {11 to 16} with a connector. After removing chamber contents and cleaning the chamber (below), flush all tubes with distilled water for several hours. Then flush and fill the tubes with 70% ethanol. To avoid evaporation, plug all tubing ends.

Recovering Perifusion Chamber Contents, Checking for Viability

Place 500 μl medium into a polypropylene tube. Disassemble the chamber assembly, remove the filter pieces from both sides of the chamber module, and place the module directly over the top of the tube. If the cells are viable, they attach to and link the beads to form a solid plug, often in less than 24 hr. Whether this plug includes endothelial cell-lined channels through which the medium flows remains to be determined. Using a wooden applicator stick, gently push the plug of beads and cells into the medium. After gentle agitation, carefully remove the stick from the chamber hole. Clean the sides of the chamber hole with a pipette. Gently break up the plug by repeatedly pulling it into and expelling it from the pipette tip. When thoroughly dispersed, transfer 100 μl to a fresh tube. Add 100 μl of 0.4% trypan blue and 800 μl of PBS. Agitate gently, and assess viability using a microscope with a hemocytometer. Freeze the remaining 900 μl for determination of DNA or other cellular contents.

Lysing Cells

To prepare cell lysates, add lysis buffer (50 mM Na_2CO_3, 2 mM EDTA, 90 U/ml bacitracin, pH 8.5) to each cell sample. Then freeze and thaw the samples twice. Collect and store the cell lysates at −20°C (27).

Disassembly and Thorough Cleaning of Rheodyne Valves

These inert valves require periodic disassembly and thorough cleaning. If this is not done, small shavings from a worn valve surface can clog the fluid tube. Before disassembly, draw a line along the top of the complete valve assembly, to orient parts during reassembly. Remove all three machine screws and carefully remove the outer metal plate, the Teflon valve head, and the valve body housing (caution: Teflon is very soft and is easily scratched). Observe the two metal pins and the order and orientation of each piece as it is removed. Immerse the valve body housing in distilled water, and gently wipe with laboratory tissue paper. Observing the valve body under a microscope, gently remove any bits of loose Teflon with the sharpened end of a wooden applicator stick. Carefully clean the three grooves, taking care to avoid damaging them. Wipe the cleaned surface with laboratory tissue paper and distilled water. Reassemble these two parts. Under the microscope, clean the valve head with the stick, laboratory tissue paper, and distilled water. This piece is especially fragile. Take care to avoid marring the surface. Reassemble the valve in the opposite order. Replace the screws into the back plate sufficiently to start the nuts. Do not tighten the nuts until the proper torque has been applied to all three machine screws. All three screws should be tightened equally and just sufficiently so as to prevent leakage. Excessive tightening can cause undue wear on the valve surface or prevent the valve from turning. (When in doubt, undertighten, test, and tighten slightly if a leak develops). Finally, tighten the nuts. Test the valve for leaks and proper switching.

Performance

Assessment of Flow Dynamics

Characterization

At 69 μl/min flow and a tubing diameter of 0.3 mm, convective (nondiffusional) velocity is 16.3 mm/sec. The larger diameter (2.7 mm) of an empty chamber represents an 80-fold greater cross-sectional area, and convective velocity would be slowed to an average of 0.2 mm/sec. However, flow in the chamber is complicated by the nonporous beads which occupy approximately 17 of the 32 μl capacity chamber. Because 1×10^6 cells occupy approximately 1 μl, the volume remaining in the chamber is close to 14 μl. This gives a vertical, convective velocity closer to 0.6 mm/sec, a value less than the mean capillary convective velocity *in vivo* of 1 mm/sec. This suggests that shear effects are likely minimal. If 1×10^6 cells are used with a mass approximating 1 mg, the flow per mass of cells in the chamber is 69 μl/mg/min. In comparison with diffusional transcapillary flow *in vivo* of 3 μl/mg/min [300,000 μl/100 gm/min; (11)], these calculations suggest that the flow rate could be cut considerably without depriving the cells of oxygen and nutrients. This could prove advantageous in reducing the washout of locally produced paracrine agents. However, flow rate reduction would lead to greater dispersion of the output signal. The calculations also suggest that the flow rate could be increased without generating turbulence or shear forces greater than those found in capillaries. Increasing flow rate would reduce dispersion (improve signal integrity) but at the price of reducing the concentration of analytes to be measured.

Laminar Flow and Turbulence

The system operates under conditions free of turbulence because, when the Reynold's number is calculated separately for the tubing, the connecting channel, and the cell chamber filled with beads and at a higher flow rate than we use, 100 μl/min, values less than 10 are obtained. These values are far less than 2000, the number at which laminar flow converts to turbulence (24).

Mean Residence Time

The mean residence time, τ (τ = volume/flow rate), describes the length of time an average medium-component particle remains in the system. Thus, τ is directly proportional to the volume and inversely proportional to the flow rate. Table I gives volumes for various components of the system using the two different electrode modules together with calculated values of τ at a flow rate of 69 μl/min and chambers containing beads.

TABLE I   Volumes for Various System
Components Using Two Different
Electrode Modules with Their
Respective Calculated τ Values[a]

Component	Volume (μl)	Tau (sec)
pH electrode unit		
Mixing point to electrode	1.7	1.0
Electrode module	4.0	3.5
Electrode module to chamber	6.7	5.9
Chamber	15.0	13.0
Electrode module	4.0	3.5
Electrode module to outlet	44.0	38.2
	75.4	65.1
Amperometric electrode unit		
Mixing point to electrode	1.72	1.5
Electrode module	2.8	2.5
Chamber	15.0	13.0
Electrode module	2.8	2.5
Electrode module to outlet	44.0	38.2
	66.3	57.7

[a] At a flow rate of 69 μl/min and in chambers containing beads.

Disruptive Effects

To assess the effects of system components on dispersion, the computer-controlled, continuous flow, microsyringe pump was used with a dual pH electrode chamber assembly to monitor at intervals of 1 sec the responses to a variety of pulse shapes, pulse frequencies, system volumes, and flow rates. As illustrated in Fig. 5, these included two, trapezoidal, constant flow rate input signals, each 8 sec in duration, separated by 10 sec (change in HBSS at pH 7.3 and 6.8) and applied at a constant flow rate of 64 μl/min to a specially configured system with no chamber or with chambers of 10 or 30 μl capacity (24). The results from each of the different trials are superimposed in Fig. 5. To interpret the resulting changes in millivolts in terms of the amount of each buffer, the analysis of these experiments required considering the capacity of the buffer. For this, the two solutions of HBSS were initially mixed stepwise. The relationship between measured millivolts versus percent pH 7.3 (with 100% HBSS at pH 7.3 assigned as 1 and 100% HBSS at pH 6.8 assigned as 0) was used to transform mV to buffer fraction. This allowed all results to be expressed as values between 0 (pH 6.8) and 1 (pH 7.3).

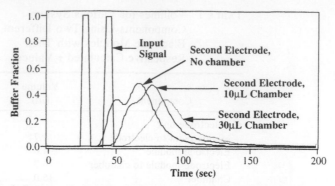

FIG. 5   The influence of chamber volume on signal dispersion. Shown are responses measured at 1-sec intervals at the second electrode to the indicated, duplicate, trapezoidal-shaped changes in pH (between 6.8 and 7.3) in the presence of either no cell chamber or an added dead volume of 10 μl or 30 μl. The input signal represents the waveform applied at the pumps. See text for explanation of buffer fraction. (Reprinted from Ref. 24 with permission of the *American Journal of Physiology*.)

With no cell chamber, a very small system volume and a mean residence time approximating the duration of the individual pulses (τ = 8.4 sec), two peaks were discernible in the measured signal. When the system volume was increased by 10 μl (τ = 17.8 sec, approximately the duration of the two pulses combined), the first pulse was only discernible as a shoulder. After adding a 30 μl dead volume (τ = 36.5 sec, a duration longer than the combined pulses), dispersion was so great that even the shoulder was lost and only one pulse could be discerned. As the duration of a pulse approaches or surpasses τ, sequential pulses can no longer be easily discerned as separate entities.

Monitoring Cellular Responses to Energy Perturbation

Changes in Energy Metabolism

Figure 6 illustrates the rapidity with which cellular responses can occur and the precision with which they can be monitored with pH electrodes (28). In this examination of proton release in response to changes in metabolism of HeLa cells, bicarbonate-free HBSS was used to amplify changes in pH and to provide a single energy source (10 mM glucose). The buffering capacity of the medium was again taken into account and all data were converted from millivolts to femtomolar change in H^+/cell-sec. Because the electrodes can measure changes as small as 0.0005 pH units, preparing test and control

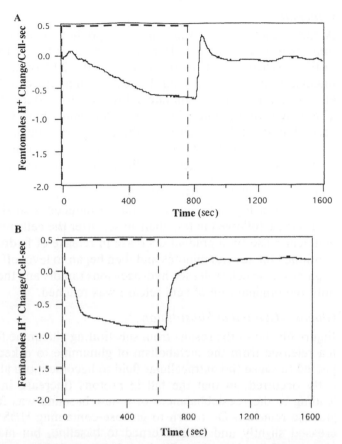

FIG. 6   Derived hydrogen ion secretory profiles in response to deprivation or substitution of extracellular energy sources. Approximately 10^5 HeLa cells were perifused with bicarbonate-free HBSS containing only glucose (10 m*M*) as an energy source until the times indicated by the broken lines, when glucose was either removed for 13 min (A) or replaced by 4 m*M* glutamine for 10 min (B). HBSS with 1.0 g/liter glucose but without bicarbonate, pH 7.2–7.4 (H1387, Sigma, St. Louis, MO), was used as the basal medium. This was chosen because of its minimal buffering capacity, single energy source, and ability to sustain cells in culture for more than 8 hr. (Reprinted from Ref. 28 with permission of Wiley-Liss, Inc. Copyright © 1994 Wiley-Liss, a division of Wiley & Sons, Inc.)

solutions with an indistinguishable pH was impossible. Thus, in addition to the effects of cellular metabolism, the pH measured downstream of the cells reflects these solution-dependent differences in pH as well as their convolution with the dispersive effects of system components.

Convolution

A convolution analysis was utilized to predict what the profile of hydrogen ions measured at the first electrode would look like at the second electrode in the absence of contributions of the cells (29). Briefly, after deriving the impulse response, describing the relationship between the first and second electrodes, the convolution integral for the applied input signal was solved. The convolution predicted the signal that would have been measured at the second (postcellular) electrode in the absence of cellular input. This predicted response was subtracted from the measured response at the second electrode to give the derived secretory profile depicted in Fig. 6.

Glucose Removal and Replacement

As shown in Fig. 6A, glucose removal induced a small initial increase in acid release followed in less than 40 sec after the cells received the glucose-deficient pulse by a gradual decrease in release of hydrogen ions. This fall continued for several minutes and then began to level off. On reintroduction of glucose, the cells released hydrogen ions rapidly and then gradually slowed until the original rate of acid release was reached.

Glucose–Glutamine Substitution

Figure 6B shows the results from substituting glutamine for glucose. Ammonia released from the metabolism of glutamine to α-ketoglutarate was expected to cause the extracellular fluid to become more alkaline. This apparently occurred, in that the fall in protons (increase in ammonia release) leading to the reestablishment of a new baseline was far faster than with glucose removal. On return to glucose-containing HBSS, acid release decreased slightly and then returned to baseline, but at a rate lower than observed following reintroduction of glucose to glucose-free HBSS.

These examples indicate that on-line monitoring of hydrogen ion release in a continuously flowing perifusion system can be a powerful method for examining short-term metabolic perturbations. By including pre- and postcellular electrodes and the technique of convolution, influences from introducing test solutions can be removed. The method is strengthened by using a medium with a single energy source, an approach that also simplifies data interpretation. The results indicate that the system has the resolution to reveal changes in energy metabolism within seconds, including the conversion between glucose- and glutamine-dominated metabolism.

Monitoring Neuroendocrine Responses

Response of Ovine Anterior Pituitary Cells to GnRH

Figure 7 illustrates the response of dispersed ovine anterior pituitary to a 4-min pulse of GnRH when monitored both by radioimmunoassay for ovine

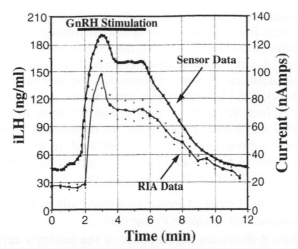

FIG. 7 Comparison of estimates of immunoreactive LH in samples obtained every 35 sec before, during, and after administering a 4-min pulse of GnRH (bar) to perifused ovine anterior pituitary cells and an amperometric difference profile recorded by cyclic voltammetry every 4 sec from electrodes located before and after the cell chamber. See text for experimental details and interpretation. The points bracketing the hormone concentrations are standard errors of the radioimmunoassay estimates. (From a doctoral thesis by Hal C. Cantor, University of Michigan, Ann Arbor, 1994.)

LH in samples collected approximately every 35 sec (6 drops/fraction) and by amperometry using cyclic voltammetry to record changes in current at selected voltages approximately every 7 sec. In these experiments, dispersed cells were first cultured with the glass beads for 4 days in growth medium (DMEM) with 10% fetal calf serum and then loaded into the chamber. To reduce possible problems with fouling of electrodes by proteins in the medium and achieve lower limits of detection with fast response times, after 2 hr of perifusion at 52 μl/min, the medium was switched to HBSS with only HEPES and L-glutamine added. A voltage-dependent peak in current flow, known to be proportional to concentration of analyte, was detected at 500 mV and recorded in the running voltammogram. The raw cyclic voltammograms recorded at the prechamber and postchamber electrodes revealed that, at 300 mV, GnRH is electroactive. To remove the contribution of GnRH input from the output signal, the input signal was convolved to the profile predicted at the output, and this convolved pattern, representing the profile given by GnRH after dispersion in the chamber, was subtracted from the measured output signal.

The resulting difference profile (Fig. 7) represents the release of molecule(s) from the pituitary and was remarkably similar to profile of LH measured by radioimmunoassay in discrete samples obtained less frequently (around every 35 sec). Because ovine LH is not electroactive under these conditions (it oxidizes at higher potentials), the likelihood is that we are measuring one or more cosecreted molecules. Because LH is known to be stored and secreted with sacrificial, protective molecules, the chromogranins and secretogranins (20–22), it is likely that these are what we measure. Note that both profiles were characterized by an early, fast response followed in turn by a rapid partial fall, a plateau lasting for the duration of the input GnRH pulse, and a slow subsequent decline. As indicated by the few values shown prior to the input of GnRH and by the suggested leveling at the end, interpulse secretion of LH continues at low levels.

Response of Ovine Anterior Pituitary Cells to Corticotropin-Releasing Hormone and Arginine Vasopressin

Figure 8 illustrates the response of the pituitary cells, presumably cortico-tropes, to 4-min pulses either of corticotropin-releasing hormone (CRH; Fig. 8, left) or arginine vasopressin (AVP; Fig. 8, right). The experimental conditions and design were as described for GnRH and LH release in Fig. 7 except that the samples were subjected to radioimmunoassay for β-lipotro-pin/β-endorphin, a product formed in a 1 : 1 relationship with corticotropin on processing of the proopiocorticotropin precursor. Contrary to LH, β-endorphin is electroactive and thus changes in its concentrations are presumably included in the cyclic voltammogram from the postchamber electrode. Because, like GnRH, CRH and AVP are each electroactive, the same convolutional approach was applied to remove the contributions of these secretagogues from the measured output signal. The resulting difference profiles are shown in Fig. 8. Again, a high degree of correspondence between the measured concentrations of β-endorphin and the sensor-detected profile, both following CRH (Fig. 8A) and especially following AVP (Fig. 8B). Note that the release of β-endorphin is slower in response to CRH than to AVP. The faster response to AVP is followed by a fast decline to a minor plateau and then decline to basal secretion. CRH not only stimulated a slower release of β-endorphin with a later peak amplitude, but was also followed by a more prolonged release phase. These kinetic differences in response profiles support the conclusions of Watanabe and Orth that the release of β-endorphin may involve different intracellular mechanisms (9).

Summary

This description of a miniature perifusion system illustrates the importance of design elements necessary to achieve high resolution, the advantages and power of monitoring cellular responses in real time, and the potential of

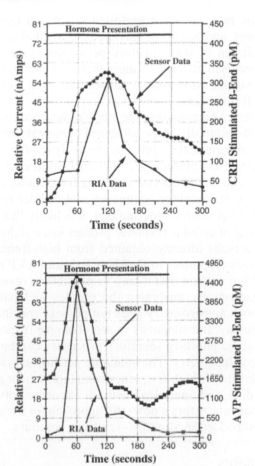

FIG. 8   Concentrations of β-endorphin in samples obtained every 35 sec from dispersed ovine anterior pituitary cells in response either to CRH (left panel) or AVP (right panel) and concomitantly monitored amperometric difference profiles recorded every 7 sec by electrodes before and after the cell chamber during and following administration of secretagogue for the time indicated by the bar. See text for experimental details and interpretation. (From a doctoral thesis by Hal C. Cantor, University of Michigan, Ann Arbor, 1994.)

applying signal processing to extract additional meaning from measured results. The flow and performance characteristics of a perifusion system are clearly dependent on all of its components: the diameter, length, and composition of the tubing; shape and volume of the cell chamber; the selection of suspending matrix for the cells; method of pumping; and effect of any incorporated device such as a sensor. Relatively small differences in volume or flow rate can have profound effects on assessment of cellular responsiveness, and attention must be paid to all elements to minimize dispersion. The design of the miniature perifusion system and on-line recording has allowed us to observe changes that could not have been measured in systems with significantly greater dispersion or in systems with far slower sampling frequencies.

Neuroendocrinology, the study of neuronal control of endocrine systems, is built on rapidly responding and complexly interacting systems. From the study of hormonal secretion to the analysis of synaptic transmission, if history has taught us anything, it is that analysis at ever smaller intervals has reaped dividends. Conventional sampling frequencies are much slower than times required for receptor occupancy and cellular response. In this dynamically rich field of neuroendocrinology, it is likely that beneath our current understanding of cellular response profiles lies a richness of subordinate information. Results already obtained from high-frequency measurements of the secretory responses to GnRH, CRH, and AVP not only bear witness to this belief but suggest that it may be possible to reveal the nature of intracellular mechanisms by analyzing the profiles and patterns of cellular response. Although amperometry does not possess specificity comparable to radioimmunoassay, the similarity in secretagogue response profiles suggests that at least some pituitary secretory processes can be monitored in perifusion by cyclic voltammetry. The miniature perifusion system described here, coupled with means for high-frequency sampling, sensor arrays for continuous on-line recording, and analytical methods for deconvolving the causative structural components, offers a powerful approach for untangling the kinetic complexities embedded in stimulus–response contours.

Acknowledgments

The authors thank Rane Curl for an introduction to and instruction in many of the principles of chemical engineering that guided this development of the system and Alice Rolfes-Curl and Thomas Valiquett for assistance and advice during the early phases of development. This work was performed as part of the NICHD's National Cooperative Program on Infertility Research and supported by Grants U54 HD29184, R01 HD18018, MH45232, T32 HD07048, and T32 GM07315.

References

1. F. E. Yates, *Biol. Reprod.* **24,** 73 (1981).
2. D. Green and R. Perlman, *Anal. Biochem.* **110,** 270 (1981).
3. M. Herrera, L. S. Kao, D. J. Curran, and E. W. Westhead, *Anal. Biochem.* **144,** 218 (1985).

4. J. Lankelma, E. Laurensse, and H. M. Pinedo, *Anal. Biochem.* **127,** 340 (1982).
5. J. E. McIntosh, R. P. McIntosh, and R. J. Kean, *Med. Biol. Eng. Comput.* **22,** 259 (1984).
6. R. P. McIntosh and J. E. McIntosh, *Endocrinology (Baltimore)* **117,** 169 (1985).
7. A. Negro-Vilar and M. D. Culler, *in* "Methods in Enzymology" (P. M. Conn, ed.), Vol. 124, p. 67. Academic Press, Orlando, Florida, 1986.
8. J. W. Parce, *et al., Science* **246,** 243 (1989).
9. T. Watanabe and D. N. Orth, *Endocrinology (Baltimore)* **121,** 1133 (1987).
10. A. J. Vander, J. H. Sherman, and D. S. Luciano, Human Physiology. The Mechanisms of Body Function. *In* "Circulation. Human Physiology," Chapt. 13. McGraw-Hill, New York, 1990.
11. R. M. Berne and M. N. Levy, Cardiovascular Physiology. *In* "The Microcirculation and Lymphatics," Chapt. 6. Mosby, St. Louis, Missouri, 1981.
12. A. Sahai, L. A. Cole, D. L. Clarke, and R. L. Tannen, *Am. J. Physiol.* **256,** C1064 (1989).
13. R. I. Woodruff and W. H. Telfer, *Nature (London)* **286,** 84 (1980).
14. E. L. Cussler, "Diffusion: Mass Transfer in Fluid Systems." Cambridge Univ. Press, New York, 1984.
15. G. Verhoeven, J. Cailleau, B. Van der Schueren, and J. J. Cassiman, *Endocrinology (Baltimore)* **119,** 1476 (1986).
16. Z. Rekasi and A. V. Schally, *Proc. Natl. Acad. Sci. U.S.A.* **90,** 2146 (1993).
17. G. Taylor, *Proc. R. Soc. London A* **219,** 186 (1953).
18. A. V. Oppenheim and R. W. Shafer, "Digital Signal Processing." Prentice-Hall, Englewood Cliffs, New Jersey, 1975.
19. E. L. Zuch, "Data Acquisition and Conversion Handbook, A Technical Guide to A/D–D/A Converters and Their Applications." Datel, 1979.
20. M. Bassetti, W. B. Huttner, A. Zanini, and P. Rosa, *J. Histochem. Cytochem.* **38,** 1353 (1990).
21. R. Fischer-Colbrie, T. Wohlfarter, K. W. Schmid, M. Grino, and H. Winkler, *J. Endocrinol.* **121,** 487 (1989).
22. R. Weiler, H. J. Steiner, R. Fischer-Colbrie, K. W. Schmid, and H. Winkler, *Histochemistry* **96,** 395 (1991).
23. A. Mosca *et al., Anal. Biochem.* **112,** 287 (1981).
24. R. B. Brand, M. N. Ghazzi, A. Rolfes-Curl, H. C. Cantor, and A. R. Midgley, *Am. J. Physiol. Endocrinol. Metab.* **266,** E739 (1994).
25. J. Van der Spiegel, *Med. Instrum.* **19,** 153 (1985).
26. W. L. Miller, M. M. Knight, H. J. Grimek, and J. Gorski, *Endocrinology (Baltimore)* **100,** 1306 (1977).
27. J. A. Flaws and D. E. Suter, *Biol. Reprod.* **48,** 1026 (1993).
28. R. B. Brand, R. H. Lyons, and A. R. Midgley, *J. Cell. Physiol.* **160,** 10 (1994).
29. R. B. Brand, "Analysis of Dynamic Changes in HeLa Cell Metabolism with an On-Line Microperifusion System." Doctoral thesis, University of Michigan, Ann Arbor, 1992.

[9] Realistic Emulation of Highly Irregular Temporal Patterns of Hormone Release: A Computer-Based Pulse Simulator

Martin Straume, Michael L. Johnson, and Johannes D. Veldhuis

Introduction

A computational strategy is described for generating synthetic hormone concentration time series that accurately reproduce both (i) the types of temporal patterning observed in empirical studies examining pulsatile hormone release as well as (ii) the distribution and magnitudes of uncertainties encountered in experimental determinations of hormone concentrations in unknown samples. The model is constructed in a manner to facilitate embodying the types of quantitative relationships frequently used to characterize endocrine secretory activity.

Because endocrine systems appear to communicate information by intermittent secretory bursts rather than direct modulation of continuous secretory activity (1–6), analysis and interpretation of experimentally determined hormone concentration time-series profiles often require quantitative consideration of complex temporal patterns. The inherent uncertainty encountered when experimentally estimating unknown hormone concentrations further confounds profiles of concentration time series, requiring accommodation during analysis of considerable contributions from noise. A variety of computational algorithms have been developed to identify peaks in hormone concentration time-series profiles (7–12), as well as to quantitatively characterize in greater detail particular properties of underlying secretory events and associated elimination kinetics (13–22). However, application of even very sophisticated, theoretically sound interpretive algorithms to real data sets requires some effort at validation of the analytical protocol before resulting conclusions can be accepted with confidence. Extensive testing, by comparative analyses, modeling efforts, and the analysis of simulations, has become an important ingredient in efforts to validate procedures and to determine expected false-positive and false-negative rates at peak detection (23–30).

The motivation for constructing a pulse simulator in the particular manner described here was to provide a mechanism for recreating as realistically as possible the temporal patterning of peak locations and amplitudes of hormone concentration time series observed in actual empirical studies. The model is thus defined within the context of the same quantifiable properties commonly elucidated from analysis of real data that are used to interpret time-dependent behavior of hormone secretory activity (1, 2, 13, 15, 17, 21, 22). The pulse simulator described here additionally has incorporated within it a specific mechanism for superimposing variability on and providing uncertainty estimates for individual concentration time points that realistically reflect the actual distribution and magnitude of uncertainty expected from hormone concentration determinations carried out in the experimental setting (31). The structure of the model is described in detail and its performance is demonstrated with reference to experimentally determined growth hormone secretory activity in the normal human male model.

Defining the Model

Temporal Distribution of Secretory Pulses

The relative distribution of secretory pulses in time is based on a relationship of cumulative probability versus the base-10 logarithm of interpulse interval as given by

$$P_{cum} = \frac{A - D}{\{1 + [\log(II)/C]^B\}^E} + D.$$

Here, P_{cum} is the cumulative probability of a secretory event occurring within time interval II (the interpulse interval in minutes), and the parameters A, B, C, D, and E define an empirically determined sigmoidal dependence of P_{cum} on $\log(II)$. Parameter A corresponds to the cumulative probability at zero $\log(II)$, D corresponds to the cumulative probability at infinite $\log(II)$, and parameters B, C, and E define the curvature and positioning along the $\log(II)$ axis of this sigmoidally shaped function. Independent cumulative probability relationships are defined for day and night secretory periods to account for any empirically observed differences in pulse frequency as a function of day/night. Interpulse interval results obtained from analysis of experimentally determined hormone concentration time series are accumulated and tabulated as discrete cumulative probabilities as a function of log of observed interpulse intervals. Interpulse intervals are ordered in increasing magnitude such that

$$\log(II_1) < \log(II_2) < \log(II_3) < \ldots < \log(II_n),$$

and discrete values of corresponding cumulative probabilities are assigned as

$$P_i = \frac{i - 0.5}{n}; \qquad i = 1, 2, 3, \ldots, n.$$

These empirical data then serve as the basis for estimating maximum likelihood parameter values for the above theoretical functional form by nonlinear least-squares analysis (32, 33). The resulting theoretical cumulative probability functions are subsequently used to estimate interpulse intervals between adjacent secretory pulses in simulation of synthetic hormone concentration time-series profiles.

Distribution of Relative Secretory Pulse Amplitudes

The distribution of relative secretory pulse amplitudes is considered in terms of a linear relationship of the base-10 logarithm of pulse amplitude ratio versus the base-10 logarithm of interpulse interval (in minutes) as given by

$$\log(PAR) = c_0 + c_1 \log(II).$$

Here, PAR is pulse amplitude ratio whereas c_0 and c_1 are parameters that define a linear relationship between $\log(PAR)$ and $\log(II)$. Values of PAR and corresponding II are assigned from analysis of experimental data as

$$PAR_i = \frac{SR_{i+1}}{SR_i}; \qquad II_i = t_{i+1} - t_i,$$

where SR_i is the experimentally evaluated secretory rate at pulse time t_i. Tabulated results of $\log(PAR)$ versus $\log(II)$ are then used to produce maximum likelihood estimates for c_0 and c_1 (by least-squares analysis) as well as the standard deviation of this linear fit [i.e., the square root of the variance of fit to uniformly, unit-weighted $\log(PAR)$ values]. The value of the SD of fit serves as a quantitative measure of the highly variable nature of the dependence of $\log(PAR)$ on $\log(II)$. Independent linear relationships are defined for day and night secretory periods to account for possible differences as a function of day/night. The day and night dependencies of $\log(PAR)$ on $\log(II)$ as given by respective parameter values for c_0 and c_1 are used in conjunction with corresponding standard deviations of fit to provide estimates of relative secretory pulse amplitudes among adjacent secretory pulses separated in time by given interpulse intervals when simulating synthetic hormone concentration time-series profiles.

Secretory Rate and Profile Parameters

Magnitudes for maximal allowable day and night secretory rates, SR_{max}, are explicitly specified to the model, as is the full width at half-maximum for the presumed Gaussian-shaped secretory pulse profile, $FWHM_{pulse}$, such that pulse secretory rate profiles, $SR_i(t)$, are given relative to pulse times, t_i, as

$$SR_i(t) = SR_{max} \exp\left[-\frac{(t - t_i)^2}{2\sigma_{pulse}^2} \right]; \qquad \sigma_{pulse} = \frac{FWHM_{pulse}}{2(2 \ln 2)^{1/2}}.$$

Additionally, the capacity for a finite rate of continuous secretory activity can be accommodated in a manner that allows basal secretion that is either constant, linearly variable with time, and/or exhibits variability as a function of day/night.

Circadian Timing Parameters

The model embodies the capability to vary either continuously (by way of two half-sine waves) or discontinuously (by way of a square wave), the transitions between day and night. A switch variable is set to define either the square or half-sine wave functional forms. Two additional parameters define the overall duration of the daytime component as well as the time at which the daytime component is maximal (or, in the case of square wave behavior, the time at which the midpoint of the daytime component occurs).

Sleep Onset Trigger

Three parameters are included in the model to permit sleep onset to act as a trigger mechanism for inducing hormone secretory activity. One parameter defines the time of sleep onset, t_{on}, another the median time after sleep onset of the first post-sleep-onset secretory pulse, Δt, and the third defines the half-width-at-half-maximum, $t_{1/2}$, of the half-Gaussian probability profile on the high-time side of the median time point. The probability distribution function for the first pulse after sleep onset, P_{fp}, is thus defined according to

$$P_{fp}[t \leq (t_{on} + \Delta t)] = \frac{1}{\sigma_{lo}(2\pi)^{1/2}} \exp\left[-\frac{1}{2}\left(\frac{t - (t_{on} + \Delta t)}{\sigma_{lo}}\right)^2\right]; \qquad \sigma_{lo} = \Delta t/3,$$

$$P_{fp}[t > (t_{on} + \Delta t)] = \frac{1}{\sigma_{hi}(2\pi)^{1/2}} \exp\left[-\frac{1}{2}\left(\frac{t - (t_{on} + \Delta t)}{\sigma_{hi}}\right)^2\right]; \qquad \sigma_{hi} = \frac{t_{1/2}}{(2\ln 2)^{1/2}}.$$

Here, $P_{fp}[t \leq (t_{on} + \Delta t)]$ is the half-Gaussian probability distribution function for pulses occurring on or before $t_{on} + \Delta t$ (the median time for a first pulse post sleep onset) and $P_{fp}[t > (t_{on} + \Delta t)]$ is the half-Gaussian probability distribution function for pulses occurring after $t_{on} + \Delta t$. The two half-Gaussians are considered independently in implementation of the model depending on the value of t. The standard deviation terms, σ_{lo} and σ_{hi}, are defined as indicated. The highly unlikely condition that a first post-sleep-onset pulse might be modeled to occur prior to t_{on} (probability equivalent to that of greater than three standard deviations) is nevertheless checked for and rigidly enforced to never occur in implementation of the model.

Hormone Elimination

Elimination of hormone, $E(t)$, can be accommodated by the model as either a single ($n = 1$) or double ($n = 2$) exponential process as

$$E(t) = \sum_{i=1}^{n} \alpha_i \exp(-k_i t); \qquad k_i = \frac{\ln 2}{t_{1/2,i}},$$

where α_i is the fractional elimination by process i with rate constant k_i as defined by the half-life for elimination, $t_{1/2,i}$.

Hormone Concentration Uncertainty

Uncertainty estimates are provided for each individual calculated hormone concentration in a manner consistent with the procedures and output generated by a data reduction procedure (reported in Ref. 31). The hormone concentrations produced and reported by the simulation procedure also exhibit random variability, the magnitude of which is scaled to be consistent with that predicted by the data reduction procedure (31). The simulated hormone concentration time-series profiles therefore contain both random variability in reported values as well as individual estimates of uncertainty assigned to each time point.

Start and Stop Times for Simulation and Sampling Frequency

The final information requested by the pulse simulator regards the start and stop times for the time period to be simulated as well as the desired sampling frequency to be modeled.

Implementation

The pulse simulator presents the user with a menu-oriented interface providing options for manual entry and/or modification of parameter values for those parameters referred to in the Defining the Model section above, as well as an option for input from an appropriately structured script input file, the values of which may be subsequently altered, if desired.

After initialization of all parameter values, simulation begins by calculation of the time of a first pulse after sleep onset. The sleep onset time, t_{on}, chosen for the simulations presented here, is 11 PM with median probability of a post-sleep-onset pulse set to occur 20 min thereafter (i.e., a Δt of 20 min). The half-width at half-maximum on the high-time side of the time of median pulse probability, $t_{1/2}$, is selected as 20 min. A unit standard deviation normal deviate is then calculated (as the sum of 12 pseudorandom numbers in the range from 0 to 1 from which the value of 6 is subsequently subtracted). If less than or equal to 0, this normal deviate is scaled as σ_{lo} and is used as the value for $[t - (t_{on} + \Delta t)]$ in the probability distribution function $P_{fp}[t \leq (t_{on} + \Delta t)]$ to assign the time of a first post-sleep-onset pulse [requiring, however, that the time of the first pulse not occur prior to t_{on} (see Sleep Onset Trigger section above)]; otherwise, the normal deviate is scaled as σ_{hi} and is used as the value for $[t - (t_{on} + \Delta t)]$ in the probability distribution given by $P_{fp}[t > (t_{on} + \Delta t)]$.

The relationship of time of pulse relative to the day/night circadian maximum probability is next identified. The proportion of day-like (f_{day}) versus night-like (f_{night}) character is defined such that the sum of f_{day} and f_{night} is equal to 1. The time-dependence of a two-half-sine-wave circadian component is used in the simulations presented here. It is selected such that a daylike component of 16-hr duration is centered at 2 PM and an 8-hr nightlike component is centered at 2 AM. This circadian profile is depicted graphically in Fig. 1. The secretory rate assigned to the first post-sleep-onset pulse, SR_{fp}, is then specified as

$$SR_{fp} = f_{day}SR_{max}(day) + f_{night}SR_{max}(night),$$

where $SR_{max}(day)$ and $SR_{max}(night)$ are the maximal allowable day and night secretory rates that are provided as input parameters to the pulse simulator (see Secretory Rate and Profile Parameters section above). The parameters SR_{max}(day) and SR_{max}(night) are assigned values of 0.42 and 1.0 ng/ml/min, respectively, for the simulations presented here [as based on results obtained from deconvolution analysis (13, 15) of 17 growth hormone time-series profiles obtained from normal human males]. The values for f_{day} and f_{night} depend on the particular time of occurrence of the pulse, as provided by $P_{fp}[t \le (t_{on} + \Delta t)]$ or $P_{fp}[t > (t_{on} + \Delta t)]$, and on the circadian profile (see Fig. 1).

Pulses occurring subsequent to this first post-sleep-onset pulse, but prior to the sleep onset time of the following day, are assigned times of occurrence according to the theoretical cumulative probability functions (one for daylike and another for nightlike character) described in Temporal Distribution of Secretory Pulses section above. Figure 2 shows graphically the day and night cumulative probability profiles relative to $\log(II)$ generated on the basis of deconvolution results from analysis of growth hormone time-series profiles obtained from 17 normal human males. A single pseudorandom number in the range from zero to one, referred to as P below (in reference to cumulative probability), is selected and applied through a composite cumulative probability function (defined in terms of a linear combination of the daylike and nightlike cumulative probabilities in proportion to the f_{day} and f_{night} that corre-

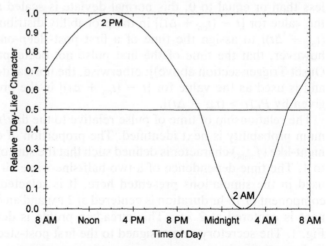

FIG. 1   The two-half-sine-wave circadian profile used in simulating growth hormone secretory activity, presented as the fraction of daylike character as a function of time of day. Maximum daylike character is modeled to occur at 2 PM (in the middle of a 16-hr "day") and maximum nightlike character at 2 AM (in the middle of an 8-hr "night").

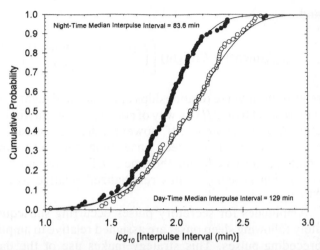

FIG. 2 Cumulative interpulse interval probabilities plotted as a function of the base-10 logarithm of experimentally determined interpulse interval (in minutes) for both the daytime (open circles) and nighttime (filled circles). Lines correspond to theoretical best-fit cumulative probability functions determined by nonlinear least-squares estimation (day: $A = 0$, $B = 7.004089$, $C = 5.052197$, $D = 1$, $E = 314.4444$; night: $A = 0$, $B = 9.505480$, $C = 2.271160$, $D = 1$, $E = 3.716779$). [Note: Values for A and D are fixed at 0 and 1, respectively, to constrain the theoretical cumulative probability functions to assume zero probability ($A = 0$) at an interpulse interval of 1 min, i.e., $\log(II) = 0$, and to never exceed unit probability ($D = 1$).] See Temporal Distribution of Secretory Pulses section for a description of the theoretical form of the cumulative probability function.

spond to the time of the current pulse). Specifically, the time of the next pulse, t_{i+1}, is determined relative to the time of the current pulse, t_i, as

$$t_{i+1} = t_i + II_i,$$

such that the interpulse interval, II_i, is given by

$$II_i = f_{\text{day}} 10^{[P_{\text{cum}}^{-1}(\text{day})]} + f_{\text{night}} 10^{[P_{\text{cum}}^{-1}(\text{night})]},$$

where

$$P_{\text{cum}}^{-1}(\text{day}) = C(\text{day}) \left[\left(\frac{A(\text{day}) - D(\text{day})}{P - D(\text{day})} \right)^{1/E(\text{day})} - 1 \right]^{1/B(\text{day})},$$

and

$$p_{\text{cum}}^{-1}(\text{night}) = C(\text{night}) \left[\left(\frac{A(\text{night}) - D(\text{night})}{P - D(\text{night})} \right)^{1/E(\text{night})} - 1 \right]^{1/B(\text{night})},$$

refer to the inverse relationships of the day and night cumulative probabilities with respect to log(II) by way of corresponding daytime and nighttime parameters A, B, C, D, and E. If, however, the calculated time for the next pulse, t_{i+1}, crosses over the sleep onset time, the pulse simulator reverts to the procedure for calculating the time of a first pulse post sleep onset. The pulse generation process is thus reinitialized relative to sleep onset after each 24-hr period.

Amplitudes for secretory pulses occurring subsequent to those immediately following sleep onset are assigned relative to amplitudes of immediately preceding pulses. This strategy makes use of the daytime and nighttime relationships of pulse amplitude ratio to interpulse interval [i.e., log(PAR) versus log(II) referred to in Distribution of Relative Secretory Pulse Amplitudes]. Figure 3 shows graphically the relationships obtained from plotting

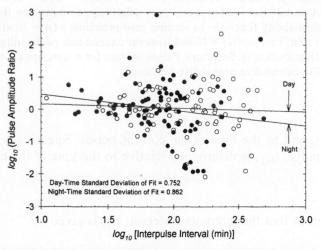

FIG. 3   Dependence of base-10 logarithms of experimentally determined pulse amplitude ratios on the base-10 logarithm of corresponding interpulse intervals (in minutes) for both the daytime (open circles) and nighttime (filled circles). Lines correspond to theoretical least squares best-fit linear dependencies (day: $c_0 = 0.2950822$, $c_1 = -0.1229449$, $\sigma_{\text{day}} = 0.75243$; night: $c_0 = 0.9625468$, $c_1 = -0.4915976$, $\sigma_{\text{night}} = 0.86187$). See Distribution of Relative Secretory Pulse Amplitudes section for definition of parameters.

log(PAR) versus log(II) as based on results of deconvolution analysis of growth hormone concentration time series obtained from 17 normal human males. Corresponding standard-deviations-of-fit are also indicated in the graph. In conjunction with values for f_{day} and f_{night} corresponding to the current pulse time, as well as a randomly generated unit standard deviation normal deviate (referred to as Z below), subsequent pulse amplitudes are calculated relative to immediately preceding pulses as

$$PA_i = PA_{i-1}\{f_{\text{day}}10^{[c_0(\text{day})+c_1(\text{day})\log(t_i-t_{i-1})+Z\sigma_{\text{day}}]} + f_{\text{night}}10^{[c_0(\text{night})+c_1(\text{night})\log(t_i-t_{i-1})+Z\sigma_{\text{night}}]}\},$$

where f_{day}, f_{night}, and t_i are as defined previously, PA_i is the pulse amplitude of the desired current pulse, Z is a unit standard deviation normal deviate, $c_0(\text{day})$, $c_1(\text{day})$, $c_0(\text{night})$, and $c_1(\text{night})$ are the parameters linearly relating log(PAR) and log(II) for the daytime and nighttime, respectively, and σ_{day} and σ_{night} are the corresponding daytime and nighttime standard deviations of fit, respectively. Once calculated in this way, the magnitudes of all pulse amplitudes are compared with the maximally allowed secretory rate [as given by $PA_{\text{max}} = f_{\text{day}} SR_{\text{max}}(\text{day}) + f_{\text{night}} SR_{\text{max}}(\text{night})$]. If any PA_i is found to be greater than PA_{max}, PA_i is set equal to PA_{max}.

The procedures for assigning times and amplitudes of pulses are conducted such that at least 2 days of secretory activity are simulated prior to beginning to record those which will be reported by the simulation. This 2-day warm-up of the simulator provides a mechanism for establishing a prior (unknown) secretory history to mimic the kinds of effects often seen in experimental settings where hormone secretory activity prior to initiation of sample collection is evident.

After having specified a sufficient distribution of pulse times and associated pulse amplitudes to satisfy the requested duration of simulation, profiles of secretory rate versus time are generated next. Explicit assignments of secretory rate are made versus time at 1-min intervals to account for contributions from both secretory pulses as well as any basal secretory activity that is being modeled (as described in the Secretory Rate and Profile Parameters section above). Secretory pulses in the examples presented here are modeled as Gaussian profiles in time, each with a full width at half-maximum of 25 min. A small amount of constant basal secretion is also included (0.44 pg/ml/min) in accordance with the average basal secretory rate estimated from deconvolution analysis of the growth hormone time-series data which serve as the basis for these simulations. An example of a simulated growth hormone secretory rate profile is shown in Fig. 4.

Calculation of the hormone concentration time-series profile, $C(t)$, is next

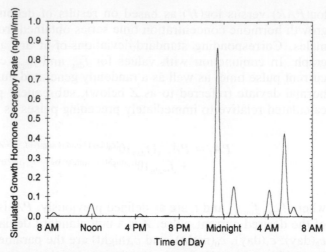

FIG. 4 Simulated growth hormone secretory rate profile showing locations and amplitudes of Gaussian-shaped secretory pulses (full width at half-maximum amplitude of 25 min). Nine simulated secretory pulses occur at the following times with corresponding maximal secretory rates:

Time (to nearest minute)	Secretory rate at maximum (pg/ml/min)
8:32 AM	18.3
11:57 AM	62.3
4:24 PM	12.5
6:50 PM	3.3
11:25 PM	863.3
12:55 AM	154.1
4:07 AM	134.3
5.29 AM	424.0
6:18 AM	47.8

There is also a constant level of 0.44 pg/ml/min of basal secretion (imperceptible at the scale of the figure). Sleep onset is set at 11 PM with the median time until a post-sleep-onset first pulse set to 20 min. See text for further details.

accomplished by convolution of the secretory rate profile, $SR(t)$, with that of the specified kinetics of hormone elimination, $E(t)$, as

$$C(t) = \int_{t_0}^{t} SR(\tau)E(t - \tau)\,d\tau,$$

where t_0 is the earliest time for which information about secretory activity is available. [Operationally, this is equivalent to integrating from $-\infty$ to t under conditions such that $SR(\tau) = 0$ for $\tau < t_0$.] Elimination in the simulations produced here is modeled as a single exponential process characterized by an elimination half-life of 18.5 min. The simulated, noise-free growth hormone concentration time series corresponding to the secretory rate profile of Fig. 4 is depicted in Fig. 5.

After the noise-free concentration profile, $C(t)$, is calculated, information regarding hormone concentration uncertainty is next superimposed. A set of 14 chemiluminescence growth hormone standard curves (34) that exhibited a high degree of reproducibility were chosen as the basis for realistically approximating magnitudes of expected experimental errors. A synthetic set of growth hormone standard curve data was generated that reflected both the average properties exhibited by the 14 chemiluminescence standard curve data sets as well as the experimental protocol employed in producing the original standards (i.e., quadruplicates at each of 10 reference growth hormone concentrations: 0, 0.0025, 0.01, 0.04, 0.12, 0.4, 1.33, 4.6, 13.3, and 45 ng/ml). This synthetic standard curve data set was then analyzed (31) and used for assigning variability and uncertainties in simulated growth hormone concentration-time series. Figure 6 shows the synthetic standard curve data set (along with the best-fit theoretical standard curve relationship) and the inset depicts the resulting discrete variance profile conjointly estimated ac-

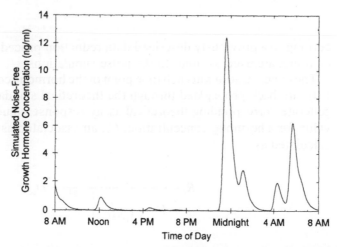

FIG. 5   Simulated growth hormone concentration time-series profile corresponding to the secretory rate profile of Fig. 4 after convolution with an 18.5-min half-life single exponential hormone elimination function.

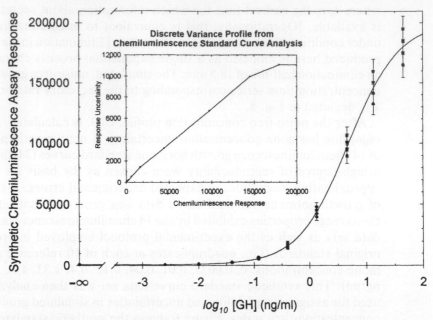

FIG. 6 Synthetic growth hormone chemiluminescence standard curve data (based on data from Ref. 34) are plotted with error bars along with the best-fit theoretical four-parameter logistic standard curve (see Ref. 31 for a detailed description of the data reduction algorithm). *Inset:* Discrete variance profile produced by the data reduction procedure and used for estimation of response uncertainties.

cording to a previously described data reduction procedure (31), the results of which are used as input to the pulse simulations.

The concentrations at each time point of the hormone concentration profile, $C(t)$, are back-propagated through the theoretical standard curve (Fig. 6) to generate corresponding theoretical assay response values. Thus, given any value for a hormone concentration, C_i, an associated assay response, R_i, is calculated as

$$R_i = \frac{A_{sc} - D_{sc}}{1 + \left[\dfrac{C_i}{10^{\log C_{sc}}}\right]^{10^{\log B_{sc}}}} + D_{sc},$$

where A_{sc}, $\log B_{sc}$, $\log C_{sc}$, and D_{sc} are the best-fit parameter values defining the theoretical four-parameter logistic standard curve as estimated by application of an iterative, variably-weighted nonlinear least-squares procedure to an appropriate set of reference hormone standards (31). Response uncertainties, $\sigma_{R,i}$, are then associated with each of the calculated responses, R_i, as estimated from the discrete variance profile generated by the data reduction procedure (see inset to Fig. 6) (31).

Noise in hormone concentration time series is generated by randomly varying the theoretical response in accordance with its corresponding uncertainty estimate and recalculating the concentration that corresponds to the altered response value. Specifically, the calculated response uncertainty, $\sigma_{R,i}$, is multiplied by a unit standard deviation normal deviate, Z, the result of which is added to the original calculated response, R_i, to produce a randomly varied response, $\overline{R}_i$, as

$$\overline{R}_i = R_i + Z\sigma_{R,i}.$$

This new response value, $\overline{R}_i$ (along with its associated uncertainty estimate, $\overline{\sigma}_{R,i}$), is then propagated through the standard curve to produce the randomly varied concentration value, $\overline{C}_i$, as well as its corresponding estimate of concentration uncertainty, $\overline{\sigma}_{C,i}$. From $\overline{C}_i$ is calculated a noise component, ε_i, relative to the original calculated concentration, C_i, as

$$\varepsilon_i = \overline{C}_i - C_i.$$

The value of ε_i is that which is actually stored by the pulse simulator, as opposed to the value of $\overline{C}_i$ itself. Figure 7 presents graphically the variability, $\varepsilon(t)$, calculated for the simulated concentration profile given in Fig. 5 (the inset is a higher resolution view of the same variability profile showing more clearly the smaller amplitude variations associated with regions of low hormone concentration).

The pulse simulator provides the means for each simulated concentration time point to have associated with it an uncertainty estimate, the magnitude of which is related to the uncertainty expected in the corresponding chemiluminescence assay response. The response uncertainty is propagated through the error inherent in estimating the theoretical hormone standard curve to arrive at the uncertainty in calculated concentration (31). Figure 8 shows the result of superimposing the concentration variability, $\varepsilon(t)$, depicted in Fig. 7 on the noise-free profile, $C(t)$, shown in Fig. 5 thus producing the noise-containing concentration time series, $\overline{C}(t)$. The magnitudes of associated standard deviations per time point, the $\overline{\sigma}_{C,i}$, are shown in the inset to Fig. 8.

The pulse simulator offers the user the opportunity to graphically review, at the conclusion of a simulation, (i) hormone concentration profiles of either the noise-free or noise-containing total simulated concentration or the pulsatile

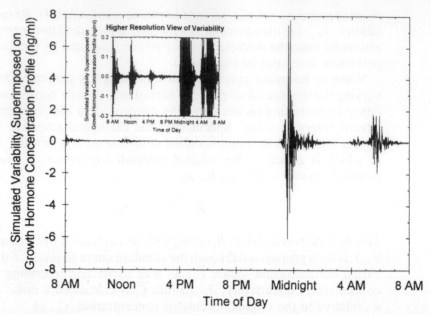

FIG. 7   Simulated random concentration variability profile, $\varepsilon(t)$, for the growth hormone concentration time series of Fig. 5. This profile is predicted in accordance with the standard curve of Fig. 6 by considering all uncertainties associated with assay responses as well as best-fit standard curve parameter values (31). *Inset:* Higher resolution view.

and basal contributions independently, (ii) secretory rate profiles of either the total secretory rate or the pulsatile and basal contributions independently, (iii) the circadian profile versus time, or (iv) the concentration variability and uncertainty profiles. Options also exist for permanently saving the above information, as well as the times at which all pulses occurred and their corresponding rates of maximal secretion, to user-specified output files.

Discussion

The highly variable and irregular temporal patterning of peak locations and amplitudes often exhibited by experimentally observed hormone concentration time-series profiles presents a challenge for quantitative analysis, particularly (as is frequently the case) when also confounded by considerable experimental uncertainty on individual time points. For these reasons, proce-

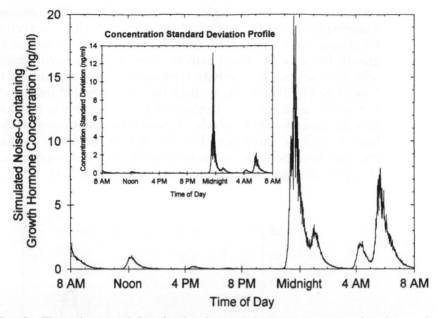

FIG. 8   The noise-containing simulated growth hormone concentration time series, $\bar{C}(t)$, corresponding to the noise-free profile, $C(t)$, of Fig. 5 after addition of the concentration variability profile, $\varepsilon(t)$, of Fig. 7. *Inset:* Time-series profile of the magnitudes of the standard deviations, $\bar{\sigma}_C(t)$, associated with each of the noise-containing concentration time points.

dures used for quantitatively analyzing such data must be carefully scrutinized regarding their ability to reliably characterize the secretory activity (and elimination kinetics) under study. However, validation efforts, to be rigorously effective, require precise knowledge of the true underlying secretory activity responsible for generating particular concentration profiles. Such information, of course, is not available from any experimentally determined time series. A means for creating realistic hormone concentration profiles for which all underlying secretory and elimination properties are known is therefore desired.

The pulse simulation algorithm described here is capable of generating synthetic hormone concentration time-series profiles that embody both the types of patterning observed in the experimental setting as well as the magnitude of variability and uncertainty individual time points are expected to exhibit based on the performance of contemporary hormone assay methods. The specific examples considered here regard growth hormone secretory activity of normal human males. Seventeen 24-hr profiles obtained at 10-min sampling intervals, three of which are shown in Fig. 9, served as the basis for the simulations presented. As is characteristic of the normal human male, growth hormone secretory activity is more pronounced during the night, with considerably less activity typically demonstrated during the daytime hours (see Fig. 9 and Ref. 1). Both the amplitude and frequency of growth hormone release appear to be elevated at night relative to the day. However, both among and within individuals, 24-hr growth hormone profiles can show high degrees of variability. For that reason, distinctly probabalistic approaches have been applied in interpreting growth hormone secretory profiles, pulsatility patterns, and elimination kinetics (1, 2, 13, 15, 17, 21, 22). The

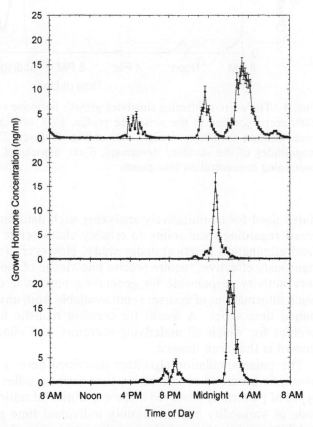

FIG. 9   Experimentally determined growth hormone concentration profiles of three normal human male subjects. Data are 24-hr records collected at 10-hr sampling intervals.

pulse simulator described here uses these same concepts as the framework for generating synthetic hormone concentration time-series profiles.

The temporal distribution of secretory pulses as well as the distribution of relative secretory pulse amplitudes are both simulated with respect to empirically based probability distributions. Secretory activity is modeled to occur as distinct secretory pulses (which are assumed to exhibit Gaussian temporal profiles) with or without contribution from constant and/or time-dependent basal secretion (35), whereas elimination kinetics are required to conform to either single or double exponential processes. The secretory and elimination properties are determined from analysis of empirical growth hormone concentration profiles by deconvolution analysis (13–17), the cumulative results of which serve to guide the simulation process. The empirical recognition that hormone secretory properties may exhibit circadian variability (36–38) is reflected in the pulse simulator, as well, through the potential for influencing pulse frequencies and amplitudes as well as basal secretory activity. The strong experimentally observed correlation between sleep and growth hormone secretory activity (1, 39, 40) is embodied into the pulse simulator as an explicit sleep onset trigger for growth hormone release. In the implementation in the pulse simulator, sleep onset serves as a reinitializing event such that the history of secretory activity prior to sleep onset effects no influence on subsequent secretory activity. However, prior secretory events may have influence on concentration time-series profiles as a consequence of the elimination of previously released growth hormone. In addition to properties directly related to secretory activity and elimination kinetics, the influences of experimental uncertainty and variability in the estimation of concentrations of unknown hormone samples are also explicitly dealt with by the pulse simulator (31).

The pulse simulator is currently capable of producing synthetic hormone concentration time-series profiles to a resolution of 1-min intervals. The effects on the nature and reliability of analytical conclusions due to varying the experimental sampling frequency in collecting hormone concentration time-series data have been considered in numerous studies on a number of hormones (41–45). The (relatively) high-temporal-density output generated by the pulse simulator will facilitate assessment of the effects of sampling frequency by permitting direct comparisons among analyses of data sets considered at any desired sampling frequency for which all underlying secretory and elimination properties are known. Figure 10 shows four different time-resolution depictions of the same synthetic hormone concentration profile as in Fig. 8, except uncertainty estimates (as error bars) are included, as well.

The high degree of variability exhibited in the temporal patterning of growth hormone release observed among and within individuals is also reproduced

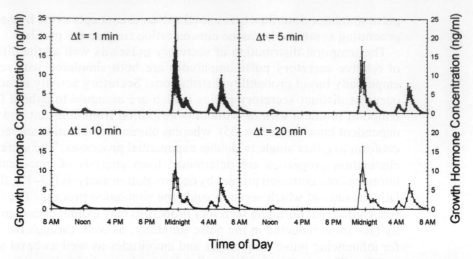

FIG. 10   The noise-containing simulated growth hormone concentration profile of Fig. 8 is presented, along with associated uncertainties (as error bars), at each of four different sampling intervals (1, 5, 10, and 20 min).

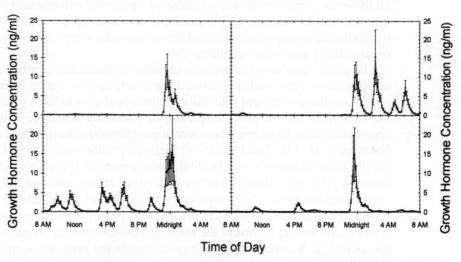

FIG. 11   Four simulated noise-containing growth hormone concentration profiles (all reported at 10-min sampling intervals), each of which were produced with the same set of system parameter values as for the simulation depicted in Fig. 10.

by the pulse simulator operating with unchanged system-defining parameters. Figure 11 shows four additional simulated growth hormone concentration time-series profiles (at a 10-min sampling frequency) generated with the same system parameters. Thus, although experimental observations indicate that temporal profiles of growth hormone secretory activity are sometimes significantly altered under conditions of fasting (46), as a function of age and sex (47), and in certain pathological settings (48, 49), even a constant set of system parameters, within the context of the structure of the pulse simulator described here, can give rise to quite highly variable growth hormone concentration profiles.

Acknowledgments

Support has been provided by the National Science Foundation Center for Biological Timing (DIR 8920162), an NIH NICHD RCDA award to J.D.V. (1K04HD00634), the NIH NICHD Reproduction Research Center at the University of Virginia Health Sciences Center (1P30HD28934-01A1; J.D.V.), the NIH Center for Fluorescence Spectroscopy of the University of Maryland at Baltimore (RR-08119; M.L.J.), and NIH Grant GM35154 (M.L.J.).

References

1. Hartman, M. L., Faria, A. C. S., Vance, M. L., Johnson, M. L., Thorner, M. O., and Veldhuis, J. D. (1991). Temporal structure of *in vivo* growth hormone secretory events in humans. *Am. J. Physiol.* **260** (*Endocrinol. Metab.* **23**), E101–E110.
2. Butler, J. P., Spratt, D. I., O'Dea, L. S., and Crowley, W. F. (1986). Interpulse interval sequence of LH in normal men essentially constitutes a renewal process. *Am. J. Physiol.* **250** (*Endocrinol. Metab.* **13**), E338–E340.
3. Hellman, L., Nakada, F., Curti, J., Weitzman, E. D., Kream, J., Roffwarg, H., Ellman, S., Fukushima, D. K., and Gallagher, T. F. (1970). Cortisol is secreted episodically by normal man. *J. Clin. Endocrinol.* **30**, 411–422.
4. Greenspan, S. L., Klibanski, A., Schoenfield, D., and Ridgway, E. C. (1986). Pulsatile secretion of thyrotropin in man. *J. Clin. Endocrinol. Metab.* **63**, 661–668.
5. Brabant, G., Ranft, U., Ocran, K., Hesch, R. D., and von zur Mülen, A. (1986). Thyrotropin—an episodically secreted hormone. *Acta Endocrinol.* **112**, 315–322.
6. Isgaard, J., Carlsson, L., Isaksson, O. G. P., and Jansson, J. O. (1988). Pulsatile intravenous growth hormone (GH) infusion to hypophysectomized rats increases insulin-like growth factor I messenger ribonucleic acid in skeletal tissues more effectively than continuous GH infusion. *Endocrinology* (*Baltimore*) **123**, 2605–2610.
7. Santen, R. J., and Bardin, C. W. (1973). Episodic luteinizing hormone secretion

in man. Pulse analysis, clinical interpretation, physiologic mechanisms. *J. Clin. Invest.* **52,** 2617–2628.

8. Clifton, D. K., and Steiner, R. A. (1983). Cycle detection: A technique for estimating the frequency and amplitude of episodic fluctuations in blood hormone and substrate concentration. *Endocrinology (Baltimore)* **112,** 1057–1064.

9. Veldhuis, J. D., and Johnson, M. L. (1986). Cluster analysis: A simple, versatile and robust algorithm for endocrine pulse detection. *Am. J. Physiol.* **250** (*Endocrinol. Metab.* **13**), E486–E493.

10. Oerter, K., Guardabasso, V., and Rodbard, D. (1986). Detection and characterization of peaks and estimation of instantaneous secretory rate for episodic pulsatile hormone secretion. *Comput. Biomed. Res.* **19,** 170–191.

11. Urban, R. J., Evans, W. S., Rogol, A. D., Johnson, M. L., and Veldhuis, J. D. (1988). Contemporary aspects of discrete peak detection algorithms. I. The paradigm of the luteinizing hormone pulse signal in man. *Endocr. Rev.* **9,** 3–37.

12. Kushler, R. H., and Brown, M. B. (1991). A model for the identification of hormone pulses. *Statistics in Medicine* **10,** 329–340.

13. Veldhuis, J. D., Carlson, M. L., and Johnson, M. L. (1987). The pituitary gland secretes in bursts: Appraising the nature of glandular secretory impulses by simultaneous multiple-parameter deconvolution of plasma hormone concentrations. *Proc. Natl. Acad. Sci. U.S.A.* **84,** 7686–7690.

14. O'Sullivan, F., and O'Sullivan, J. (1988). Deconvolution of episodic hormone data: An analysis of the role of season on the onset of puberty in cows. *Biometrics* **44,** 339–353.

15. Veldhuis, J. D., and Johnson, M. L. (1992). Deconvolution analysis of hormone data. *In* "Methods in Enzymology" (L. Brand and M. L. Johnson, eds.), Vol. 210, pp. 539–575. Academic Press, San Diego.

16. De Nicolao, G., and Liberati, D. (1993). Linear and nonlinear techniques for the deconvolution of hormone time-series. *IEEE Trans. Biomed. Eng.* **40,** 440–455.

17. Johnson, M. L., and Veldhuis, J. D. (1995). This volume, Chapter 1.

18. Hendricks, C. M., Eastman, R. C., Takeda, S., Asakawa, K., and Gorden, P. (1985). Plasma clearance of intravenously administered pituitary growth hormone: Gel filtration studies of heterogenous compartments. *J. Clin. Endocrinol. Metab.* **60,** 864–867.

19. Veldhuis, J. D., Fraioli, F., Rogol, A. D., and Dufau, M. L. (1986). Metabolic clearance of biologically active luteinizing hormone in man. *J. Clin. Invest.* **77,** 1122–1128.

20. Lopez, F. J., and Negro-Vilar, A. (1988). Estimation of endogenous adrenocorticotropin half-life using pulsatility patterns: A physiological approach to the evaluation of secretory episodes. *Endocrinology (Baltimore)* **123,** 740–746.

21. Faria, A. C. S., Veldhuis, J. D., Thorner, M. O., and Vance, M. L. (1989). Half-time of endogenous growth hormone (GH) disappearance in normal man after stimulation of GH secretion by GH-releasing hormone and suppression with somatostatin. *J. Clin. Endocrinol. Metab.* **68,** 535–541.

22. Hindmarsh, P. C., Matthews, D. R., Brain, C. E., Pringle, P. J., Di Silvio, L.,

Kurtz, A. B., and Brook, C. G. D. (1989). The half-life of exogenous growth hormone after suppression of endogenous growth hormone secretion with somatostatin. *Clin. Endocrinol.* **30**, 443–450.

23. Merriam, G. R., and Wachter, K. W. (1982). Algorithms for the study of episodic hormone secretion. *Am. J. Physiol.* **243** (*Endocrinol. Metab.* **6**), E749–E759.

24. Veldhuis, J. D., Rogol, A. D., and Johnson, M. L. (1985). Minimizing false-positive errors in hormonal pulse detection. *Am. J. Physiol.* **248** (*Endocrinol. Metab.* **11**), E475–E481.

25. Guardabasso, V., De Nicolao, G., Rocchetti, M., and Rodbard, D. (1988). Evaluation of pulse-detection algorithms by computer simulation of hormone secretion. *Am. J. Physiol.* **255** (*Endocrinol. Metab.* **18**), E775–E784.

26. Van Cauter, E. (1988). Estimating false-positive and false-negative errors in analysis of hormone pulsatility. *Am. J. Physiol.* **254** (*Endocrinol. Metab.* **17**), E786–E794.

27. Urban, R. J., Kaiser, D. L., Van Cauter, E., Johnson, M. L., and Veldhuis, J. D. (1988). Comparative assessments of objective peak detection algorithms. II. Studies in men. *Am. J. Physiol.* **254** (*Endocrinol. Metab.* **17**), E113–E119.

28. Veldhuis, J. D., and Johnson, M. L. (1988). A novel general biophysical model for simulating episodic endocrine gland signaling. *Am. J. Physiol.* **255** (*Endocrinol. Metab.* **18**), E749–E759.

29. Urban, R. J., Johnson, M. L., and Veldhuis, J. D. (1989). Biophysical modeling of the sensitivity and positive accuracy of detecting episodic endocrine signals. *Am. J. Physiol.* **257** (*Endocrinol. Metab.* **20**), E88–E94.

30. Urban, R. J., Johnson, M. L., and Veldhuis, J. D. (1989). *In vivo* biological validation and biophysical modeling of the sensitivity and positive accuracy of endocrine peak detection: I. The LH pulse signal. *Endocrinology (Baltimore)* **124**, 2541–2547.

31. Straume, M., Veldhuis, J. D., and Johnson, M. L. (1994). Model-independent quantification of measurement error: Empirical estimation of discrete variance function profiles based on standard curves. *In* "Methods in Enzymology" (M. L. Johnson and L. Brand, eds.), Vol. 240, pp. 121–150. Academic Press, San Diego.

32. Johnson, M. L., and Frasier, S. G. (1985). Nonlinear least squares analysis. *In* "Methods in Enzymology" (C. H. W. Hirs and S. N. Timasheff, eds.), Vol. 117, pp. 301–342. Academic Press, San Diego.

33. Straume, M., Frasier-Cadoret, S. G., and Johnson, M. L. (1991). Least-squares analysis of fluorescence data. *In* "Topics in Fluorescence Spectroscopy," (J. R. Lakowicz, ed.), Vol. 2: Principles, pp. 177–240. Plenum, New York.

34. Chapman, I. M., Hartman, M. L., Straume, M., Johnson, M. L., Veldhuis, J. D., and Thorner, M. O. (1994). Enhanced sensitivity growth hormone (GH) chemiluminescence assay reveals lower postglucose nadir GH concentrations in men than women. *J. Clin. Endocrinol. Metab.* **78**, 1312–1319.

35. Hartman, M. L., Veldhuis, J. D., Vance, M. L., Faria, A. C. S., Furlanetto, R. W., and Thorner, M. O. (1990). Somatotropin pulse frequency and basal concentrations are increased in acromegaly and are reduced by successful therapy. *J. Clin. Endocrinol. Metab.* **70**, 1375–1384.

36. Plotnick, L. P., Thompson, R. G., Kowarski, A., De Lacerda, L., Migeon, C. J., and Blizzard, R. M. (1975). Circadian variation of integrated concentration of growth hormone in children and adults. *J. Clin. Endocrinol. Metab.* **40,** 240–247.

37. Carnes, M., Brownfield, M. S., Kalin, N. H., Lent, S., and Barksdale, C. M. (1988). Pulsatile ACTH secretion: Variation with time of day and relationship to cortisol. *Peptides* **9,** 325–331.

38. Veldhuis, J. D., Iranmanesh, A., Lizarralde, G., and Johnson, M. L. (1989). Amplitude modulation of a burstlike mode of cortisol secretion subserves the circadian glucocorticoid rhythm. *Am. J. Physiol.* **257** (*Endocrinol. Metab.* **20**), E6–E14.

39. Takahashi, Y., Kipnis, D. M., and Daughaday, W. H. (1968). Growth hormone secretion during sleep. *J. Clin. Invest.* **47,** 2079–2090.

40. Kerkofs, M., Van Cauter, E., Van Onderbergen, A., Caufriez, A., Thorner, M. O., and Copinschi, G. (1993). Sleep-promoting effects of growth hormone-releasing hormone in normal men. *Am. J. Physiol.* **264** (*Endocrinol. Metab.* **27**), E594–E598.

41. Shin, S. H. (1982). Detailed examination of episodic bursts of rGH secretion by high frequency blood sampling in normal male rats. *Life Sci.* **31,** 597–602.

42. Veldhuis, J. D., Evans, W. S., Rogol, A. D., Drake, C. R., Thorner, M. O., Merriam, G. R., and Johnson, M. L. (1984). Performance of LH pulse-detection algorithms at rapid rates of venous sampling in humans. *Am. J. Physiol.* **247** (*Endocrinol. Metab.* **10**), E554–E563.

43. Veldhuis, J. D., Evans, W. S., Johnson, M. L., Wills, M. R., and Rogol, A. D. (1986). Physiological properties of the luteinizing hormone pulse signal: Impact of intensive and extended venous sampling paradigms on its characterization in healthy men and women. *J. Clin. Endocrinol. Metab.* **62,** 881–891.

44. Evans, W. S., Faria, A. C. S., Christiansen, E., Ho, K. Y., Weiss, J., Rogol, A. D., Johnson, M. L., Blizzard, R. M., Veldhuis, J. D., and Thorner, M. O. (1987). Impact of intensive venous sampling on characterization of pulsatile GH release. *Am. J. Physiol.* **252** (*Endocrinol. Metab.* **15**), E549–E556.

45. De Nicolao, G., Guardabasso, V., and Rocchetti, M. (1990). The relationship between rate of venous sampling and visible frequency of hormone pulses. *Comput. Methods Prog. Biomed.* **33,** 145–157.

46. Ho, K. Y., Veldhuis, J. D., Johnson, M. L., Furlanetto, R., Evans, W. S., Alberti, K. G. M. M., and Thorner, M. O. (1988). Fasting enhances growth hormone secretion and amplifies the complex rhythms of growth hormone secretion in man. *J. Clin. Invest.* **81,** 968–975.

47. Ho, K. Y., Evans, W. S., Blizzard, R. M., Veldhuis, J. D., Merriam, G. R., Samojlik, E., Furlanetto, R., Rogol, A. D., Kaiser, D. L., and Thorner, M. O. (1987). Effects of sex and age on the 24-hour profile of growth hormone secretion in man: Importance of endogenous estradiol concentrations. *J. Clin. Endocrinol. Metab.* **64,** 51–58.

48. Asplin, C. M., Faria, A. C. S., Carlsen, E. C., Vaccaro, V. A., Barr, R. E., Iranmanesh, A., Lee, M. M., Veldhuis, J. D., and Evans, W. S. (1989). Alterations in the pulsatile mode of growth hormone release in men and women

with insulin-dependent diabetes mellitus. *J. Clin. Endocrinol. Metab.* **69,** 239–245.

49. Hartman, M. L., Pincus, S. M., Johnson, M. L., Matthews, D. H., Faunt, L. M., Vance, M. L., Thorner, M. O., and Veldhuis, J. D. (1994). Enhanced basal and disorderly growth hormone secretion distinguish acromegalic from normal pulsatile growth hormone release. *J. Clin. Invest.* **94,** 1277–1288.

[10] Simulation of Peptide Prohormone Processing and Peptidergic Granule Transport and Release in Neurosecretory Cells

Daniel K. Hartline, Robert Whitney Newcomb, and Robert Wayne Newcomb

Introduction

Peptidergic secretion offers a potential degree of subtlety and complexity not possible with classic small molecule neurotransmitters and hormone systems. Bioactive peptides are synthesized as part of larger amino acid chains, which may contain a number of bioactive amino acid sequences. The prohormone is packaged into secretory granules, which serve for the storage and release of the peptides. Within the secretory granules, enzymatic cleavage of the prohormone into active peptides occurs [for reviews, see Douglass *et al.* (1); Hokfelt (2)]. Consequently, such cells release multiple chemical messengers in response to a single stimulus, potentially permitting a single neural or endocrine cell to control a coordinated response to its secretion (Fig. 1).

It is experimentally observed that the degree of proteolytic cleavage (and thus the spectrum of peptides potentially released) often varies with cellular location and physiological state (3–6). Both the complexity of the chemical mixture as well as its changing character with physiological state are potentially important features of peptidergic secretion. These properties contrast with classic transmitter systems [for review, see Kanner and Schuldiner (7)], whose behavior is, despite kinetic complexity, limited to an increase or decrease in the rate of secretion of one or several bioactive compounds. Schematic representations contrasting these two types of bioactive compounds are summarized in Fig. 2. Given the complexity of the peptidergic cell, methods which can describe its dynamic properties should help in elucidating the mechanisms by which generation and release of mixtures of bioactive peptides are controlled. This in turn will help further the understanding of how individual peptide-secreting cells function in neural and endocrine networks.

To facilitate the quantitative analysis of these complexities, this chapter

Methods in Neurosciences, Volume 28

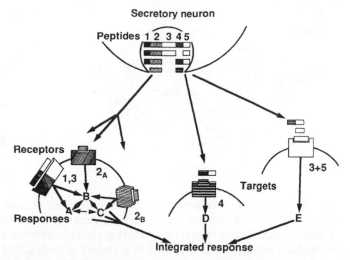

FIG. 1 Potential complexity of the response to peptidergic secretion. The partial proteolytic cleavage of a propeptide with multiple similar and distinct sequences is illustrated. Interaction of the released mixture with multiple receptors on distinct cell types is diagrammed.

develops several computer models and mathematical methods which provide the following capabilities. First, they provide analysis of the kinetics of the proteolytic cleavage of a prohormone. This will include the determination of the rate constants of proteolytic processing from biochemical measurements. Although outside of the scope of this chapter, there are a number of newer experimental approaches to the study of prohormone processing which, in combination with classic pulse–chase experiments, should be amenable to the type of quantitative analysis developed here (8–12). Second, the models provide simulation of the behavior of a peptidergic secretory system (i.e., computer models of peptide synthesis, processing, transport, storage, and release).

Kinetics of Proteolytic Cleavage of Prohormone

When a prohormone is processed into peptide fragments, there is in general a set of specific potential cleavage sites, each acted on by one or more specific enzymes. As the prohormone and its fragments are progressively attacked, more cleavage sites may become exposed, and/or the affinity of previously exposed sites for enzymes (and hence the rate constant for cleav-

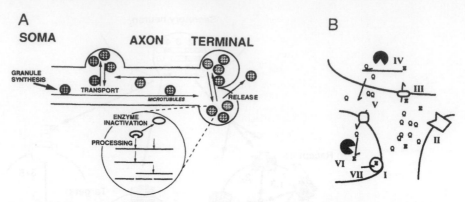

FIG. 2 Schematic comparison of cellular properties of peptidergic secretion and classic secretion of small molecule neurotransmitters. (A) Peptidergic system, with transport, storage, and release of secretory granules in different axonal compartments, proteolytic cleavage of propeptides in secretory granules, and potential changes in enzymatic activity within secretory granules. Further detail is provided in the text [modified from Hartline and Newcomb (13)]. (B) Classic system: small molecule transmitters. Glutamate (E) is used as a specific example. The neurotransmitter is released into the synaptic cleft from storage vesicles (I), where it acts on its receptors (II) and is then removed from the cleft by a specific transporter (III). The active molecule is enzymatically converted (IV) to the inactive compound glutamine (Q) in nonneuronal cells. Glutamine is released into the extracellular fluid and transported into the neuron (V), where it is enzymatically converted back to glutamate (VI) and transported back into the vesicle (VII).

age) may change. In this section we discuss how this situation may be represented mathematically so that the time course of rise and fall of various peptide products and intermediates may be calculated.

1. A prohormone is represented as a chain of peptide segments representing the fragments produced by complete cleavage at all potential sites on the prohormone. Cleavage sites between segments are numbered 1 through N, starting at the amino-terminal end (left end in the figure schematics). Segments are identified by the number of the cleavage site to their right (the final segment is designated $N + 1$). This situation is diagrammed in Fig. 3.

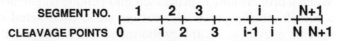

FIG. 3 Schematic representation of a prohormone showing the cleavage sites, peptide segments, and numbering system used for the cleavage model.

2. $P_{i,j}$ will represent a peptide fragment (or the amount or concentration of such) including that of segments i through j. $P_{i,i}$ (also abbreviated P_i) is the single ith segment.

3. $K_{i,j,m}$ will be the first-order nonnegative rate constant of cleavage of fragment $P_{i,j}$ at site m (higher order reactions are not considered in this treatment).

4. A crucial assumption is that the system initially starts, at $t = 0$, with $P_{1,N+1}$ of nonzero concentration P_0 and all others zero.

Basic Equations

For each fragment, $P_{i,j}$, the amount present is determined by a combination of reactions which (1) generate that fragment from precursors (two categories of reaction do this: cleavages at site $i - 1$ of all fragments ending at site j and cleavages at site j of all fragments beginning at site $i - 1$) and (2) degrade it by cleavage at any internal site m (with $i \leq m < j$). For first-order reactions this means (6)

$$dP_{i,j}/dt = \sum_{m=1}^{i-1} K_{m,j,i-1} P_{m,j} + \sum_{m=j+1}^{N+1} K_{i,m,j} P_{i,m} - \left(\sum_{m=1}^{j-1} K_{i,j,m} \right) P_{i,j},$$

$$\text{for } i < j \text{ over } 1, 2, ..., N + 1. \tag{1}$$

Stoichiometry constrains this set of equations so that, whereas there are $(N + 1)(N + 2)/2$ peptides (including the precursor) in the mixture (13), only $N(N + 1)/2$ are independently variable. The sum of all peptides containing a particular segment must equal the amount of prohormone originally present, whence

$$P_{i,i} = P_0 - \sum_{m=1}^{i-1} P_{m,i} - \sum_{j=i+1}^{N+1} P_{i,j}. \tag{2}$$

Note that Eq. (2) reflects the fact that the final concentration of 1-segment products, $P_{i,i}$, approaches the initial concentration of the prohormone, since all of the multisegment intermediates are degraded over time and hence $P_{i,j} \rightarrow 0$ for i not equal to j.

Solutions to Basic Equations for N = 3

We now derive the mathematical form describing the time course of buildup and decay of the remaining peptides of Eq. (1). To express the concepts efficiently, matrix notation will be used, for which an excellent reference is the text of Searle (14). The equations describing the cleavage processes are expressed as ordinary differential equations for which an outstanding and geometrically oriented reference is the text of Arnold (15).

To best illustrate the state of affairs let us choose a three-cleavage-site case ($N = 3$). First we write out Eq. (1) and collect the results into the following matrix form using a convenient ordering of the $P_{i,j}$ into a vector **P**:

$$\frac{d}{dt}\begin{bmatrix} P_{1,4} \\ P_{1,3} \\ P_{1,2} \\ P_{2,4} \\ P_{2,3} \\ P_{3,4} \end{bmatrix} = \begin{bmatrix} -K_{1,4,1} - K_{1,4,2} - K_{1,4,3} & 0 & 0 & 0 & 0 & 0 \\ K_{1,4,3} & -K_{1,3,1} - K_{1,3,2} & 0 & 0 & 0 & 0 \\ K_{1,4,2} & K_{1,3,2} & -K_{1,2,1} & 0 & 0 & 0 \\ K_{1,4,1} & 0 & 0 & -K_{2,4,2} - K_{2,4,3} & 0 & 0 \\ 0 & K_{1,3,1} & 0 & K_{2,4,3} & -K_{2,3,2} & 0 \\ K_{1,4,2} & 0 & 0 & K_{2,4,2} & 0 & -K_{3,4,3} \end{bmatrix} \begin{bmatrix} P_{1,4} \\ P_{1,3} \\ P_{1,2} \\ P_{2,4} \\ P_{2,3} \\ P_{3,4} \end{bmatrix}. \quad (3a)$$

By simply renaming the elements of the large matrix as A_{ij}s we see that this last equation has the nice lower triangular structure

$$\frac{d}{dt}\begin{bmatrix} P_{1,4} \\ P_{1,3} \\ P_{1,2} \\ P_{2,4} \\ P_{2,3} \\ P_{3,4} \end{bmatrix} = \begin{bmatrix} A_{11} & 0 & 0 & 0 & 0 & 0 \\ A_{21} & A_{22} & 0 & 0 & 0 & 0 \\ A_{31} & A_{32} & A_{33} & 0 & 0 & 0 \\ A_{41} & 0 & 0 & A_{44} & 0 & 0 \\ 0 & A_{52} & 0 & A_{54} & A_{55} & 0 \\ A_{61} & 0 & 0 & A_{64} & 0 & A_{66} \end{bmatrix} \begin{bmatrix} P_{1,4} \\ P_{1,3} \\ P_{1,2} \\ P_{2,4} \\ P_{2,3} \\ P_{3,4} \end{bmatrix}, \quad (3b)$$

whereas the relabeling is recorded in the 6×6 matrix quality

$$\mathbf{A} = \begin{bmatrix} -K_{1,4,1} - K_{1,4,2} - K_{1,4,3} & 0 & 0 & 0 & 0 & 0 \\ K_{1,4,3} & -K_{1,3,1} - K_{1,3,2} & 0 & 0 & 0 & 0 \\ K_{1,4,2} & K_{1,3,2} & -K_{1,2,1} & 0 & 0 & 0 \\ K_{1,4,1} & 0 & 0 & -K_{2,4,2} - K_{2,4,3} & 0 & 0 \\ 0 & K_{1,3,1} & 0 & K_{2,4,3} & -K_{2,3,2} & 0 \\ K_{1,4,2} & 0 & 0 & K_{2,4,2} & 0 & -K_{3,4,3} \end{bmatrix}, \quad (4)$$

that is, $A_{11} = -(K_{1,4,1} + K_{1,4,2} + K_{1,4,3})$, etc., as is seen by equating the coefficients in Eq. (3b) with those in Eq. (3a).

The solution to Eqs. (3a) and (3b) may be obtained using convolution, the fundamental operation of linear time-invariant systems theory (16), for which an extensive mathematical treatment exists (17). Because of the nice form of the state equations, an explicit form for the solution can be given using the convolution integral, this latter, denoted as $a * b$, being defined for two time functions $a(\cdot)$ and $b(\cdot)$ by the rather messy looking but extremely useful integral

$$a(\cdot) * b(\cdot)(t) = \int_0^t a(t - \tau)b(\tau)\, d\tau \qquad \text{for } t > 0. \tag{5}$$

In our case all convolutions essentially reduce to those of two exponentials so the result is quite convenient, being

$$e^{\alpha t} * e^{\beta t} = \begin{cases} \dfrac{e^{\beta t} - e^{\alpha t}}{\beta - \alpha} & \text{if } \beta \neq \alpha \\[2ex] t e^{\alpha t} & \text{if } \beta = \alpha. \end{cases} \tag{6}$$

Further, in proceeding down the rows of Eq. (3a) or (3b) after the first row, we will only need to solve equations of the form

$$\frac{dx}{dt} = f(t) + \alpha x \qquad x(0) = 0, \tag{7}$$

for which the convolution solution is

$$x(t) = e^{\alpha x} * f(t). \tag{8}$$

In the case of $N = 3$, $P_{1,4}$ is first found by direct integration of the first row of Eq. (3b), then used in the convolutions to obtain the other fragments, as follows:

$$P_{1,4} = e^{A_{11}t}P_0,$$

$$P_{1,3} = e^{A_{22}t} * A_{21}P_{1,4} = e^{A_{22}t} * A_{21}e^{A_{11}t}P_0 = \begin{cases} \dfrac{e^{A_{22}t} - e^{A_{11}t}}{A_{22} - A_{11}} A_{21}P_0 & \text{if } A_{22} \neq A_{11} \\[2ex] t e^{A_{11}t}A_{21}P_0 & \text{if } A_{22} = A_{11} \end{cases},$$

$$P_{1,2} = e^{A_{33}t} * (A_{31}P_{1,4} + A_{32}P_{1,3}) = e^{A_{33}t} * (A_{31} + A_{32}e^{A_{22}t}A_{21}) * e^{A_{11}t}P_0, \tag{9}$$

$$P_{2,4} = e^{A_{44}t} * A_{41}P_{1,4} = e^{A_{44}t} * A_{41}e^{A_{11}t}P_0,$$

$$P_{2,3} = e^{A_{55}t} * (A_{52}P_{1,3} + A_{54}P_{2,4}) = e^{A_{55}t} * (A_{52}e^{A_{22}t}A_{21} + A_{54}e^{A_{44}t}A_{41}) * e^{A_{11}t}P_0, \tag{10}$$

$$P_{3,4} = e^{A_{66}t} * (A_{61}P_{1,4} + A_{64}P_{2,4}) = e^{A_{55}t} * (A_{61} + A_{64}e^{A_{44}t}A_{41}) * e^{A_{11}t}P_0. \tag{11}$$

As seen by these equations, the convolution approach views the time course of a particular intermediate as the result of a simple first-order drain in the peptide concentration at an aggregate rate constant equal to the sum of all of the first-order rate constants for cleavage of that particular intermediate [the last term in Eq. (1)]. This simple exponential is driven by (convolved with) the, in general, complex, time course of production of that intermediate from all of its various immediate precursors [Eq. (8)]. Because of the nice form of the original cleavage equation, Eq. (1), Eqs. (9)–(11) are readily extended for any N.

An example of the kinetics just described, which shows as well some of the difficulties in the interpretation of the results of experiments which look at prohormone processing, is illustrated in Fig. 4. This figure considers two different kinetic situations which might arise in the processing of a

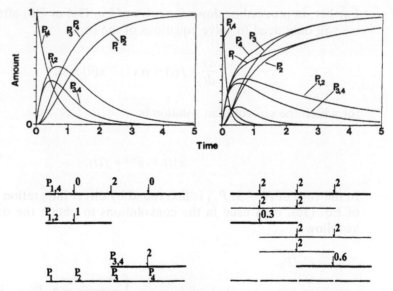

FIG. 4   Simulated processing of model prohormones by sequential (left) and simultaneous (right) mechanisms. Numbers indicate cleavage rate constants in arbitrary reciprocal time units. Intermediates which accumulate significantly are emphasized with thick lines in the diagrams at bottom.

prohormone containing three internal cleavage sites. In the graph at the left, an ordered processing reaction involving peptide-length-dependent cleavage sites is shown. In this example, all rate constants are set to 0, except for $K_{1,4,2} = 2.0$ on $P_{1,4}$, $K_{3,4,3} = 2.0$ on $P_{3,4}$, and $K_{1,2,1} = 1.0$ on $P_{1,2}$, where rate constants are in arbitrary time units (e.g., days^{-1}). The decay of the prohormone from this mechanism is contrasted on the right-hand side of Fig. 4 to a situation in which multiple processing pathways occur. In the example shown, rapid cleavage (K terms of 2.0) occurs at all but several cleavage sites on selected intermediates ($K_{1,2,1} = 0.3$ on $P_{1,2}$, and $K_{3,4,3} = 0.6$ on $P_{3,4}$). In this latter case, much of the biosynthesis of the final product peptides occurs through a variety of intermediates that do not accumulate to a significant extent. However, both this and the length-dependent cleavage rate situation involve the accumulation of a limited number of intermediates in significant quantity. Qualitative description of prohormone processing, particularly if coupled with low resolution methods of chemical analysis, will not be able to distinguish between these two different situations.

General Solution to the Basic Equations

Equation (1) represents linear time-invariant ordinary differential equations with no input except the one initial condition P_0. Consequently they can be rewritten in state-vector form as

$$\frac{d\mathbf{P}}{dt} = \mathbf{A} \cdot \mathbf{P}, \tag{12}$$

where $\mathbf{P}$ is a column vector of the $P_{i,j}$, terms, called the state variable in systems theory nomenclature, and $\mathbf{A}$ a square matrix with its entries being linear combinations of the $K_{i,j,m}$, as the above example illustrates. The solution of Eq. (12) is readily obtained and is (see p. 106 in Ref. 15):

$$\mathbf{P(t)} = [\exp(\mathbf{A} \cdot t)] \cdot \mathbf{P(0)}, \tag{13}$$

where various methods are available for evaluating the matrix exponential (see pp. 123, 137, and 167 in Ref. 15); see also Eqs. (9)–(11) above for the form of most use to the equations on hand. Although this solution [Eq. (13)] is simply expressed, the simplicity is somewhat deceptive as it shows that estimation from this form of the solution involves nonlinear (exponential) functions of the K terms. For this reason we attempt to take advantage of the structure inherent in the problem.

By properly ordering the entries in the state vector $\mathbf{P}$ a particularly nice form for $\mathbf{A}$ results, this being lower triangular and conveniently partitioned if the first $N + 1$ entries of $\mathbf{P}$ go with $i = 1$ and inversely ordered numerically with j, the second N entries going with $i = 2$, and again inversely ordered numerically with j, the third $N - 1$ entries going with $i = 3$, etc. The column vector $\mathbf{P}$ is thus chosen as [written as a row vector through use of the matrix transpose (see p. 38 in Ref. 14), which changes a row vector into a column vector]

$$\mathbf{P} = [P_{1,N+1}, P_{1,N}, ..., P_{1,2}; P_{2,N+1}, ..., P_{2,3}; ...; P_{N-1,N+1}, P_{N-1,N}; P_{N,N+1}]^{\text{transpose}},$$
$$\mathbf{P(0)} = [P_0, 0, ..., 0]^{\text{transpose}}, \tag{14}$$

which agrees with the number given above (15). From this we see that the number of entries in $\mathbf{P}$ is $N(N + 1)/2 = N + (N - 1) + ... + 1$ terms total, which is also the number of rows and columns in $\mathbf{A}$.

We note from our example above that the $\mathbf{A}$ matrix is indeed lower triangular, the advantage of which is that each row is dependent only on the rows above it and not those below. This allows us to start a solution from the top row and work down to the bottom row, finding the unknown entries of $\mathbf{A}$ of a given row at each step using the convolution on all but the first row, as shown in the example above.

Finding Rate Constants

There are a number of means available for finding the rate constants $K_{i,j,k}$ from $\mathbf{P(t)}$ though not all of them are equally effective, primarily due to numerical inaccuracies and other sources of calculation and measurement noise. For those interested, there are commercial programs which can be generally used though usually requiring considerable learning to be effective. For those interested in writing their own program, by which probably more insight can be gained, the curve fitting method based on the exact solution via convolution is recommended [see Eqs. (21)–(25)]. The method is actually presented last since it relies on a number of developments including least-squares and related fittings. Indeed almost all practical techniques seem to require some sort of least-squares fitting, so that aspect is emphasized. For convenience the techniques presented here are summarized in Table I.

Since most parameter determination methods will first find the A terms we need to determine the K terms given the A's. We again use the example for $N = 3$, for which the 14 nonzero equations of Eq. (4) (only 10 of which

TABLE I Summary of Techniques

	Technique; subequations	Comments
Main equations	$\mathbf{K} = \mathbf{L_K A}$	Linear transformation of $\mathbf{A}$; see (15)
1. $\mathbf{A} = \Theta[d\Theta/dt]^{-1}$	Θ represents columns of $\mathbf{P}(t)$	Numerically inaccurate
2. $\mathbf{A} = I\Theta[\Theta]^{-1}$	$I\Theta$ is the numerical integral of Θ	Slightly improved
3. Symbolic solution $[\mathbf{A}t = \ln \mathbf{P}(t) - \ln \mathbf{P}(0)]$		Exact; needs precise data
Least-squares fits		
4. $\mathbf{P}_e(t) = e^{\mathbf{A}_e t} \cdot \mathbf{P}(0)$		Good; messy equations
5. $\mathbf{P}_e(t) = \mathbf{A}_e I\mathbf{P}(t) + \mathbf{P}(0)$	$I\mathbf{P}$ is the numerical integral of $\mathbf{P}$; see Eq. (16)	Satisfactory
6. $\mathbf{P}_e(t) = \mathbf{B}_e E\mathbf{P}(t)$ $\mathbf{A}_e = \mathbf{L_A B}_e$	Curve fitting using convolution exponentials; EP represents columns of exponentials; see Eq. (24) for $\mathbf{L_A}$	Accurate for experimental data
Commercial programs SCIENTIST	Least-squares fit to ordinary differential equation $d\mathbf{P}/dt = \mathbf{A(K)} \cdot \mathbf{P}$	Accurate, reasonably fast, easy to use
MatLab	Matrix solutions of all of above	Good; need to learn its language
Matlab system identification toolbox	General parameter estimation	Generic estimation package for MatLab language

are independent) are solved for the K's in terms of the A's, resulting in the following transformation (which can also be written in matrix form as the transformation is linear, but which is avoided here because it requires introducing other complications such as an appropriate numbering of the K's):

$$
\begin{aligned}
&K_{1,4,1} = -A_{11} - A_{21} - A_{31} = A_{41}, \qquad K_{2,4,2} = -A_{44} - A_{55} = A_{64}, \\
&K_{1,4,2} = A_{31}, \qquad\qquad\qquad\qquad\qquad\quad K_{2,4,3} = A_{45}, \\
&K_{1,4,3} = A_{21}, \qquad\qquad\qquad\qquad\qquad\quad K_{2,3,2} = -A_{55}, \\
&K_{1,3,1} = -A_{22} - A_{32} = A_{52}, \qquad\qquad K_{3,4,3} = -A_{66}. \\
&K_{1,3,2} = A_{32}, \\
&K_{1,2,1} = -A_{33},
\end{aligned}
\tag{15}
$$

It should be noticed that there is dependence among the A_{ij} terms, though the $K_{i,j,m}$ are themselves generally independent.

The most obvious technique is to use $\mathbf{P(t)}$ at $N \times (N + 1)/2$ different times using them as columns in a square matrix Θ; placing their derivatives in $d\Theta/dt$ allows $\mathbf{A}$ to be solved via $\mathbf{A} = \Theta[d\Theta/dt]^{-1}$ where the superscript -1 denotes the matrix inverse (see p. 83 in Ref. 14). Unfortunately, the numerical formation of $d\Theta/dt$ is a very noisy process, something which can be circumvented by numerically integrating $\mathbf{P(t)}$; using the integral $I\mathbf{P(t)}$ leads to the integral $I\Theta$ of Θ, in which case, again using Eq. (13), $\mathbf{A} = I\Theta[\Theta]^{-1}$. Because of the lower triangular structure of $\mathbf{A}$ one can proceed in blocks, one block for each i of the $P_{i,j}$, something which allows some alleviation of the numerical problems of forming the inverse. But the results have not proven to be very satisfying in practice.

A more satisfying solution is to use least-squares estimation techniques (see p. 228 in Ref. 14) on Eq. (3a) rewritten in integral form, $\mathbf{P} = \mathbf{A} \cdot I\mathbf{P} + \mathbf{P(0)}$. In this technique we take an estimated $\mathbf{A}$ (call it $\mathbf{A_e}$), and using $I\mathbf{P}$, obtain an estimated $\mathbf{P}$ (call it $\mathbf{P_e}$), as $\mathbf{P_e} = \mathbf{A_e} \cdot I\mathbf{P} + \mathbf{P(0)}$. Then iterations are used on $\mathbf{A_e}$ to minimize the least-squares error

$$E(\mathbf{A_e}) = \|\mathbf{P_e}(\mathbf{A_e}) - \mathbf{P}\|^2, \tag{16}$$

where $\|\cdot\|^2$ denotes the square of the norm; any reasonable norm can be used. One common norm is the Euclidean norm (see p. 98 in Ref. 15), this being the square root of the sum of the squares of the components of the vector for which the norm is being taken, that is, $\|x\|^2_{\text{Euclidean}} = x^{\text{transpose}} \cdot x$. However, in our case the Euclidean norm would vary with time and, thus, a theoretically more useful norm is the square root of the integral over the total time of the square of the Euclidean norm, that is, $\|x\|^2 = 1/T[\int_0^T \sum_{i=1}^n x_i(t)^2 \, dt]$ when x is an n-vector. As a practical matter with experimental data, though, it is most convenient to replace the integral by a sum over the data points, that is, for $d\#$ the number of data points, to use

$$\|x\|^2 = \frac{1}{d\#} \sum_{k=1}^{d\#} \sum_{i=1}^n x_i(t_k)^2. \tag{17}$$

In using this least-squares technique, iterative changes are performed on the estimate $\mathbf{A}$ matrix, $\mathbf{A_e}$, using the gradient, ∇E, of $E(\mathbf{A_e})$ with respect to $\mathbf{A_e}$. Writing the square matrix $\mathbf{A_e}$ as a vector of components A_{ei} we update the nth iteration of $\mathbf{A_e}$, $\mathbf{A_e}(n)$, to the $(n + 1)$th iteration by

$$A_e(n + 1)_i = A_e(n)_i - \lambda \cdot \nabla E_i, \tag{18}$$

in which λ is a free parameter used to insure convergence (normally $\lambda < 1$) and the gradient is defined as the vector of partial derivatives of E, of Eq.

(16), with respect to the corresponding components of the vector $\mathbf{A}_e$. In doing so, Eq. (18) essentially is the weight update formula for a neural network (18) in which case neural network theory can be used to an advantage in improving the performance [e.g., by adding momentum terms to Eq. (18)].

In any event the number of iterations needed becomes very large, generally in the 10,000 range, in order to achieve a reasonably small error, taking perhaps 15 min on a 50 MHz personal computer (PC). One can write a relatively simple program in something like QuickBASIC to find a suitable $\mathbf{A}_e$, or there are commercial programs for such purposes. Indeed we found that the program SCIENTIST (19) is reasonably efficient for carrying out this type of least-squares fit. Because it too uses least square fitting it may take some time. Other available programs include MatLab with its System Identification toolbox (20), though in practice it may be harder for nonengineers to use than SCIENTIST.

A very viable alternate is to use the precise form of the solution using convolution, as given in Eqs. (5)–(8). Because the convolution framework involves only $\mathbf{P}$ and $\mathbf{A}_e$, but not the integral $I\mathbf{P}$ or the derivative of $\mathbf{P}$, it offers an improved means of proceeding numerically. As shown in the case of $N = 3$ above, by using Eqs. (7) and (8) and working down through the lower triangular state-space equations we can express the results in terms of the initial condition P_0 and the unknowns A_{rs}. In fact if the $P_{i,j}(t)$ are precisely known, these equations [Eqs. (9)–(11) for $N = 3$] can be solved precisely, showing that given N and $P_{i,j}(t)$ the K's are uniquely determined (this is also borne out by the fact that the exponentials are linearly independent and the convolution of functions on $[0,t]$ has no zero divisors). However, it should be pointed out that because of measurement noise and numerical error there could be a number of solutions for the K's that satisfactorily yield the proper cleavage results. On the other hand, by the example shown in Fig. 4, it also appears that in some cases small changes in the K's can yield significantly different cleavage pathways.

The convolution equations put us in a position to obtain an exact solution for $\mathbf{A}$, assuming exact data. As an example of the exact determination, from the first of Eq. (9)

$$P_0 = P_{1,4}(0),$$

$$A_{11} = \frac{\ln[P_{1,4}(t)] - \ln[P_0]}{t} \qquad \text{for any } t > 0. \tag{19}$$

To obtain A_{21} and A_{22} we look at the derivative of $P_{1,3}$

$$\frac{dP_{1,3}}{dt} = \frac{A_{22}e^{A_{22}t} - A_{11}e^{A_{11}t}}{A_{22} - A_{11}} A_{21}P_0, \tag{20}$$

which, when evaluated at $t = 0$ yields A_{21} and then Eq. (9) gives A_{22}. For larger i, j we have to solve nonlinear (nondifferential) equations but these have unique solutions. However, experimental data are not exact and it is better to proceed by curve fitting the given data, $P_{i,j}(t)$, to the known form of the solutions, that is, to Eqs. (9)–(11) in the case of $N = 3$.

For the curve fitting procedure we note that the solutions are linear combinations of $\exp(A_{ii}t)$ where only those exponentials are present which go with rows at or above the $P_{i,j}$ being considered. Thus, we can use the diagonal entries of $\mathbf{A}$ to form an exponential vector

$$EP(t) = [e^{A_{11}t}, e^{A_{22}t}, e^{A_{33}t}, \ldots,]^{\text{transpose}}. \tag{21}$$

Then the convolution solution evaluates to the form

$$\mathbf{P(t)} = \mathbf{B} \cdot EP(t), \tag{22}$$

where $\mathbf{B}$ and $EP(t)$ are functions of the entries of $\mathbf{A}$ with $\mathbf{B}$ being a lower triangular square matrix. Assuming distinct A_{ii}, for our $N = 3$ case we find $\mathbf{B}$ directly from Eqs. (9)–(11) to be 6×6 with the 36 entries

$$
\begin{aligned}
&B_{11} = P_0; \qquad B_{12} = B_{13} = B_{14} = B_{15} = B_{16} = 0, \\
&B_{21} = A_{21}P_0/(A_{11} - A_{22}) = -B_{22}; \qquad B_{23} = B_{24} = B_{25} = B_{26} = 0, \\
&B_{31} = [A_{31}(A_{11} - A_{22} + A_{21}A_{32})]P_0/[(A_{11} - A_{22})(A_{11} - A_{33})], \\
&B_{32} = -A_{21}A_{32}P_0/[(A_{11} - A_{22})(A_{22} - A_{33})], \\
&B_{33} = [A_{31}(A_{22} - A_{33} - A_{21}A_{32})]P_0/[(A_{11} - A_{33})(A_{22} - A_{33})], \\
&B_{34} = B_{35} = B_{36} = 0, \\
&B_{41} = A_{41}P_0/(A_{11} - A_{44}) = -B_{44}, \\
&B_{42} = B_{43} = B_{45} = B_{46} = 0, \\
&B_{51} = [A_{21}A_{52}(A_{11} - A_{44} + A_{41}A_{52})(A_{11} - A_{22})]P_0/ \\
&\qquad\quad [(A_{11} - A_{22})(A_{11} - A_{44})(A_{11} - A_{55})], \\
&B_{52} = -A_{21}A_{52}P_0/[(A_{11} - A_{22})(A_{44} - A_{55})], \\
&B_{54} = -A_{41}A_{54}P_0/[(A_{11} - A_{44})(A_{44} - A_{55}), \\
&B_{55} = [A_{21}A_{52}(A_{44} - A_{55} + A_{41}A_{54})(A_{22} - A_{55})]P_0/ \\
&\qquad\quad [(A_{11} - A_{55})(A_{22} - A_{55})(A_{44} - A_{55})], \\
&B_{53} = B_{56} = 0, \\
&B_{61} = [A_{61}(A_{11} - A_{44} + A_{41}A_{64})]P_0/[(A_{11} - A_{44})(A_{11} - A_{66})], \\
&B_{64} = -A_{41}A_{64}P_0/[(A_{11} - A_{44})(A_{44} - A_{66})], \\
&B_{66} = [A_{41}A_{64} - A_{61}(A_{44} - A_{66})]P_0/[(A_{11} - A_{66})(A_{44} - A_{66})], \\
&B_{62} = B_{63} = B_{65} = 0.
\end{aligned}
\tag{23}
$$

Although these equations look formidable for finding $\mathbf{A}$ in terms of $\mathbf{B}$, once the A_{ii} terms are known (from the exponentials in time) the equations in Eq. (23) are linear in the other A_{ij} and, so, $\mathbf{A}$ is readily obtained from $\mathbf{B}$ and with $\mathbf{A}$, the K's, using the other linear transformation of Eq. (15). Specifically, assuming that the A_{ii} are found independently (as they will be via estimation of the exponentials) direct calculations on the above equations for $\mathbf{B}$ give

$$A_{21} = B_{21}(A_{11} - A_{22})/P_0,$$

$$A_{31} = [B_{31}(A_{11} - A_{33}) + B_{32}(A_{22} - A_{33})]/P_0,$$

$$A_{32} = -\frac{B_{32}}{B_{21}}(A_{22} - A_{33}),$$

$$A_{41} = B_{41}(A_{11} - A_{44})/P_0,$$

$$A_{52} = -\frac{B_{52}}{B_{21}}(A_{44} - A_{55}),$$
(24)

$$A_{54} = -\frac{B_{54}}{B_{41}}(A_{44} - A_{55}),$$

$$A_{61} = [B_{61}(A_{11} - A_{66}) + B_{64}(A_{44} - A_{66})]/P_0,$$

$$A_{64} = -\frac{B_{64}}{B_{41}}(A_{44} - A_{66}).$$

The curve fitting procedure is then to evaluate Eq. (22) with an estimated $\mathbf{B}$, call it $\mathbf{B_e}$, that is, $\mathbf{P_e(t)} = \mathbf{B_e}E\mathbf{P(t)}$, evaluate the norm, $\|\mathbf{P_e} - \mathbf{P}\|$, and update the $\mathbf{B}$ by using the gradient of the error, exactly as at Eq. (18) (with $\mathbf{A_e}$ replaced by $\mathbf{B_e}$). Once the error is small enough, $\mathbf{A_e}$ is found from $\mathbf{B_e}$ and from $\mathbf{A_e}$ the K's. One could also proceed directly to get the $\mathbf{A}$ matrix without the intermediary of the $\mathbf{B}$ matrix, but the gradient with respect to the entries of $\mathbf{A}$ appears to be numerically more noisy than that for the entries of $\mathbf{B}$ due to the convolution placing entries of $\mathbf{A}$ into denominator terms.

Because each row of $\mathbf{P}$ is independent of the lower rows and only introduces one new exponential, it is expedient to do a least-squares fit one row at a time. In that case the (transposed) parameter vector to be updated at each iteration becomes $[A_{ii}; B_{i1}, B_{i2}, ..., B_{ii}]$, and the gradient column vector, using the norm defined by Eq. (17), becomes

$$\nabla E_i = \frac{1}{d\#} \sum_{k=1}^{d\#} 2\{P_{ei}(t_k) - P(t_k)\}[t_k B_{ii} e^{A_{ii}t_k}; e^{A_{11}t_k}, e^{A_{22}t_k}, ..., e^{A_{ii}t_k}]^{\text{transpose}}, \quad (25)$$

where the subscript i on P_{ei} denotes the entry of $\mathbf{P_e}$ which goes with the ith row of $\mathbf{A}$.

By curve fitting the experimental data $P_{i,j}(t)$ to these formulas as indicated (using a least-squares fit), values for the unknown entries of **A** can be obtained in reasonably accurate form. Alternatives are deconvolution techniques, as available in MatLab or direct solution of Eqs. (9)–(11).

Some comments are in order. The number of cleavage sites, N, is assumed known but may vary from process to process. Thus, if the equations are programmed for a given N they can be used for all smaller N by constraining the appropriate rows of **A** to be 0. In the latter methods of this section we have assumed that no two diagonal entries of **A** are identical, which is probably almost always true (even though some K's may be the same). If there are equal A_{ii}'s this will show up in a divide by zero error and can be corrected by using the time multiplied exponential form of the convolution for the offending terms. If one does not have full data on all of the P's then one may not be able to fully determine the K's. This can occur when using experimental data if some of the P's are buried in noise and appear to be 0; in such a case several different sets of K's may appear to give the same end result P's. Because of this it may be best to do estimation including the end results, the P_i's, rather than relying solely on the $P_{i,j}$'s as was done here, because the measurement of the P_i's is usually assured.

In Table I the various alternatives for finding the K's are listed; in Table I $\mathbf{L_K}$ and $\mathbf{L_A}$ are matrices for the linear transformations of Eqs. (15) and (24), respectively.

Granule Transport and Release

The cleavage processes described above occur separately (and for our purposes, independently) in each granule, but they are (again by assumption) the same in all granules. Thus to assess the mix of bioactive peptides released under given circumstances, we need only know the ages of the granules released. This depends on how long it takes following synthesis for granules to be transported to release sites, how long they spend in storage, and what the temporal characteristics are for release from the cell. A model including these processes has been developed as follows:

Representation

The schema which we will represent is as follows.

1. Synthesis of a secretory granule occurs in the main body of a secretory cell (e.g., the soma of a neuron), including prohormone and processing enzymes.

2. Transport of the granule to various parts of the cell for storage and release occurs (e.g., along an axon and into a terminal). In the model, the different physical regions are represented by multiple, diffusionally connected compartments (21–23). With different physiological situations, diffusional rates between compartments may conceivably vary.

3. Granules may be released from certain of the compartments at rates which depend on physiological conditions (21, 24).

4. Growth or change in the shape of the terminal must be taken into account in some cases, for example, in a growing organism (25, 26).

Movement of Granules within Terminal

Cell organelles, including secretory granules, are actively and directionally transported along microtubule tracts via specific attachment proteins (23, 27). Providing secretory granule concentrations are not high enough to be near saturating the microtubule transport system, it may be assumed that the migration of granules from one part of the cell (compartment) into another will occur at a rate proportional to the concentration of the granules in the compartment of origin. A first-order differential equation may thus be used to represent the process, so the secretory cell can be modeled as a set of compartments and connecting pathways using a multicompartmental anisotropic diffusion algorithm. The equations governing granule transport may then be written as (6):

$$dn_i/dt = \sum_{j \neq i} k_{ij} n_j / V_j - n_i \left(\sum_{j \neq i} k_{ji} + R_i \right) \bigg/ V_i + S_i, \tag{26}$$

where n_i is the number of granules in the ith compartment, k_{ij} the rate constant for transport into ith compartment from jth (≥ 0), R_i the release rate constant from the ith compartment (≥ 0), S_i the rate of direct synthesis into ith compartment (≥ 0), and V_i the volume of the ith compartment.

Certain assumptions in the formulation in Eq. (26) must be noted, as they may be violated in real situations. First, granule concentration [n/V in Eq. (26)] is assumed to be without limits. Real granules are not point objects, and only a finite number can occupy a fixed volume. A corollary of this is that compartment volumes are assumed to be independent of granule content. Thus the standard diffusion model may not be valid at high concentrations. Second, granule transport and release rate constants are assumed to be independent of other model variables and time, including factors such as internal content, changing external molecular coats, second-messenger modulation of transport parameters, and size changes with age. The only form of time dependence we have included is for release probability to depend optionally on granule age.

If the coefficients in Eq. (26) are time independent, the equation can be written in matrix notation in a manner equivalent to Eq. (1):

$$d\mathbf{n}/dt = \mathbf{kn} + \mathbf{S},\tag{27}$$

where bold letters indicate the column vector or square matrix equivalents of the terms in Eq. (26). In matrix notation the solution for the particular case of constant synthesis is

$$\mathbf{n} = e^{\mathbf{k}t} \cdot \mathbf{n(0)} + e^{\mathbf{k}t} * \mathbf{S},\tag{28}$$

where $\mathbf{n(0)}$ is a column vector of initial conditions and $\mathbf{S}$ is a column vector of synthesis into each compartment. This gives for each compartment a time course of material concentration consisting of summed decaying exponentials no greater in number than the number of compartments [provided that all compartments have at least indirect access to release sites; Peikari (28)]. Each exponential in general differs in amplitude from compartment to compartment, but the rate constants of decay are shared by all compartments. Geometry can reduce the number of distinct exponentials in the solution. Because this case is too restrictive for many physiological situations of interest, we will leave this line of development and pursue a more general approach based on simulation.

Growth and Volume Changes

The granule transport model makes provision for a simple type of growth in which the geometry of the cell is assumed to expand uniformly over time. The rate constant for intercompartmental granule transport is assumed proportional to the area of the channel between the compartments. A doubling of the cross-sectional area produces a doubling of the transport apparatus and hence a doubling of the flux of material. Such an assumption implies that the rate-limiting step in transport is binding of vesicles to the transport system. The area grows as the square of the linear dimension and the volume as the cube. Thus, for a given number of granules in each compartment, transport and release rate constants are inversely proportional to the linear dimension:

$$dn_i/dt = \left[\sum_{j \neq i} k_{i,j} n_j / V_{oj} - n_i \left(\sum_{j \neq i} k_{j,i} + R_i \right) \middle/ V_{oi} + S_i \right] \middle/ v^{1/3}, \qquad (29)$$

where v is the volume expansion factor giving the proportional increase in volume from a reference time, and V_{oi} and V_{oj} represent compartment volumes at the reference time. The latter serve to fix the geometrical ratios of volumes. The model also expands the synthetic apparatus of the cell in direct proportion to cell volume.

In the present configuration of the model, the volume factor, v, has a linear dependence (including one with 0 slope) on time, although other growth paradigms could be implemented easily. Thus

$$v = v_0 + k_g t, \qquad (30)$$

where t is time, v_0 is the volume factor at $t = 0$, and k_g is a growth rate constant.

Age-Dependent Release

It has been observed in some secretory systems that newly synthesized granules are released preferentially. One possible mechanism would be for affinity of the release mechanism to decrease with granule age. We have included a simple age dependency as an option in our model. This modifies the probability that a granule, once targeted for release in the model, will actually be released. The probability decreases exponentially toward 0 starting from a value of 1.0 at an age of 0:

$$p_{rel} = e^{-k_{rel} t}, \qquad (31)$$

where p_{rel} is the probability-modification factor and k_{rel} is the rate constant of probability decay.

Implementation of Model

The implementation of this model consists of two programs, named GRANULES and PROCON. GRANULES computes the age distributions of granules released from a model cell according to the schema described in the previous section. PROCON utilizes this age distribution to calculate the mixture of peptide products according to the processing scheme described in the first section. Flow charts for the two programs are shown in Fig. 5.

In the GRANULES model (version 1.1; Fig. 5A), granules are represented by two one-dimensional arrays: one stores compartment-assignment integers for the granule (numbered consecutively from 1 to N_c), and the other the time of granule synthesis (birth date). Initializations include reading in an

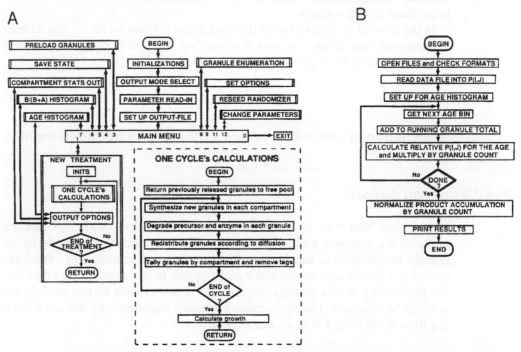

FIG. 5 Flow diagram for GRANULES model and peptide product calculations. (A) GRANULES. Program commences at BEGIN and follows the arrows. Various options are selected at the Main Menu, and then New Treatment is selected to initiate a model run. Calculations proceed by repeated calls to the subroutine ONE CYCLE's CALCULATIONS which performs simulations of granule synthesis and redistribution according to the compartmental and kinetic structure specified in an input-parameter file, as well as daily growth recalculations of model parameters. (B) PRO-CON program for computing peptide products from granule age distributions. The program reads a data file giving the time courses of occurrence of each of the peptide fragments in a granule, then convolves it with an age histogram file generated by GRANULES. This produces a net product distribution as a function of time for all peptides in a compartment or terminal (see text). Symbols are defined as follows: ovals, entry and exit points; double end-bar boxes, subroutine calls; diamonds, branch points or loops. [Modified from Hartline and Newcomb (13)].

ASCII format parameter file. The main menu gives options such as saving and restoring the state of the model on disk, changing parameters, and outputting results in various formats. When New Treatment is selected, the model is run for a user-specified period. A loop is entered to compute changes one cycle at a time (the time allotted to a cycle is arbitrary, being used only to determine the frequency of tallies and growth increments). Euler integration is applied in integrating first-order differential equations, meaning that time is divided into small (compared to the fastest kinetics) increments. At each time increment, the number of new granules to be added to each compartment is calculated from the compartment-specific synthesis rate multiplied by the volume factor. Empty slots are found in the granule arrays and filled with compartment assignment and birth date. Fractional granules for each compartment are held over and summed until a value greater than one accumulates, whereon an extra granule is synthesized for that compartment. Granules already in compartments are randomly selected, in numbers proportional to total compartment occupancy and intercompartmental diffusion constants, for transfer to connected compartments. Newly transferred granules are tagged and not allowed to move twice in the same time increment. A compartment with no back diffusion is used for granules released from the terminal. At the end of each time increment, tallies of granules in each compartment are made, and tags are removed. At daily intervals, ages of released granules are tallied, and histograms for granule age and extent of peptide processing (see next) are generated for outputting. Array-space for released granules is returned to the unassigned pool prior to initiation of the next days calculations.

To facilitate certain simple applications, our implementation of this model includes an option (available if fewer than 6300 granules are used) for two special granule arrays, one containing the remaining amount of prohormone in each granule and the other the rate constant of the processing enzymes. This may be used for simulation of one-cleavage-site processing, and includes the simulation of physiologically induced changes in both initial peptide amounts and processing activity in newly synthesized granules. For more complex situations, these quantities can be calculated from the granule age distributions.

The PROCON calculation of peptide fragment release concentrations as a function of time proceeds according to the flow diagram in Fig. 5B (PROCON version 1.0). After various initializations, a data file containing the stored time courses of peptide fragment kinetics is read into a fragment abundance array, computing normalized fragment distributions for each of the times corresponding to midpoints of the age histogram bins generated by GRANULES. Then the histogram of granule ages is read in one cycle at a time and one bin at a time. The number of granules in each bin is multiplied by the fragment abundance for each fragment for that age. After the histogram of each time step has been processed, the number is normalized to (divided by) the total granule count for the histogram, stored on disk, and then the program goes to the next time step.

Model Applications

A single example is provided here of the combined operation of the GRAN-ULES model and the peptide processing mathematics (Fig. 6). Additional examples may be found elsewhere (6, 13, 26). In the situation shown in Fig. 6, a prohormone containing three cleavage sites has fixed cleavage rate constants at each site, regardless of the length of the peptide containing the site (Fig. 6A). Cleavage at the middle site ($K_{i,j,2} = 0.02$) is 15 times slower

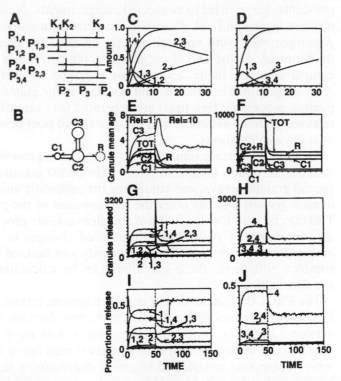

FIG. 6 Simulated output of a model peptidergic neuron using the GRANULES model. The model simulates a neuron with two compartments, one (upper in diagram) for granule storage; one for release.

than that at the two end sites ($K_{i,j,1} = K_{i,j,3} = 0.3$). The kinetic changes in the concentration of the various peptides in a granule as a function of time after synthesis are shown in Fig. 6C,D. Such granules are synthesized into compartment C1 of the cell, as diagrammed in Fig. 6B. The synthesis compartment communicates directly with a release compartment (C2), which in turn is connected to a storage compartment (C3). The rate constants for transport between all compartment pairs are equal and about the same as that for peptide degradation at the fast sites (=0.33). The model is started at $t = 0$ and run until a steady state in the granule content of the cell is reached, in which release balances synthesis (Fig. 6F). At $t = 10$ (time units are arbitrary), the release rate constant is increased 10-fold. Figure 6F shows the resulting sharp drop in the number of granules in each compartment (arrows labeled C1, C2, C3) and in total granule content of the cell. The rate of granule release (arrow R) increases sharply, then decreases as the granule supply in the cell is reduced (synthesis is held constant). Figure 6E shows the mean age of granules in each compartment under this protocol. Note that with the increase in release there is an initial increase in age of granules in the storage compartment (C3) as the supply of young granules in C2 is depleted. Figure 6G,H and 6I,J show the time courses for the amounts and proportions, respectively, of the different peptides released under the conditions modeled. The potential for physiologically significant effects not only from changes in amounts of bioactive peptides released (as would be expected with any transmitter system) but from changes in the proportions released, may be imagined.

Neuroendocrine Secretory Structures

The computer models presented here have been used to construct working hypotheses for secretory structures in which there are comparatively simple processing schemes [Table II; for details, see Newcomb *et al.* (6, 26)]. They have allowed us to make testable predictions of cellular or kinetic mechanisms.

First, the carboxyl-terminal truncation of the oxytocin neurophysin in the rat pituitary gland (29) has served as a model processing system in which experimental determination of *in vivo* processing rate constants under a variety of secretory conditions is feasible (26). The kinetic analysis of this allowed us to ask what actual *in vivo* processing rate constants are, at what rate they inactivate with time in secretory granules, and how the rate of propeptide conversion varies with changes in secretory activity. By showing a constancy of processing rate with changes in secretion rate, we can attribute changes in gland peptide content with physiological activity to changes in

TABLE II   Summary of Quantitative Modeling of Peptide Processing
in Neurosecretory Structures

Preparation	Measured quantities	Hypothesis tested	Ref.
Vertebrate neurohypophysis	Neurophysin conversion rate constants	Secretory granule turnover time and gland volume change with growth	Newcomb et al. (26)
	Changes in neurophysin content with age		
	Neurophysin conversion rate constants	Various hypotheses of secretory granule compartmentation and release	
	Changes in neurophysin content with in vivo depletion of gland content		
	Measurement of stored versus released neurophysin content in vitro		
Crustacean sinus gland	Stored peptide amounts in different animals	Biosynthetic mechanism (from stoichiometry)	Newcomb et al. (6)

secretory granule turnover times (as opposed to changes in processing enzyme activity) (26).

Analysis of gland oxytocin neurophysin content with growth and depletion of the neurohypophysis, in combination with various analyses of release and subcellular fractions, allows us to test the behavior of models for movement, storage, and release of secretory granules in the nerve endings of the posterior pituitary. In this approach, simulations of the peptide content of the neural lobe which include various rate constants for processing and transport kinetics, were compared with experimental measurements. Comparison of the model and experiments then suggested agreement, or specific experiments which will test the validity of models of cellular compartmentation (26).

Second, for the two-site internal cleavage of peptide H of the land crab sinus gland (8), modeling has shown how the approach can be used with a more complex prohormone in a case where the endocrine and cell biology of secretory structure is less completely understood. Application of the stoichiometric equations of processing allowed verification that a proposed biosynthetic mechanism does occur. Further use of the model showed how different cell biological situations can all give rise to experimentally observed variations in sinus gland peptide content; experimental determination of the kinetics of propeptide conversion might be used to distinguish between these variations (6).

Future Directions

The mathematical and computer-simulation tools we have presented have so far been used only in relatively simple situations. Their potential power extends to substantially more complex situations. Such applications have been limited in the past by technical difficulties in biochemically monitoring concentrations of all relevant peptide products in a prohormone cleavage family. With improvement in biochemical techniques, more complex systems should be amendable to the type of analysis developed here.

As indicated in the development above, not all intermediates and final products need be measured in order to deduce the rate constants of cleavage. For simplicity, we have chosen to eliminate the final products from the calculations. Because these compounds are frequently the more accessible to measurement, it would be useful (and should be possible) to develop a formulation dropping some of the less accessible intermediates instead.

We have also indicated some of the technical difficulties with the numerical procedures we have applied. Finding an optimal procedure, especially one which is appropriate for noisy experimental data, is another challenge for the future.

The approach begun here can be extended to more complex situations still. Thus, issues dealing with the different effects of granule aging, targeting of granules to different release sites, and the modulatory control of many potentially susceptible sites in the synthesis, processing, transport and release system, can be addressed in a straightforward way in building more sophisticated models to further our understanding of endocrine systems. (Note: Copies of the simulation programs running on IBM PC-compatible microcomputers are available from the authors.)

Acknowledgments

Past work leading to this chapter has been supported by National Institues of Health Grants NS15314 to D.K.H., NS-24739 to R.Wh.N. and RCMI Grant RR-03061 to the University of Hawaii.

References

1. J. Douglass, O. Civelli, and E. Herbert, Polyprotein gene expression: Generation of diversity of neuroendocrine peptides. *Annu. Rev. Biochem.* **53,** 665–715 (1984).

2. T. Hokfelt, Neuropeptides in perspective: The last ten years. *Neuron* **7,** 867–879 (1991).

3. P. A. Rosa, P. Policastro, and E. Herbert, A cellular basis for the differences in the regulation of synthesis and secretion of ACTH/endorphin peptides in anterior and intermediate lobes of the pituitary. *J. Exp. Biol.* **89,** 215–237 (1980).

4. S. Zakarian and D. G. Smyth, Beta-endorphin is processed differently in specific regions of the rat pituitary and brain. *Nature* (*London*) **296,** 250–252 (1982).

5. C. Evans, D. L. Hammond, and R. C. Frederickson, The opioid peptides. *In* "The Opiate Receptor" (G. W. Pasternak, ed.), pp. 23–71. Humana Press, Clifton, New Jersey, 1988.

6. R. W. Newcomb, D. K. Hartline, and I. M. Cooke, Changes in information content with physiological history in secretory neurons. *Curr. Top. Neuroendocrinol.* **9,** 151–184 (1988).

7. B. I. Kanner and S. Schuldiner, Mechanism of transport and storage of neurotransmitters. *Crit. Rev. Biochem.* **22,** 1–38 (1987).

8. R. W. Newcomb, Amino acid sequences of neuropeptides in the sinus gland of the land crab *Cardisoma carnifex:* A novel neuropeptide proteolysis site. *J. Neurochem.* **49,** 574–583 (1987).

9. J. Bourdais, A. P. Pierotti, H. Boussetta, N. Barre, G. Devilliers, and P. Cohen, Isolation and functional properties of an arginine-selective endoprotease from rat intestinal mucosa. *J. Biol. Chem.* **266,** 23386–23391 (1991).

10. C. Evans, N. Maidmen, and R. Newcomb, Reversed phase liquid chromatography of biological peptides: Technical aspects and applications. IBRO Handbook Series, Vol. 15, "High Performance Liquid Chromatography in Neuroscience Research" (B. Holman, A. J. Cross, and M. H. Joseph, eds.), pp. 263–294. Wiley, New York, 1993.

11. K. G. Galanopoulou, S. N. Rabbani, N. G. Seidah, and Y. C. Patel, Heterologous processing of prosomatostatin in constitutive and regulated secretory pathways. *J. Biol. Chem.* **268,** 6041–6049 (1993).

12. C. S. Konkoy and T. P. Davis, Time-course administration of neuroleptics decreases regional neurotensin metabolism in intact brain slices. *Soc. Neurosci. Abstr.* **19,** 1364 (1993).

13. D. K. Hartline and R. W. Newcomb, Simulation of peptide processing, compartmentation and release in neurosecretory cells. *Neurochem. Int.* **19,** 281–296 (1991).

14. S. R. Searle, "Matrix Algebra for the Biological Sciences." Wiley, New York, 1966.

15. V. I. Arnold, "Ordinary Differential Equations." MIT Press, Cambridge, Massachusetts, 1978.

16. R. W. Newcomb, "Concepts of Linear Systems and Controls," Chapters 3 and 6. Brooks/Cole Publ., Belmont, California, 1968.

17. J. Mikusinski, "Operational Calculus," 2nd ed., Pergamon, New York, 1983.

18. S. Bhama and H. Singh, Single layer neural networks for linear system identification using gradient descent technique. *IEEE Transactions on Neural Networks* **4**, 884–888 (1993).

19. SCIENTIST, version 2.0, for DOS, MicroMath, Inc., Salt Lake City, Utah, 1993.

20. L. Ljung, "System Identification Toolbox for Use with MatLab." The Math Works, Natick, Massachusetts, 1993.

21. J. J. Nordmann, Hormone content and movement of neurosecretory granules in the rat neural lobe during and after dehydration. *Neuroendocrinology* **40**, 25–32 (1985).

22. J. F. Morris, D. B. Chapman, and H. W. Sokol, Anatomy and function of the classic vasopressin-secreting hypothalamus–neurohypophyseal system. *In* "Vasopressin" (D. B. Gash and G. J. Boer, eds.), pp. 1–90. Plenum, New York, 1987.

23. M. P. Scheetz, R. Vale, B. Schnapp, T. Schroer, and T. Reese, Movements of vesicles on microtubules. *Ann. N.Y. Acad. Sci.* **493**, 409–416 (1987).

24. J. J. Nordmann and G. Dayanithi, Release of neuropeptides does not only occur at nerve terminals. *Biosci. Rep.* **8**, 471–483 (1988).

25. G. I. Hatton, Cellular reorganization in neuroendocrine systems. *Curr. Top. Neuroendocrinol.* **9**, 1–28 (1988).

26. R. W. Newcomb, D. K. Hartline, J.-G. Lorentz, A. Depaulis, and J. J. Nordmann, Quantitative analysis and computer simulation of oxytocin–neurophysin processing in the rat neurohypophysis. *Neurochem. Int.* **19**, 297–312 (1991).

27. I. R. Gibbons, Dynein ATPases as microtubule motors. *J. Biol. Chem.* **263**, 15837–15840 (1988).

28. B. Peikari, "Fundamentals of Network Analysis and Synthesis." Prentice-Hall, Englewood Cliffs, New Jersey, 1974.

29. R. W. Newcomb and J. J. Nordmann, Quantative analysis of rat neurophysin processing. *Neurochem. Int.* **11**, 229–240 (1987).

[11] Systems-Level Analysis of Physiological Regulatory Interactions Controlling Complex Secretory Dynamics of the Growth Hormone Axis: A Dynamical Network Model

Lubin Chen, Johannes D. Veldhuis, Michael L. Johnson, and Martin Straume

Introduction

Time-series data of growth hormone (GH) concentrations in the circulation of humans and of other animal models clearly demonstrate pulsatile patterns indicative of both rhythmic and episodic release from the pituitary gland (1–5). These patterns sometimes exhibit quite regular time dependence, as demonstrated by the approximately 3.3-hr periodicity observed in the male rat (2), but more often are irregular and sometimes highly variable, as in humans (1), sheep (4), and the female rat (3). Systemic GH concentration time series vary not only as a consequence of species and sex differences, but their temporal patterns, as well as the responses elicited from the GH neuroendocrine axis by perturbing stimuli, vary in relation to different healthy and pathological conditions.

Understanding the physiological mechanisms that control the complex secretory dynamics exhibited by the GH neuroendocrine axis requires detailed appreciation of a highly regulated phenomenon that is dependent on functional interactions among a large number of processes comprising an exceedingly complex network of stimulatory and inhibitory feedback. The primary regulatory species responsible for mediating pituitary GH release are two hypothalamic peptides, a stimulatory factor, GH-releasing hormone (GHRH), and an inhibitory tetradecapeptide, somatostatin (SRIH) (6–21). GHRH and SRIH are secreted from the median eminence of the hypothalamus into the hypophysial portal blood to be transported to the pituitary gland, where their influence on GH production and secretion is exerted. Growth hormone, after being released into the systemic circulation, stimulates production of insulin-like growth factor-I (IGF-I) in numerous peripheral tissues, including liver, bone, and muscle (22–25). It is

Methods in Neurosciences, Volume 28

IGF-I that serves the role of primary mediator of GHs peripheral effects. The GH neuroendocrine axis is a closed-loop system because both GH and IGF-I exert regulatory feedback on the hypothalamus in the form of decreasing GHRH and increasing SRIH signaling (26–38). Reciprocal intrahypothalamic interactions between GHRH and SRIH neurons provide an additional means for regulation at the hypothalamic level (39–51). Negative feedback is also exerted at the pituitary level by IGF-I, which can suppress net GH output (26, 52). The existence of two GH binding proteins (GHBPs) (53–57) and six IGF-I binding proteins (IGFBPs) (58, 59) provides a capacity for dynamic buffering of free concentrations of peripheral GH and IGF-I (56), whereas the kinetics of hypophysial and systemic distribution as well as elimination of each of these species provide additional means for modulating the dynamic behavior of the GH neuroendocrine axis (1, 56, 60–64).

The descriptive information summarized above was accumulated over decades of anatomical, biochemical, and physiological research that elucidated what have become recognized as the principal individual component processes comprising the GH regulatory network. This information, although exceedingly detailed in some cases (often to the level of molecular interactions in specific signal transduction pathways), largely addresses qualitative aspects of regulation of individual system components. Little explicit attention has been paid to the dynamics (i.e., time-dependent properties) of these regulatory processes, and even less to dynamic systems-level coupling among them. To achieve a mechanistic understanding of how GH secretory dynamics are regulated, knowledge is required not only about the individual time-dependent behavior of each regulatory process, but also about how dynamic interplay among the various stimulatory and inhibitory elements within the network produces the emergent functional properties of regulated GH secretory dynamics.

The complex and sometimes highly irregular nature of GH concentration time series has long presented challenges for interpretation by classic mathematical/analytical approaches. Advances have been made, however, through the development of methods for deconvolution analysis to rigorously quantify secretory events in terms of rates and durations of release as well as of half-lives of elimination (65). Results from such analyses have been used to construct statistical correlates of intra- and interburst interpulse intervals versus secretory burst amplitudes as a function of time-of-day (1). However, statistical interpretations imply underlying stochastic (i.e., random, and therefore not predictable) system properties as opposed to deterministic (i.e., cause-and-effect, and therefore predictable) relationships (66–69). Stochastic contributions to system dynamics cannot be ruled out, but deterministic physiological relationships have clearly been demonstrated and characterized, as mentioned above, albeit primarily at the qualitative and descriptive level. Statistical analyses can provide probabilistic descriptions of complex system dynamics but do not directly address mechanisms of regulation, which can be achieved by interpretations guided explicitly by knowledge of underlying physiological control processes.

Progress in the theory of nonlinear dynamical systems has been producing a new level of understanding of the dynamics of intensively observed biological systems (66–73). It is clear that variable and highly irregular temporal behavior can be produced from even quite simple deterministic relationships that involve no stochastic component. Such system output, however, requires that the system possess appropriate nonlinear properties (i.e., properties that do not scale linearly with the magnitudes of at least some system parameters). In biological systems, such nonlinear properties can be imparted by common phenomena such as ligand–receptor–effector cooperativity, thresholds, saturable binding, and end product inhibition. Coupling among kinetic processes possessing nonlinear dynamical properties and the presence of temporal delay (74, 75) provide additional means whereby systems may exhibit self-organizing behavior, produce characteristic emergent dynamic patterns, and abruptly and aggressively respond to changing (internal and/or external) conditions (66–73). Adaptability to minimally predictable change is a necessary and characteristic quality of life and is not readily achieved in systems devoid of nonlinear character.

Developing an understanding of nonlinear biological systems is generally difficult because of the considerable inherent challenges presented by direct analysis (66–70). Direct analytical approaches designed for reconstructing attractors (referring to the tendency of a nonlinear dynamical system for a particular region or trajectory in state space) are not feasible for neuroendocrine time-series data, because such data are typically relatively short in duration, sparse in point density, and can contain considerable amounts of noise. To generate reliable conclusions from direct nonlinear dynamical analytical methods, approximately infinitely long, noise-free time series are imperative (66–70). The alternative approach of attempting to model mathematically the dynamical behavior of the GH neuroendocrine axis in a physiological context offers the potential for understanding the dynamics of this system in functionally mechanistic terms. One fundamental hypothesis to be addressed by such an approach is whether the pulsatile temporal patterns of GH seen in humans and other animal models emerge as a consequence of nonlinear dynamical coupling among already identified physiological processes or whether explicit involvement of endogenous pulse generators and/or ultradian pacemakers is required.

Construction of the Model

Growth Hormone Axis as a Regulatory Network

Abundant experimental data exist regarding the neuroendocrine GH axis and serve as the basis for constructing a regulatory network. Specifically, GHRH stimulates GH synthesis (20, 21) and release (10, 11, 19) as well as SRIH release (39, 40, 43). SRIH inhibits GH release (12, 13), its own release (50, 51), as well as GHRH synthesis (47–49) and release (41, 42, 44–46). GH stimulates SRIH synthesis (30, 36, 37) and release (30, 32–34), inhibits GHRH synthesis (35, 37, 38) and release (31, 33), and stimulates IGF-I synthesis (22–24) and release (22). IGF-I stimulates SRIH synthesis (38) and release (26, 33), inhibits GHRH synthesis (38) and release (26, 33), and inhibits GH synthesis (52) and release (26, 52). These multiple experimentally determined regulatory interactions, together with binding proteins for GH and IGF-I, and the distribution and elimination kinetics for each of these species, provide the basis for the GH regulatory network depicted in Fig. 1. The numbering convention used throughout refers to GHRH as species 1, SRIH as species 2, GH as species 3, and IGF-I as species 4. S_i refers to the mass of hormone in a storage pool, H_i refers to the concentration of released hypophysial (for $i = 1,2$) or systemic (for $i = 3,4$) hormone, and H_{ih} refers to the mass of intrahypothalamically released hormone (for $i = 1,2$).

Regulatory interactions that are explicitly considered in the initial implementation of the model discussed herein are referred to in Fig. 1 as the modifier functions labeled as F_{iXj}^+ and F_{iXj}^-. The $+/-$ signs refer to up- or down-regulation, respectively, and the iXj refer to regulation imposed by species i on process X acting on species j. Two types of processes are operationally considered in the current formulation, $X = R$ for release, and $X = S$ for synthesis. Thus, for example, F_{1R3}^+ refers to the stimulatory effect of hypothalamically released hypophysial portal GHRH on pituitary GH release. Not explicitly depicted graphically in Fig. 1, but considered during mathematical implementation, are the elimination processes that serve to reduce levels of all released hormone within pools.

Primary Processes

Expressions for synthesis, release, and elimination of the network species outlined above are developed as primary system processes. Distribution is dealt with separately (see Temporal Delay section). These act as the physiological processes on which regulatory control is exerted. Implicit as-

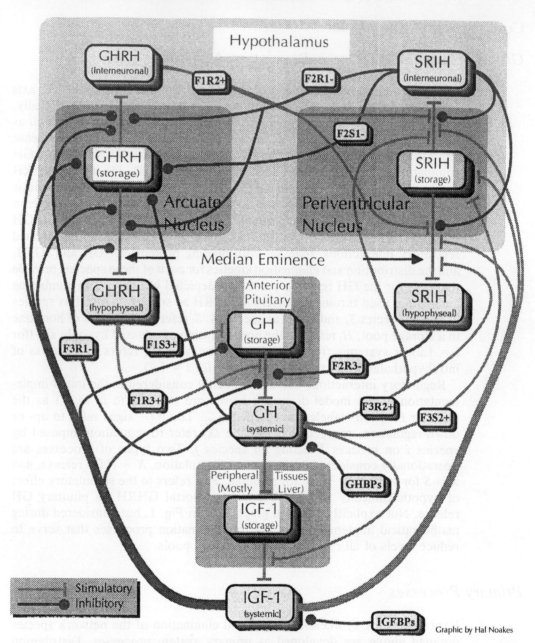

FIG. 1 Network of physiological regulatory interactions involved in controlling secretory activity of the growth hormone (GH) neuroendocrine axis. Up-regulatory interactions are indicated by lines terminating in T endings; down-regulatory interactions are indicated by lines terminating in ball endings. See text for a detailed description of this network. (Graphic by Hal Noakes.)

sumptions in the current formulation are the following: (a) hormone synthetic rates are proportional to differences between single-compartment storage pool maxima, S_i^m, and current storage pool levels, S_i, as $(S_i^m - S_i)$; (b) hormone secretory rates are proportional to current storage pool levels, S_i; (c) on secretion, hormones instantaneously attain their final concentration values (i.e., as if they were immediately diluted throughout their respective distribution volumes, V_i) (temporal delay, see below, allows for latencies introduced by such factors as distribution and admixture); (d) independent hypophysial portal and intrahypothalamically released pools exist for GHRH and SRIH (i.e., H_i and H_{ih}, respectively, for $i = 1, 2$); and (e) irreversible hormone elimination includes any degradation contributing to removal and occurs by first-order processes (i.e., monoexponential decay) that are proportional to current levels of released pools, H_i and H_{ih}. Conservation of mass and the lack of postsecretory activation or interconversion of the hormone species are also assumed.

Translating these assumptions into mathematical descriptions for intrinsic time rates of change of stored and released pools produces the following differential equations:

$$\frac{d}{dt} S_i = k_{Si}(S_i^m - S_i) - k_{Ri}S_i - k_{Rih}S_i, \qquad i = 1, 2,$$

$$\frac{d}{dt} S_i = k_{Si}(S_i^m - S_i) - k_{Ri}S_i, \qquad i = 3, 4,$$

$$\frac{d}{dt} H_i = \frac{k_{Ri}S_i}{V_i} - k_{Ei}H_i, \qquad i = 1, 2, 3, 4,$$

$$\frac{d}{dt} H_{ih} = k_{Rih}S_i - k_{Eih}H_{ih}, \qquad i = 1, 2.$$

The variables S_i, H_i, and H_{ih} are dynamic variables (i.e., they exhibit explicit time dependence). The terms $k_{Si}(S_i^m - S_i)$, $k_{Ri}S_i$, $k_{Rih}S_i$, $k_{Ei}H_i$, and $k_{Eih}H_{ih}$ correspond to intrinsic rates of synthesis, release, and elimination as defined by the first-order rate constants k_{Si}, k_{Ri}, k_{Rih}, k_{Ei}, and k_{Eih}. The V_i terms refer to hormonal distribution volumes of the hypophysial portal blood (for $i = 1, 2$) or the systemic circulation (for $i = 3, 4$). In this formulation of primary processes only, the dynamic variables S_i, H_i, and H_{ih} transiently approach steady-state values due to lack of any regulatory interactions or temporal delays (in fact, these equations are linear and can be solved exactly). However, these primary processes will serve as targets for regulatory intervention, as developed below.

Regulatory Modifier Functions

Regulatory processes are incorporated into the model by defining dose-responsive modifier functions, $F_{iXj}^{+/-}$, that exert influence on the previously defined primary system processes. The general mathematical form used for these modifier functions is

$$F_{iXj}^{+/-}(D) = \frac{(u - v)}{1 + (D/D_t)^p} + v.$$

From the point of view of physiological regulation, four important properties are embodied in the above mathematical representation: (a) minimal and (b) maximal attainable magnitudes of modifying regulatory influence (given by u, the value of F_{iXj} at $D = 0$, and v, the value of F_{iXj} at $D = \infty$), (c) a threshold level at which switching of the function occurs (given by D_t), and (d) the rate of responsiveness of the function in the vicinity of the threshold (related to p). D is a delay variable (see Temporal Delay section) corresponding to the effective concentration of hormone exerting regulatory influence after accounting for effects of temporal delay. The power, p, to which (D/D_t) is raised is a parameter permitting variation of the sharpness of the sigmoidal transition in the modifier function, as demonstrated in Fig. 2. Stimulatory regulation is elicited by the condition $u < v$ (as in the upward-tending sigmoid of Fig. 2) whereas inhibitory regulation is imposed when $u > v$ (i.e., a downward-tending sigmoid). Nonregulation occurs when $u = v$. F_{iXj} is imple-

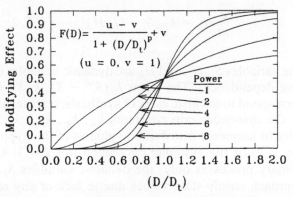

FIG. 2   Example of the dependence of an up-regulatory modifier function on the ratio of temporally delayed hormone concentration, D, to threshold, D_t, as a function of the power, p.

mented to represent the degree of fractional regulatory influence (i.e., $0 \leq u, v \leq 1$ and therefore $0 \leq F_{iXj} \leq 1$).

Each regulatory process of the network is operationally defined in this manner and has associated with it a function of this form. Operationally defining regulatory interactions in this way simplifies modeling efforts while maintaining the capability for direct physiological interpretation and estimation without need for considering, in great detail, those molecular-level processes involved in production of regulatory influence.

In the case of $u = 0$ and $v = 1$, the form of F_{iXj} reduces to that of the Hill equation (76, 77), originally derived to assess cooperative binding of oxygen to hemoglobin tetramers. In that case, the parameter p represents an index of the cooperativity of oxygen binding, that is, a measure of the nonlinear functional coupling among the four binding sites for molecular oxygen in the hemoglobin tetramer. In the current setting, the parameter p may therefore be thought of as representing some index of the degree of functional coupling among the numerous biochemical processes that underlie each operationally defined regulatory process.

Regulatory control exerted by modifier functions on target primary processes is implemented in a multiplicative manner into the differential equations that characterize the respective primary process. The present formulation of the model considers only those nine modifier functions explicitly referred to in Fig. 1 (constituting what will be referred to as the reduced core network).

Three processes are modeled as being regulated by a single modifier function: (a) synthesis of hypothalamic GHRH is down-regulated by SRIH via F_{2S1}^{-}, (b) synthesis of hypothalamic SRIH is up-regulated by GH via F_{3S2}^{+}, and (c) synthesis of pituitary GH is up-regulated by GHRH via F_{1S3}^{+}. Their effects are modeled by direct multiplication of corresponding rates of synthesis.

Three sets of primary processes are modeled as being regulated by pairs of modifier functions: (a) rates of release of hypothalamic GHRH into both hypophysial portal and intrahypothalamic interneuronally released pools are down-regulated by both SRIH and GH via F_{2R1}^{-} and F_{3R1}^{-}, respectively, (b) rates of release of hypothalamic SRIH into both hypophysial portal and intrahypothalamic interneuronally released pools are up-regulated by both GHRH and GH via F_{1R2}^{+} and F_{3R2}^{+}, respectively, and (c) the rate of pituitary GH release into systemic circulation is up-regulated by GHRH via F_{1R3}^{+} and down-regulated by SRIH via F_{2R3}^{-}. Dual regulation of primary processes is implemented mathematically in analogy to expressions that describe classic enzyme kinetics, with appropriate generalization, as discussed below.

In the event that $p = 1$ (and $u = 0$, $v = 1$), F_{iXj} takes on the same form as the Michaelis–Menten equation (78)

$$\frac{[ES]}{[E]_{tot}} = \frac{[S]}{[S] + K_m} = \frac{[S]/K_m}{1 + [S]/K_m}.$$

Here, $[ES]$ and $[E]_{tot}$ are substrate-bound and total concentrations of enzyme, respectively, $[S]$ is free substrate concentration, and K_m is the Michaelis–Menten constant. Competitive and noncompetitive inhibition in classic enzyme kinetics, in turn, are described by the following expressions:

$$\frac{[ES]}{[E]_{tot}} = \frac{[S]/K_m}{1 + [S]/K_m + [I]/K_i}, \quad \text{competitive,}$$

$$\frac{[ES]}{[E]_{tot}} = \frac{[S]/K_m}{1 + [S]/K_m} \cdot \frac{1}{1 + [I]/K_i}, \quad \text{noncompetitive.}$$

Here, $[ES]$, $[E]_{tot}$, $[S]$, and K_m are as previously defined, $[I]$ is free concentration of inhibitor, and K_i is the equilibrium dissociation constant for inhibitor. The form of these expressions is now extended to address dual-acting regulatory modifier functions. If the assumption is made that mutual up- or down-regulating influences interact in a competitive manner for the same up- or down-regulatory pathways, competitive mutual GHRH/GH up-regulation of SRIH release and competitive mutual SRIH/GH down-regulation of GHRH release can be expressed, after appropriate generalization of the expressions for inhibition in classic enzyme kinetics, as

$$\text{Mutual up-regulation of SRIH release} = \frac{(D/D_t)^p_{1R2^+} + (D/D_t)^p_{3R2^+}}{1 + (D/D_t)^p_{1R2^+} + (D/D_t)^p_{3R2^+}},$$

$$\text{Mutual down-regulation of GHRH release} = \frac{1}{1 + (D/D_t)^p_{2R1^-} + (D/D_t)^p_{3R1^-}}.$$

These expressions can be rewritten (given that the values of u and v are constrained appropriately to values of exactly 0 and 1) directly in terms of corresponding regulatory modifier functions as

$$\text{SRIH up-regulation} = \frac{F^+_{1R2}(D_{1R2}) + F^+_{3R2}(D_{3R2}) - 2 F^+_{1R2}(D_{1R2})F^+_{3R2}(D_{3R2})}{1 - F^+_{1R2}(D_{1R2})F^+_{3R2}(D_{3R2})},$$

$$\text{GHRH down-regulation} = \frac{F^-_{2R1}(D_{2R1})F^-_{3R1}(D_{3R1})}{F^-_{2R1}(D_{2R1}) + F^-_{3R_1}(D_{3R1}) - F^-_{2R1}(D_{2R1})F^-_{3R1}(D_{3R1})}.$$

Constraining the values of u and v appropriately to exactly 0 and 1 was necessary to maintain similar structural form for the equations describing inhibition of classic enzyme kinetics and that of the regulatory modifier functions. The resultant derived expressions for dual mutual up- and down-regulation made explicitly in terms of the modifier functions themselves now relieve the need for continuing to maintain a constraint on u and v.

In a similar (and much simpler) manner, if GHRH (up-regulating via F_{1R3}^+) and SRIH (down-regulating via F_{2R3}^-) interact in a noncompetitive manner (as has been reported in Refs. 14 and 15) to influence pituitary GH release (i.e., by presumably not competing for influence of a common regulatory pathway), these regulatory modifier functions can be incorporated by direct multiplicative implementation into the differential equation describing the rate of GH secretion (in analogy to the direct multiplicative form for noncompetitive inhibition seen in classical enzyme kinetics).

Temporal Delay

Responses to released regulatory species are not instantaneously manifested by the system, but rather, will require some finite amount of time for their realization. A myriad of distribution, admixture, transport, and biochemical processes can be envisioned as potentially contributing to temporal delays at numerous stages of the GH network depicted in Fig. 1. The mathematical complexity implicit in specifically accounting for multistep processes is significantly reduced by operationally defining temporal delay by way of delay variables through a delay convolution integral in terms of two parameters, response delay and lifetime of the regulatory stimulus. The form of the delay convolution integral is

$$D(t) = \frac{1}{\tau} \int_{-\infty}^{t-d} H(z) \exp\left[-\frac{(t-d)-z}{\tau} \right] dz.$$

Here, $H(z)$ is the concentration of regulatory hormone at time z, d is the amount of time by which the particular network process influenced by species H is delayed, and τ is the lifetime of the regulatory stimulus. Both D and H are in units of concentration when considering H_i ($i = 1,2,3,4$) and in units of mass when considering H_{ih} ($i = 1,2$), whereas d and τ have units of time. This integral is derived as the solution of a delay differential equation (as described in the APPENDIX at the end of this chapter). This expression can be conceptualized as $H(z)$ being composed of a series of infinitely sharp pulses at times z, each pulse inducing an equivalent response (i.e., also infinitely sharp) in D time d later, followed by decay of D with lifetime τ.

Summing these effects over time up to $(t - d)$ gives $D(t)$. The effects of d and τ on a delay variable profile are shown in Fig. 3.

Each network regulatory process has response delay operationally defined in this way. Although not directly interpretable in terms of individual biochemical processes, d and τ have physiological significance. The net delay imposed by the sequence of biochemical processes involved in generating a response induced by a particular stimulus is characterized by d. The parameter τ can be thought of as the lifetime of the (operationally defined) vehicle responsible for transducing the stimulatory or inhibitory signal into an elicited response (for example, as the duration of activation or down-regulation of an implicated second messenger pathway).

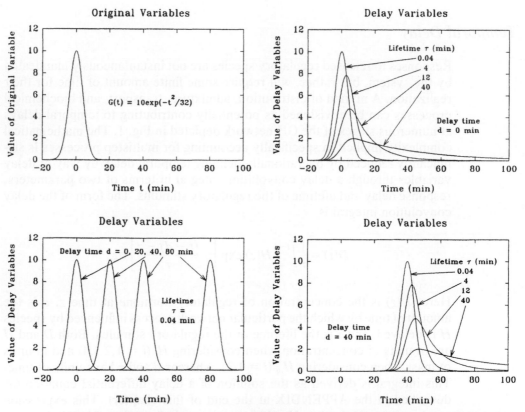

FIG. 3   Effects of delay time, d, and lifetime, τ, on profile shape and delayed temporal positioning of a secretory hormone profile in generation of delay variable profiles.

Differential Equations of the Reduced Core Growth Hormone Network

The complete set of coupled nonlinear ordinary differential equations that embody the concepts developed in the previous paragraphs is given below:

$$\frac{d}{dt} S_1 = k_{S1} F_{2S1}^-(D_{2S1})(S_1^m - S_1) - RS_1 - RS_{1h},$$

$$\frac{d}{dt} H_1 = \frac{RS_1}{V_1} - k_{E1}H_1,$$

$$\frac{d}{dt} H_{1h} = RS_{1h} - k_{E1h}H_{1h},$$

$$RS_1 = k_{R1}S_1 \frac{F_{2R1}^-(D_{2R1})F_{3R1}^-(D_{3R1})}{F_{2R1}^-(D_{2R1}) + F_{3R1}^-(D_{3R1}) - F_{2R1}^-(D_{2R1})F_{3R1}^-(D_{3R1})},$$

$$RS_{1h} = k_{R1h}S_1 \frac{F_{2R1}^-(D_{2R1})F_{3R1}^-(D_{3R1})}{F_{2R1}^-(D_{2R1}) + F_{3R1}^-(D_{3R1}) - F_{2R1}^-(D_{2R1})F_{3R1}^-(D_{3R1})},$$

$$\frac{d}{dt} S_2 = k_{S2} F_{3S2}^+(D_{3S2})(S_2^m - S_2) - RS_2 - RS_{2h},$$

$$\frac{d}{dt} H_2 = \frac{RS_2}{V_2} - k_{E2}H_2,$$

$$\frac{d}{dt} H_{2h} = RS_{2h} - k_{E2h}H_{2h},$$

$$RS_2 = k_{R2}S_2 \frac{F_{1R2}^+(D_{1R2}) + F_{3R2}^+(D_{3R2}) - 2F_{1R2}^+(D_{1R2})F_{3R2}^+(D_{3R2})}{1 - F_{1R2}^+(D_{1R2})F_{3R2}^+(D_{3R2})},$$

$$RS_{2h} = k_{R2h}S_2 \frac{F_{1R2}^+(D_{1R2}) + F_{3R2}^+(D_{3R2}) - 2F_{1R2}^+(D_{1R2})F_{3R2}^+(D_{3R2})}{1 - F_{1R2}^+(D_{1R2})F_{3R2}^+(D_{3R2})},$$

$$\frac{d}{dt} S_3 = k_{S3} F_{1S3}^+(D_{1S3})(S_3^m - S_3) - RS_3,$$

$$\frac{d}{dt} H_3 = \frac{RS_3}{V_3} - k_{E3}H_3,$$

$$RS_3 = k_{R3}S_3 F_{1R3}^+(D_{1R3})F_{2R3}^-(D_{2R3}).$$

The mathematical formulation of the reduced core GH neuroendocrine axis, as implemented in the present study, employs 9 regulatory modifier functions defined by 36 parameters, 9 delay processes described by 18 parameters, 13 rate constants accounting for synthesis, release, and elimination processes, 3 parameters defining maximal storage pool sizes, and 2 parameters defining distribution volumes, for a total of 72 system parameters.

Estimation of Parameters

An attempt was made to produce physiologically reasonable estimates for each of the model parameters by examination of published data regarding maximal storage pools for GHRH and SRIH in the hypothalamus and GH in the pituitary (35, 60, 79, 80), systemic and hypophysial portal blood volumes (56, 61, 62), and elimination rates of regulatory species (1, 56, 62–64). Estimation of physiologically reasonable ranges for delay times and durations of influence experienced by target tissues to stimuli are based on the underlying biological processes that are likely involved in the generation of the response (i.e., circulation, transport, diffusion, activation of signal transduction pathways, gene transcription, and protein translation). Intrinsic rates of synthesis and release were considered as adjustable model parameters in the sense that their magnitudes were made such that system outputs conformed with experimental observations. Similarly, parameters defining regulatory modifier functions were adjusted accordingly to bring model output into correspondence with maximal normal physiological concentrations observed for GHRH and SRIH in hypophysial portal blood and for GH in the systemic circulation (1, 2, 4, 17).

Table I presents the parameter values implemented for definition of the

TABLE I   Parameter Values Used in Defining Regulatory Modifier Functions

Modifier function	Threshold (D_t)	Power (p)	Initial value (u)	Final value (v)	Delay time (d, min)	Lifetime (τ, min)
F_{2S1}^-	0.05 pg	4	1	0	72	32
F_{2R1}^-	0.039 pg	7.2	1	0	0	0.1
F_{3R1}^-	45 ng/ml	7.2	1	0	1.5	0.8
F_{3S2}^+	49 ng/ml	4	0.25	1	158	40
F_{1R2}^+	0.05 pg	7.2	0.25	1	0	0.1
F_{3R2}^+	25 ng/ml	7.2	0.25	1	65	32
F_{1S3}^+	45 pg/ml	4	0	1	81	32
F_{1R3}^+	142 pg/ml	6	0	1	0.4	0.4
F_{2R3}^-	108 pg/ml	7.2	1	0	0.4	0.4

regulatory modifier functions and their associated delay and lifetime parameters. Values for these parameters are not available as *a priori* known or even approximate quantities. Rather, they were considered as adjustable parameters whose absolute values were modified after numerous rounds of simulations until their composite values conveyed to the network dynamic properties consistent with those observed experimentally in the male rat model. Thresholds were restricted to reside between minimal and maximal observed concentrations of each respective species. Powers of at least four for the regulatory modifier functions were found to be necessary to produce oscillatory behavior and were permitted to range between 4 and 7.2. Up-regulatory functions (a) used values of 0 for the parameter u except in the case of regulation of SRIH where u had a value of 0.25 [to reflect observations suggesting the presence of basal levels of SRIH even under conditions that provide little or no up-regulatory influence on SRIH levels (16, 17)] and (b) used values of 1 for the parameter v. Down-regulatory functions used values of 1 and 0 for the parameters u and v, respectively, under the assumption that down-regulation can be complete given sufficiently aggressive down-regulatory conditions.

Relatively long delays and lifetimes were imposed on regulatory processes affecting hormone synthesis ($\sim$1–3 hr for delays and $\sim$0.5 hr for lifetimes) to reflect the considerable cellular machinery and activity expected to be required to respond to stimuli encouraging intracellular hormone production. Reciprocal interneuronal interactions between GHRH and SRIH neurons in the hypothalamus were assigned no delay and very short lifetime (0.1 min) parameters to reflect the typically extremely rapid nature of neuronal communication. Processes involving hypothalamus-to-pituitary stimuli were assigned delay and lifetime values (0.4 min for each) to reflect the rapid transport to and departure from the pituitary of released hypothalamic factors traveling via the hypophysial portal vasculature. Processes involving GH feedback on the hypothalamus were assigned delay and lifetime values (1.5 and 0.8 min, respectively) to reflect the additional time expected to be required for a released pituitary factor to travel via dilution into and transport by the systemic circulation to influence the release of hypothalamic factors. The feedback of GH on SRIH release, however, required modification to significantly longer delay and lifetime values (65 and 32 min, respectively) to produce the approximate 3.3-hr periodicity seen in bursts of GH in the male rat. (This is an accommodation to the lack of explicit consideration of the GH → IGF-I → SRIH loop in the current implementation of the model; see Discussion section for more on this point.)

Table II presents the parameter values used to define storage pool maxima, hypophysial portal and systemic circulatory distribution volumes, and intrinsic rates of hormone synthesis, release, and elimination. Maximal storage

TABLE II   Parameter Values for Storage Pool Maxima, Distribution Volumes, and Rates of Synthesis, Release, and Elimination

Hormone	S^m	k_S (min^{-1} $\times 10^{-4}$)	k_R (min^{-1} $\times 10^{-3}$)	k_E (min^{-1})	V	k_{Rh} (min^{-1} $\times 10^{-5}$)	k_{Eh} (min^{-1})
GHRH	1.4 ng	1.68	2.70	1.386	0.82 μl	7.20	3.466
SRIH	16 ng	0.31	0.031	1.386	0.82 μl	0.583	3.466
GH	1.95 mg	4.20	2.47	0.0495	28 ml		

pools were estimated from reported total contents of GHRH, SRIH, and GH in dissected tissues (79, 80). The GH distribution volume was estimated as 7% of a 395-g body weight for an average male rat (7% being an estimate frequently cited in the literature for humans and assumed approximately the same in the rat). Estimation of the hypophysial portal distribution volume involved using the same 7% proportion, but in this case relative to the reported weight of a rat pituitary ($\sim$11.7 mg) (80). The rat GH elimination rate was assumed to be approximately the same as reported in humans (a 14-min half-life for elimination was assumed, translating to a 0.0495 min^{-1} first-order elimination rate constant; $k_E = \ln 2/t_{1/2}$). Rapid effective hypophysial portal GHRH and SRIH elimination rates (corresponding to 0.5-min half-lives) were used to reflect the relatively brief time available for exposure of hypophysial portal blood to the pituitary. Even more rapid elimination rates (corresponding to 0.2-min half-lives) were used to characterize the elimination of intrahypothalamically released GHRH and SRIH. Intrinsic rates of synthesis and release were treated as adjustable parameters that were modified to produce maximal circulating hormone concentrations in agreement with experimentally observed levels in the male rat.

Mathematical Implementation of Experimental Perturbation Studies

Two experimental techniques commonly employed to study regulatory influences on hormone secretion are (a) exogenous infusion and (b) antibody-mediated specific withdrawal of regulatory hormones. System responses to such perturbations are then observed. To test the dynamic behavior of this computer-based model in response to similar (simulated) system perturbations, mechanisms were incorporated for mathematically implementing in the model equivalent infusion and withdrawal of selected species as in the experimental setting.

An infusion function, F_{Ii}, is defined as

$$F_{Ii}(t) = \begin{bmatrix} RI_i; & T_{Ini} \le t \le T_{Ifi} \\ 0; & t < T_{Ini}, t > T_{Ifi} \end{bmatrix} + A_{Ii} \exp\left(-\frac{(t - T_{Ii})^2}{2SD_{Ii}^2}\right), \qquad i = 1, 2, 3.$$

This function can provide (a) a constant rate of infusion of species i at a rate of infusion RI_i between the infusion onset and offset times T_{Ini} and T_{Ifi} and/or (b) a Gaussian-shaped bolus infusion of species i with maximum infusion rate A_{Ii} at time T_{Ii} with standard deviation SD_{Ii} (reflecting the duration of influence). F_{Ii}, RI_i, and A_{Ii} have the same units as dH_i/dt, and T_{Ini}, T_{Ifi}, T_{Ii}, and SD_{Ii} have units of time.

Withdrawal of a particular species is implemented by increasing the rate of its elimination, as perceived by the model, through a withdrawal function, F_{Wi}, defined as

$$F_{Wi}(t) = \begin{bmatrix} k_{Wi}; & T_{Wni} \le t \le T_{Wfi} \\ 0; & t < T_{Wni}, t > T_{Wfi} \end{bmatrix}, \qquad i = 1, 2, 3.$$

Here, k_{Wi} is the withdrawal rate constant for species i with onset and offset of withdrawal given by T_{Wni} and T_{Wfi}, respectively.

Time rates of change of infused hormone are then considered as

$$\frac{d}{dt} H_{Ii} = F_{Ii} - k_{EIi} H_{Ii} - F_{Wi} II_{Ii}, \qquad i = 1, 2, 3.$$

Here, H_{Ii} is the concentration of exogenously infused hormone and k_{EIi} its elimination rate constant. It is necessary, in the cases of GHRH and SRIH, to distinguish exogenous and endogenous elimination rate constants. Endogenously secreted GHRH and SRIH require very rapid effective endogenous elimination rates to account for their rapid transit out of the hypophysial portal blood, their region of bioactive pituitary influence, and into systemic circulation. This is not the case in exogenously administered hormone which is systemically distributed and not limited only to the hypophysial portal blood. Thus, exogenous H_I is mathematically considered independently of its endogenous counterpart, but affects the system through addition to endogenous hormone in the convolution integral for calculating the delay variable as

$$D(t) = \frac{1}{\tau} \int_{-\infty}^{t-d} [H(z) + H_I(z)] \exp\left[-\frac{(t - d) - z}{\tau}\right] dz.$$

The withdrawal function F_{Wi} exerts its influence on endogenous species i as well as on any exogenously introduced hormone that may be present at the onset of withdrawal. Therefore, the effects of F_{Wi} are exerted directly at the level of elimination by addition to the endogenous elimination rate constant in the previously defined differential equations.

Numerical Integration of Differential Equations

Calculation of time-dependent system properties was performed by numerically integrating, via a fourth- fifth-order Runga–Kutta–Fehlberg method (81, 82), the set of differential equations describing the network. Integration was carried out in 0.25-min intervals. Sufficient calculational cycles (i.e., simulated time) were performed such that transient dynamic effects were absent and stable time-series behavior was produced (i.e., the system was sufficiently warmed up, typically for several days of simulated time, to avoid contamination by any decaying transient effects from initial conditions relative to which integration was initiated).

Results

The male rat represents an animal model that exhibits quite regular GH concentration time-series (2) and about which a great deal of experimental data has been obtained. It therefore provides an excellent candidate system with which to begin development of a computer-based physiological model of the neuroendocrine GH axis.

Reference System

Rhythmic temporal behavior in GH concentration profiles was readily generated by the model. Subsequent rounds of parameter adjustment were successful in producing GH concentration profiles highly analogous to those observed experimentally in the male rat with regard to both frequency of temporal patterning and concentration range.

The top graphs of Figs. 4–7 (marked as Reference) present the simulated profiles of the unperturbed system. (Note that the time axes of Figs. 4–7 are for representational purposes only; the simulated data sets were generated in terms of relative time only.) The unperturbed profile exhibits a periodicity in the appearance of bursts of GH of approximately 3.3–3.4 hr [in agreement with experimentally observed periods of approximately 3.1–3.3 hr (2, 10)].

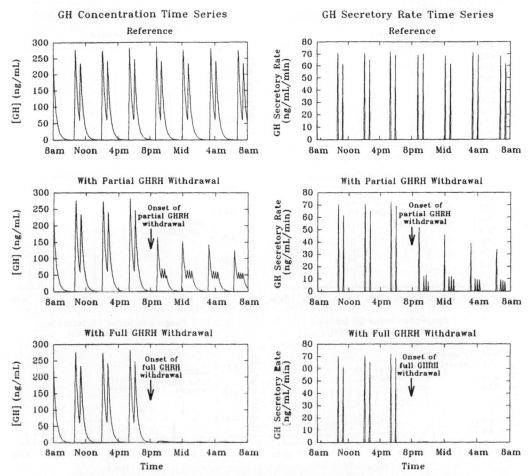

FIG. 4 Simulated effects of partial and full GHRH withdrawal on time-series profiles of GH concentration and associated secretory rate. Time axis labeling is representational only.

Each burst is, in turn, composed of a pair of GH pulses, also in analogy to experimentally determined GH profiles from the male rat (2). Peak concentrations of GH of approximately 280 ng/ml along with nadir concentrations approaching 0 also agree with experimental findings (2, 10). Additionally, peak hypophysial portal concentrations of GHRH and SRIH produced by the simulation were approximately 680 and 170 pg/ml, respectively (data not shown), which compare well with reported literature values of around 800–900 and 165 pg/ml, respectively (17).

A series of preliminary computer-based experiments was then performed

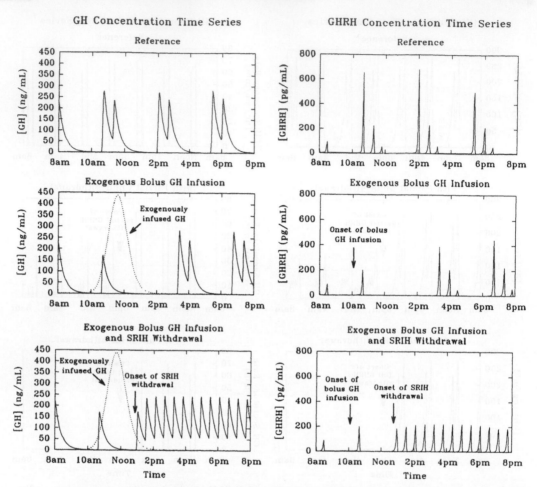

FIG. 5 Simulated effects of bolus infusions of GH, with and without subsequent withdrawal of SRIH, on GH and GHRH concentration time-series profiles. Time axis labeling is representational only.

in an attempt to quantitatively reproduce behavior observed experimentally in response to specific physiological perturbations.

Experiment 1: Partial and Full Growth Hormone-Releasing Hormone Withdrawal

The effects of partial and full withdrawal of GHRH [as accomplished experimentally by administration of increasing amounts of GHRH-specific antibody

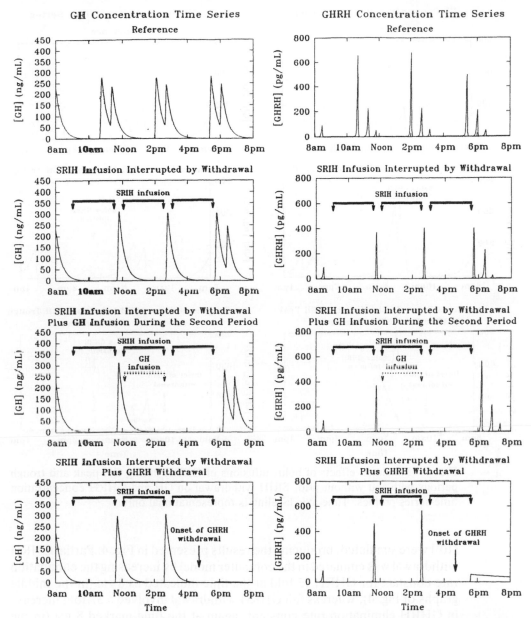

FIG. 6   Simulated effects of interrupted continuous 2.5-hr infusions of SRIH alone and in conjunction with GH infusion or subsequent GHRH withdrawal on GH and GHRH concentration time-series profiles. Time axis labeling is representational only.

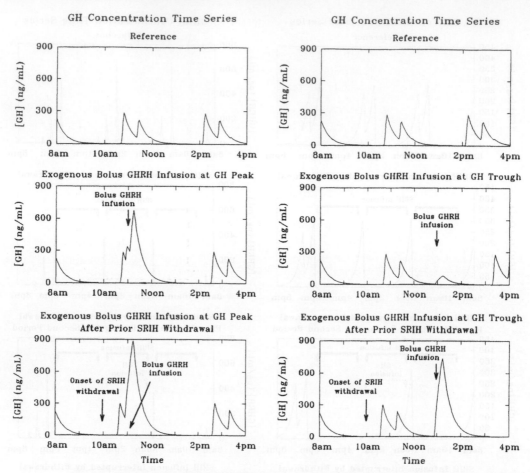

FIG. 7   Simulated effects of bolus infusions of GHRH during GH peak and trough periods with and without prior SRIH withdrawal on GH and GHRH concentration time-series profiles. Time axis labeling is representational only.

(10)] were simulated, producing the results presented in Fig. 4. Partial GHRH withdrawal was emulated in the computer model by increasing the elimination rate constant for GHRH 12-fold at the time indicated as 8 PM (in the middle graphs of Fig. 4), whereas full GHRH withdrawal involved a 37-fold increase in GHRH elimination rate constant, again at the time marked 8 PM (in the bottom graphs of Fig. 4). Simulated partial GHRH withdrawal produced lower peak GH concentration maxima, considerably accelerated within-burst pulsing, and resulted in a small decrease in interburst period, whereas full withdrawal of GHRH completely eliminated circulating GH levels, in direct analogy to behavior observed experimentally after infusion with increasing dosages of GHRH antibody (10). Also shown in Fig. 4 are the simulated GH secretory rate profiles that correspond to the associated GH concentration time series, clearly demonstrating the broadening exhibited by concentration profiles relative to GH secretory activity.

Experiment 2: Exogenous Growth Hormone Infusion with and without Subsequent Somatostatin Withdrawal

Exogenous infusion of a bolus of GH was modeled as an approximately 53-μg Gaussian-shaped infusion of GH centered at 11:25 AM on the representational time scale of Fig. 5, middle graphs, producing a maximum exogenous GH concentration of around 450 ng/ml. The elimination rate constant of the exogenously introduced GH was the same as that for endogenous GH. This perturbation to the reference condition of the model induced an approximate 4.3-hr suppression of endogenous GH and GHRH secretion followed by resumption of the normal 3.3- to 3.4-hr periodicity. The response of the model to this perturbation compares well with the experimentally observed approximate 4-hr suppression of GH (GHRH was not measured) in response to an analogous exogenous bolus infusion of GH that attained a maximum systemic concentration of approximately 250 ng/ml, was of approximately the same duration as in the simulation, and was applied at the same point relative to the GH concentration time-series (i.e., initiation of the GH bolus near the time of onset of a GH burst). Not shown in Fig. 5 is the elevation of SRIH levels produced by the model in response to the exogenous GH bolus.

The bottom graphs of Fig. 5 portray the loss of SRIH regulation in the model as induced by antibody-mediated SRIH withdrawal (implemented as a 5-fold increase in the rate of SRIH elimination at the 12:40 PM marker in the bottom graphs of Fig. 5) after an identical exogenous bolus infusion of GH as in the middle graphs. The elevated between-pulse GH levels and (slightly) increased pulse frequency (of both GH and GHRH) to approximately 30- to 35-min periodicity are in excellent agreement with the response produced experimentally in GH concentration profiles by the analogous perturbation protocol (34).

Experiment 3: Somatostatin Infusion, Growth Hormone Infusion, and Growth Hormone-Releasing Hormone Withdrawal

The response of the model to constant exogenous infusion of SRIH for successive 2.5-hr periods (interrupted by 0.5-hr periods of no SRIH infusion) is depicted in the second pair of panels in Fig. 6. Secretion of GH and

GHRH was abolished during periods of SRIH infusion, with rebound effects appearing immediately on cessation of the infusion. An analogous result was observed experimentally (83), although a quantitative difference between the response of the model and that seen experimentally is evident. The model produced GH rebound to about 305 ng/ml following 2.5-hr infusions of SRIH at 2 μg/hr (with an exogenous SRIH elimination rate constant of 0.693 min^{-1}, corresponding to a 1-min half-life). The experimental setting produced GH rebound to 334 ± 45 ng/ml after 2.5-hr SRIH infusions of 20 μg/hr, a 10-fold higher rate of SRIH infusion. Also, the dependence on SRIH dosage of the magnitude of GH rebound determined experimentally has not yet been examined in the model. Thus, although qualitative agreement between model and experimental behavior is observed, further refinement of the model will be required to achieve closer quantitative agreement.

As shown in the third pair of graphs of Fig. 6, concomitant infusion of exogenous GH (at 100 μg/hr in both model and experimental settings) and SRIH (at 2 μg/hr in the model and 20 μg/hr experimentally) prevented GH and GHRH rebound on termination of infusion. The identical result was observed experimentally regarding GH rebound (GHRH was not measured) (83).

The bottom pair of graphs of Fig. 6 demonstrate that GHRH withdrawal (modeled as a 37-fold increase in GHRH elimination rate at 4:30 PM on the representational time scale, i.e., initiated approximately midway through a 2.5-hr bout of SRIH infusion) is capable of (very nearly) eliminating all signs of a GH rebound on cessation of SRIH infusion, again in excellent agreement with experiments (83).

Experiment 4: Growth Hormone-Releasing Hormone Infusions at Peaks or Troughs in Growth Hormone Profile with and without Prior Somatostatin Withdrawal

Response of the model to bolus infusions of exogenous GHRH during periods of either peak or trough levels of circulating systemic GH is depicted in the middle graphs of Fig. 7. In both model simulations and the experimental setting (16), exogenous GHRH was capable of stimulating GH release during both peak and trough GH periods. However, the amount of GHRH-stimulated GH release was substantially greater during GH peaks than troughs. The GHRH dosage applied in the model was 1.25 μg (with a 0.347 min^{-1} exogenous GHRH elimination rate constant, i.e., a 2-min half-life) whereas 10 μg of GHRH was introduced experimentally (16).

Withdrawal of SRIH prior to introduction of a bolus of GHRH (modeled as a 1.75-fold increase in the rate of SRIH elimination) caused enhanced GHRH-induced release of GH, particularly when GHRH application occurred during a GH trough period (see bottom graphs of Fig. 7). A similar effect was observed experimentally (16).

Although model responses in the above perturbation protocols qualitatively agree with those observed experimentally, quantitative agreement was not achieved. In all likelihood, a significant contributing factor to this quantitative disagreement has to do with the 8-fold lower GHRH infusions introduced in the model simulations. This was necessary due to difficulties encountered during numerical integration of the differential equations when attempts were made to simulate larger GHRH infusions (see Discussion section for more on this). When GHRH is introduced during a GH peak period without prior SRIH withdrawal, GH concentrations were elevated to approximately 680 ng/ml in the model but rose to around 800–2000 ng/ml experimentally. Introduction of GHRH during a GH trough without prior withdrawal of SRIH saw GH concentrations rise to about 100 ng/ml in the model, whereas concentrations of around 180–360 ng/ml were observed experimentally. When SRIH was withdrawn prior to introduction of a bolus of GHRH during a GH peak, the model produced GH concentrations of approximately 900 ng/ml, whereas experimentally observed concentrations of GH rose to around 1400–3000 ng/ml. Bolus infusion of GHRH during GH troughs after prior SRIH withdrawal caused the model to produce GH concentrations up to about 740 ng/ml, although the same protocol experimentally produced GH levels as high as around 1150–2800 ng/ml. Despite these quantitative differences, the model was able to reproduce in almost identical relative manner the reduction of peak/trough-dependent differences in GH responsiveness to GHRH infusion by prior withdrawal of SRIH.

Discussion

A physiologically based mathematical model of the neuroendocrine GH axis has been developed that demonstrates an ability to simulate a number of experimentally observed dynamic system properties and responses exhibited by the male rat GH neuroendocrine axis. Although the implementation described does not incorporate all of the physiological regulatory interactions involved in controlling GH secretory dynamics to which reference is made in the network of Fig. 1, we are able to obtain qualitatively and, to a considerable extent, quantitatively accurate representations of GH concentrations, pulsatile patterns, periodicity, and responses to experimental perturbations. Thus, although further refinement is required, significant first steps toward development of an accurate, computer-based model of the GH axis have been taken in the present modeling endeavor.

First attempts at generation of reference male rat GH concentration time-series (i.e., GH pulsatility with multiple-peak bursts recurring with an ~3.3-hr interburst periodicity) made evident the requirement for incorporating delay on the order of an hour, with an approximately half-hour lifetime, into the positive regulatory feedback of GH on SRIH release (F_{3R2}^+). The sum of the delay and lifetime parameters of F_{3R2}^+ corresponds to approximately one-half the interburst period. Secretion of GH also occurs during approximately half of the approximate 3.3-hr interburst period, which is subsequently suppressed by elevated SRIH levels induced by prior high GH concentrations. After SRIH levels again fall, consequent to lack of GH up-regulation, another burst of GH secretion can occur, due to the absence of SRIH down-regulation, and the cycle repeats. Thus, in this implementation of the model, reciprocal coupling between GH and SRIH is responsible for mediating bursts of GH release about every 3.3 hr.

The direct feedback of GH on hypothalamic regulatory targets is not necessarily expected to occur by such slow dynamic coupling. Delay of the order of minutes, to account for circulatory distribution and transit of GH from the pituitary back to the hypothalamus, may be more physiologically reasonable. In fact, the higher frequency spiking of GH release within bursts is due to the more rapid dynamic coupling between GH and GHRH as imposed by the F_{1R3}^+ and F_{3R1}^- regulatory interactions. However, because effects of IGF-I are not explicitly included in this initial reduced core network model, the slow dynamic coupling implicit in F_{3R2}^+ is a required approximation in order to more accurately embody the positive GH $\rightarrow$ IGF-I $\rightarrow$ SRIH regulatory feedback loop. Dynamic responses of IGF-I to changing system conditions are reported as occurring on the order of hours (23, 60). Induction by GH of IGF-I production, either in hypothalamically distant tissue, from which it must subsequently be transported systemically to the brain, or more locally within the brain, is a physiological rationalization for the slow dynamic coupling currently being provided by F_{3R2}^+ (to compensate for the absent IGF-I intermediary loop). Future explicit consideration of the IGF-I component in this network is anticipated to obviate the need for this approximation.

Refinement of model parameter values was performed next by examining responses elicited by the model to various system perturbations that have been studied experimentally. In this manner, the dynamic responsiveness of the model was brought more closely in line with that which has been observed experimentally in response to analogous system perturbations, thus providing a means for more accurately embodying quantitative aspects of dynamic coupling between network regulatory elements. The first system response examined (experiment 1) was with respect to GHRH withdrawal.

The model was shown to be capable of exhibiting the same dynamic behavior as observed in the experimental setting (10). Under conditions of partial GHRH withdrawal, the amplitude of GH secretory bursts, although still appearing rhythmically, was diminished in proportion to the extent of GHRH withdrawal. This is consistent with a direct and primary involvement of GHRH in stimulating pituitary GH secretion. The increased frequency of intraburst pulsing and slight reduction of interburst period exhibited by model simulations under partial GHRH withdrawal were not apparent in experimental data. However, noise in experimentally determined GH concentrations, as well as relatively long experimental sampling intervals (i.e., low temporal point density), would tend to obscure any such relatively subtle system responses.

A second experimental perturbation protocol that was examined involved considering system responses to bolus infusions of GH. Resumption of endogenous GH secretion following a simulated exogenous infusion of GH was delayed by about 4 hr or more (experiment 2), directly analogous to the effect observed experimentally (34). When exogenous GH infusion was followed by SRIH withdrawal, similar GH secretory pulses appear almost coincident with the onset of SRIH withdrawal in both the model and experimental settings. These combined results are consistent with GH exhibiting an autoinhibitory effect on its own secretion via the involvement of SRIH. The rapid pulsing of GH that arises in the latter case emerges as a consequence of uncoupling the influence of SRIH from the network. Only the coupling interactions between GHRH and GH, and their respective secretory and elimination rates, remain, thus producing an altered dynamical behavior.

The rebound pulses of GH secretion observed experimentally (83), and reproduced by the model, on cessation of prolonged (2.5 hr) infusions of SRIH in experiment 3 are consistent with a diminished endogenous SRIH tone that results as a consequence of exogenous SRIH infusion. Chronically elevated levels of SRIH suppress circulating levels of GH (and GHRH), thus removing up-regulatory influences to the hypothalamus to release SRIH. Rapid elimination of the exogenous source of SRIH then freely permits GHRH and GH levels to rise because no down-regulatory pressure on these species is present (i.e., that which would have been provided by endogenous SRIH). Experimentally, and in the case of the model, this rebound effect (due to cessation of SRIH infusion) could be eliminated by two different means. Endogenous SRIH capacity could be sufficiently enhanced by simultaneous infusion of GH (to stimulate hypothalamic SRIH release) during the period of SRIH infusion so as to suppress any rebound pulse of endogenous GH. Similarly, withdrawal of GHRH prior to cessation of SRIH infusion also eliminated the appearance of GH rebound, again demonstrating the direct involvement of GHRH in stimulating pituitary GH release.

The responsiveness of the GH axis to bolus infusions of GHRH has been demonstrated experimentally to differ between periods of peak and trough systemic GH concentrations (i.e., times of high versus low circulating GH levels, respectively) (16). The model system was able to qualitatively reproduce this same differential sensitivity to GHRH, however, without quantitative agreement in absolute response magnitudes (see experiment 4 results). A partial explanation of these quantitative differences may rest in the different GHRH doses implemented in the experimental and model settings (10 μg GHRH infused experimentally versus 1.25 μg infused in the modeled perturbations). Inability to successfully numerically integrate the system of differential equations when larger infusions of GHRH were attempted required simulating this perturbation protocol using smaller dosage infusions. [Difficulties arose in the form of numerical overflow during calculation of the ratio of temporally-delayed GHRH to GHRH threshold raised to the power p, i.e., $(D/D_t)^p$, in GHRH-dependent regulatory modifier functions.] The resulting smaller amplitude system responses may therefore be, at least partially, a consequence of this fact. However, even though absolute model responses were smaller than those observed experimentally, relative effects of GHRH infusions during peak and trough GH periods, with and without prior SRIH withdrawal, qualitatively paralleled the experimentally reported system behavior.

Two significant observations regarding establishment of regulatory modifier function parameter values were made apparent by the simulations of experiment 4. The SRIH up-regulatory functions F_{1R2}^+, F_{3S2}^+, and F_{3R2}^+ required modification such that their minimal attainable values were 0.25 (i.e., the parameter u was set to 0.25 for each function). The physiological interpretation of this is that, even in the absence of any up-regulatory stimulus (i.e., GHRH for F_{1R2}^+ and GH for F_{3S2}^+ and F_{3R2}^+), some level of unstimulated basal secretion of SRIH from the hypothalamus is permitted. In the absence of a finite amount of unstimulated SRIH secretion, the exceedingly low GH (and GHRH) levels encountered during trough GH periods effectively reduced SRIH levels to zero. In such cases, no differential effects of GHRH infusions on GH release were apparent during trough GH periods as a function of prior SRIH withdrawal because no SRIH was available for withdrawal.

The second modification to regulatory modifier function parameter values was related to the stimulatory interaction of GHRH on GH release, F_{1R3}^+. The responsiveness of F_{1R3}^+ to the level of GHRH, as reflected in the power p, needed to be reduced (i.e., a smaller value for p) in order for exogenous infusion of GHRH during GH peak periods to produce any enhancement of GH secretory activity. To elicit such a GHRH-induced stimulation of GH release, the capacity for GHRH to further influence GH release must be less than saturated under conditions of relatively high GH. However, conditions of high GH occur in concordance with high levels of GHRH. Therefore, the function F^+_{1R3} must have available a high-end dynamic range to increase further under conditions of already relatively high GHRH levels. Reduction of the value of p in F^+_{1R3} (from 7.2 to 6) provided a sufficient reduction in the steepness of the sigmoidal ascent of this response function to permit the model to produce a GHRH-induced enhancement of GH secretion during a peak GH period.

These efforts to model system responses to a variety of experimental perturbation protocols demonstrate that the manner in which the current model of the neuroendocrine GH axis is formulated permits (a) highly accurate emulation of the rat animal model and (b) direct physiological implications to be associated with particular mathematical properties of the model. The ability to reliably simulate the dynamical behavior and responsiveness of a complex network as well as to establish correlates between physiology and mathematics is anticipated to provide deeper quantitative insights into the dynamic regulatory mechanisms responsible for GH secretory control and demonstrates the utility of having an accurate, computer-based model of the GH axis. Availability of such a model will offer the potential for gaining insight and testing hypotheses, through simulation and by comparison with experimental results, regarding how specific aspects of dynamic coupling among physiological regulatory interactions affect this system's overall dynamic properties and responsiveness to perturbations.

Concluding Remarks/Future Directions

The computer-based model of the neuroendocrine GH axis developed in this chapter represents an initial attempt at formulating the framework for a physiologically based network model, directly in terms of experimentally identified functional regulatory interactions, to describe GH secretory dynamics. Because the male rat exhibits regular periodicity in circulating levels of GH and because a great deal of experimental information has been obtained about this particular animal model, initial work has concentrated on this system. Empirical studies have identified the regulatory interactions depicted in Fig. 1 as relevant components of the network involved in secretory control of GH. The mathematical implementation of this dynamical network, as described herein, is less than complete. However, even the current reduced model, albeit with some necessary approximations, showed the capability to qualitatively and, to a large extent, quantitatively reproduce the system dynamics and responsiveness to perturbations observed experimentally in the male rat GH axis.

Any detailed quantitative description of the physiological processes regulating GH secretion will require accurate, explicit consideration of the dynamic coupling in which an exceedingly complex network of regulatory interactions are engaged. Empirical studies aimed at considering individual interactions within the network will be challenged to address this issue. Unless considered as a unified interactive system, emergent temporal properties exhibited by a nonlinear, dynamically coupled system may not be readily apparent from reductionist experiments examining only selected portions of the system at a time. Mathematically designing a network model that operationally defines physiological control mechanisms (as, e.g., in terms of modifier functions and convolution delay integrals) obviates the need for embodiment of excessive molecular level detail while maintaining the capability for interpretation in physiologically meaningful contexts.

Recognition of the system-wide physiological basis of the functional mechanisms underlying GH secretory control will come only by appreciation of how the individual regulatory pathways influence emergent system dynamics in conjunction with all other dynamically coupled processes. The strategy of altering, in a physiologically consistent manner, different functional aspects of selected (combinations of) pathways in an accurate model is anticipated to facilitate appreciation not only of physiological mechanisms responsible for GH-related disorders, but also the bases for species, age, and sex differences in GH secretory dynamics. In this way, insight will be gained regarding GH pathology, potential means for rational design of therapeutic interventions, and the possible consequences of different intervention strategies. As the thoroughness of an evolving model of the GH axis grows to embody more completely its essential elements, an eventual long-term goal is integration of the influences of other hormone axes, independently and cooperatively, to ultimately produce an extended, multi-hormone-axis description of neuroendocrine regulatory control.

Immediate refinements to the existing model will involve including, in totality, all of the regulatory interactions indicated in the network diagram of Fig. 1 (i.e., adding explicit consideration of IGF-I and its regulatory interactions as well as the binding proteins for GH and IGF-I). Another type of modification is suggested from a particular aspect of the performance of the current model. The involvement of synthetic processes in regulatory control could not be effectively examined in the current study. In the present formulation, maximal hypothalamic stores of GHRH and SRIH and pituitary stores of GH were modeled as single-compartment pools (i.e., pools to which synthesis contributed directly and from which release occurred). Reported hypothalamic and pituitary levels for these hormones (35, 79, 80), when used as estimates of storage pool levels, were so large relative to amounts released during secretion that no appreciable changes in pool contents could be elicited under any conditions. This inability eliminates not only the possibility of experiencing depletion effects in response to secretory activity, but also dynamic variation in system properties to which regulation of synthetic processes is directed. A modification is planned in which hypothalamic and pituitary storage pools will be considered as multiple compartments (representing, e.g., an immediately postsynthetic pool, a pool in which prerelease processing takes place, and an immediately releasable pool) between which transfer of hormone will proceed in a sequential manner as defined by regulated kinetic transfer parameters.

Although not yet at all thoroughly examined, preliminary attempts at modeling with depletable, immediately releasable pools suggest that both effects of depletion and synthetic regulation will impact on system dynamics. Considerably greater variability and irregularity in GH concentration time series have been produced under such conditions as a consequence of greater system nonlinearity arising from influences of previously unavailable dynamically coupled processes.

Generation of variable and irregular GH temporal patterns is a necessary capability that must be embodied in any accurate and useful network dynamics model of GH regulation. In practically all models other than the male rat, considerably more dynamic complexity is apparent in GH profiles (1, 3, 4). As demonstrated by the current model, episodic hormone release can occur as an emergent system property of this network without requirement of any explicit endogenous pulse generator mechanism, per se. And complex temporal patterns are anticipated to emerge as a result of incorporating depletable hormone storage pools. However, empirical studies suggest that at least part of the complexity seen in GH temporal profiles involves sleep effects and/or a circadian component (84–86). These are therefore another class of phenomena (i.e., manifested by regulatory neural input pathways) that will require attention and may be important considerations in an examination of the underlying causes of temporal GH secretory complexity.

Among other long-term objectives of this modeling effort are plans to incorporate the capability of learning, as in connectionist neural networks (87). A physiologically accurate model formulated with such capability will provide a mechanism for estimation of physiological constants on exposure and training to actual hormone concentration time-series data. Such an unconventional form of analysis of experimental data will permit interpretation not just in phenomenological terms (as rates and durations of secretory activity and hormone elimination, for example), but rather directly in terms of the dynamically coupled physiological regulatory interactions embodied in the empirically based formulation of the network. In turn, it is further anticipated that availability of such a functional connectionist network model will facilitate testing the performance of other new and existing analytical tools, as well as potentially eventually reducing the numbers of experiments needed with animals.

Appendix

Temporal delay in elicitation of the effects exerted by regulatory interactions in the network of the growth hormone neuroendocrine axis was modeled by way of operationally defined delay convolution integrals (see Temporal Delay section). Specifically, delay variables were defined in a manner that accounted for both (a) a delay in stimulus-induced response (quantified by the variable d) and (b) the effective lifetime of an operationally defined mediator responsible for response production (quantified by the variable τ). The general form of the convolution delay integral is derived here beginning from a delay differential equation.

Define $H(t)$ to be the time dependence of a given hormone concentration and $D(t)$ to be the corresponding delay variable. The time rate of change of the delay variable can be described by

$$\frac{d}{dt}D(t) = k_F H(t - d) - k_E D(t),$$

where the rate of formation of D at time t is given by $k_F H(t - d)$ and is proportional to the concentration of H some time d earlier, and the rate of elimination of D at time t is given by $k_E D(t)$ and is proportional to the current value of D.

Consider the case of $H(t)$ being held constant at some concentration H_0. The value of $D(t)$ must then also be constant at some level D_0. The time rate of change of D is therefore 0 [i.e., $dD(t)/dt = 0$], requiring that $k_F H_0 = k_E D_0$. If D_0 is required to equal H_0, then k_F must equal k_E in order for D to scale equivalently with H.

Now, consider further that the concentration of H goes to (and remains) 0 at some time $t_0 - d$. Then $dD(t)/dt = -k_E D(t)$ for $t > t_0$. The solution of this last expression is $D(t) = D(t_0) \exp(-t/\tau)$, where $\tau = k_E^{-1}$ is the monoexponential lifetime for elimination of D.

After substitution of this result and rearrangement, the original delay differential equation can be rewritten as

$$\left[\tau \frac{d}{dt} + 1\right] D(t) = H(t - d).$$

However, $H(t - d)$ is also given by the expression

$$H(t - d) = \int_{-\infty}^{+\infty} H(z) \, \delta[(t - d) - z] \, dz,$$

where δ is Kronecker's delta. A solution to the expression

$$\left[\tau \frac{d}{dt} + 1 \right] G(t - d, z) = \delta[(t - d) - z]$$

will therefore provide a solution to the original delay differential equation where $G(t - d, z)$ is the Green's function for this problem. Solution for $G(t - d, z)$ gives

$$G(t - d, z) = \begin{cases} 0 & , \quad t < z + d \\ G(z, z) \exp\left[-\frac{t - (z + d)}{\tau} \right], & t \geq z + d. \end{cases}$$

A solution for $G(z, z)$ is obtained by performing the following integration:

$$\left[\tau \frac{d}{dt} + 1 \right] \int_{z+d-\varepsilon}^{z+d+\upsilon} G(t - d, z) \, dt = \int_{z+d-\varepsilon}^{z+d+\varepsilon} \delta[(t - d) - z] \, dt,$$

where ε is some small positive quantity. Performing this integration produces

$$\tau G(z, z + \varepsilon) + \int_{z+d-\varepsilon}^{z+d+\varepsilon} G(t - d, z) \, dt = 1.$$

As $\varepsilon \to 0$, the integral in this expression goes to 0 [as long as $G(t - d, z)$ is not singular at $t = z + d$] giving the result that $G(z, z) = \tau^{-1}$. Therefore

$$G(t - d, z) = \begin{cases} 0 & , \quad t < z + d \\ \dfrac{1}{\tau} \exp\left[-\dfrac{t - (z + d)}{\tau} \right], & t \geq z + d \end{cases}$$

or, similarly

$$G(t - d, z) = \begin{cases} 0 & , \quad z > t - d \\ \dfrac{1}{\tau}\exp\left[-\dfrac{(t - d) - z)}{\tau}\right], & z \leq t - d. \end{cases}$$

Note that $G(t - d, z)$ is discontinuous at $t = z + d$, but no singularity exists there.

Finally, the following integration with respect to z provides a definition of $D(t)$:

$$\left[\tau\frac{d}{dt} + 1\right]\int_{-\infty}^{+\infty} H(z)G(t - d, z)\, dz = \int_{-\infty}^{+\infty} H(z)\delta[(t - d) - z]\, dz.$$

The right-hand side of this expression is $H(t - d)$, as shown earlier. Therefore, in concordance with the previous expression for the delay differential equation, $D(t)$ is given by

$$D(t) = \int_{-\infty}^{+\infty} H(z)G(t - d, z)\, dz = \frac{1}{\tau}\int_{-\infty}^{t-d} H(z)\exp\left[-\frac{(t - d) - z}{\tau}\right]\, dz.$$

The above convolution integral is the solution for the time dependence of the delay variable in the original delay differential equation. The physical interpretation of the above derivation is that $H(t)$ can be thought of as a continuous train of infinitely sharp pulses (as represented by an integral of δ functions). Each pulse in $H(t)$ induces an infinitely sharp pulse in $D(t)$ that is delayed by time d and scaled in magnitude to $\tau^{-1}H(t)$. Each of the pulses making up $D(t)$ decay monoexponentially with lifetime τ. The time dependence of $D(t)$ is then the sum total contribution over time of each of these exponentially decaying pulses (as given by the time integral).

It is interesting to note that the form of the delay differential equation presented above is highly analogous to that which describes first-order hormone release and elimination kinetics:

$$\frac{d}{dt} H(t) = k_R S(t) - k_E H(t).$$

In this case, S and H refer to hormone in prerelease storage pools and in circulation, respectively, and k_R and k_E are corresponding release and elimination rate constants, respectively. A fundamental difference between the delay differential equation considered above and this latter expression is that no effects of delay are explicitly considered in the latter; the time rate of change of circulating hormone is dependent only on current levels of stored and circulating hormone. The convolution integral produced as the solution for first-order hormone secretion and elimination thus possesses a function analogous to $G(t - d, z)$, except lacking the contribution of delay, d. Referred to as the pulse response function (65), it quantitatively accounts for elimination of circulating hormone in deconvolution analysis of hormone concentration time-series.

An alternative approach to modeling temporal delays has been applied to address mechanisms responsible for ultradian oscillations of insulin and glucose (88). Third-order delay was introduced in this case by a series of three consecutive delay differential equations that were applied sequentially to ultimately generate a delay variable by which regulatory influence was manifest. A parameter analogous to the τ derived above was utilized, but in the absence of any explicit consideration of delay, per se (i.e., no parameter analogous to d). In notation similar to that used above, delay was modeled as (88)

$$\frac{d}{dt} D_1(t) = 3\tau_D^{-1} [H(t) - D_1(t)],$$

$$\frac{d}{dt} D_2(t) = 3\tau_D^{-1} [D_1(t) - D_2(t)],$$

$$\frac{d}{dt} D_3(t) = 3\tau_D^{-1} [D_2(t) - D_3(t)].$$

In these equations, $H(t)$ refers to plasma insulin, τ_D to a delay time between plasma insulin and glucose production, and the $D_i(t)$ to delay variables introduced to account for the fact that the effect of insulin on glucose production involves a substantial time delay (88). The delay variable $D_3(t)$ was subsequently employed as the means through which down-regulatory influence on glucose production was exerted by insulin. Delay is implicit in this formulation, taking time to build up, for example, a sufficient level of effective signal, $D_3(t)$, in response to a step increase in $H(t)$. Attempts to use similar expressions to introduce temporal delay in the model of the GH neuroendocrine axis suggested a lack of sufficiently aggressive delay production, even when up to tenth-order processes were implemented. The current strategy, in which an explicit delay property is considered, was therefore selected instead.

The opposite extreme in implementation of the delay convolution integral would be equivalent to having the parameter τ take on a value of 0 and employ only the effect introduced by d (i.e., $D(t) = H(t - d)$). In such a case, the temporal profile of $D(t)$ would appear identical to that of $H(t)$, except its appearance would be delayed relative to $H(t)$ by time d.

Acknowledgments

Support has been provided by the National Science Foundation Center for Biological Timing (DIR 8920162), a National Institutes of Health NICHD RCDA award to J.D.V. (1K04HD00634), the NIH NICHD Reproduction Research Center at the University of Virginia Health Sciences Center (1P30HD28934-01A1; J.D.V.), the NIH Center for Fluorescence Spectroscopy of the University of Maryland at Baltimore (RR-08119; M.L.J.), and NIH Grant GM35154 (M.L.J.).

References

1. Hartman, M. L., Faria, A. C. S., Vance, M. L., Johnson, M. L., Thorner, M. O., and Veldhuis, J. D. (1991). Temporal structure of *in vivo* growth hormone secretory events in humans. *Am. J. Physiol.* **260** (*Endocrinol. Metab.* **23**), E101–E110.

2. Tannenbaum, G. S., and Martin, J. B. (1976). Evidence for an endogenous ultradian rhythm governing growth hormone secretion in the rat. *Endocrinology* (*Baltimore*) **98**, 562–570.

3. Clark, R. G., Carlsson, L. M. S., and Robinson, I. C. A. F. (1987). Growth hormone secretory profiles in conscious female rats. *J. Endocrinol.* **114**, 399–407.

4. Frohman, L. A., Downs, T. R., Clarke, I. J., and Thomas, G. B. (1990). Measurement of growth hormone-releasing hormone and somatostatin in hypothalamic-portal plasma of unanesthetized sheep. *J. Clin. Invest.* **86**, 17–24.

5. Winer, L. M., Shaw, M. A., and Baumann, G. (1990). Basal plasma growth hormone levels in man: New evidence for rhythmicity of growth hormone secretion. *J. Clin. Endocrinol. Metab.* **70**, 1678–1686.

6. Frohman, L. A., Downs, T. R., and Chomczynski, P. (1992). Regulation of growth hormone secretion. *Front. Neuroendocrinol.* **13**, 344–405.

7. Tannenbaum, G. S. (1991). Neuroendocrine control of growth hormone secretion. *Acta Paediatr. Scand. Suppl.* **372**, 5–16.

8. Davesa, J., Lima, L., and Tresguerres, J. A. F. (1992). Neuroendocrine control of growth hormone secretion in humans. *Trends Endocrinol. Metab.* **3**, 175–183.

9. Muller, E. E. (1987). Neural control of somatotropic function. *Physiol. Rev.* **67**, 962–1053.

10. Wehrenberg, W. B., Brazeau, P., Luben, R., Bohlen, P., and Guillemin, R. (1982). Inhibition of the pulsatile secretion of growth hormone by monoclonal

antibodies to the hypothalamic growth hormone releasing factor (GRF). *Endocrinology (Baltimore)* **111**, 2147–2148.

11. Thorner, M. O., Rivier, J., Spiess, J., Borges, J. L., Vance, M. L., Bloom, S. R., Rogol, A. D., Cronin, M. J., Kaiser, D. L., Evans, W. S., Webster, J. D., MacLeod, R. M., and Vale, W. (1983). Human pancreatic growth-hormone-releasing factor selectively stimulates growth-hormone secretion in man. *Lancet* **1/8**, 24–28.

12. Brazeau, P., Vale, W., Burgus, R., Ling, N., Butcher, M., River, J., and Guillemin, R. (1973). Hypothalamic polypeptide that inhibits the secretion of immunoreactive pituitary growth hormone. *Science* **179**, 77–79.

13. Brazeau, P., Rivier, J., Vale, W., and Guillemin, R. (1974). Inhibition of growth hormone secretion in the rat by synthetic somatostatin. *Endocrinology (Baltimore)* **94**, 184–187.

14. Vale, W., Vaughan, J., Yamamoto, G., Spiess, J., and Rivier, J. (1983). Effects of synthetic human pancreatic (tumor) GH releasing factor and somatostatin, triiodothyronine and dexamethasone on GH secretion *in vitro*. *Endocrinology (Baltimore)* **112**, 1553–1555.

15. Brazeau, P., Ling, N., Bohlen, P., Esch, F., Ying, S.-Y., and Guillemin, R. (1982). Growth hormone releasing factor, somatocrinin, releases pituitary growth hormone *in vitro*. *Proc. Natl. Acad. Sci. U.S.A.* **79**, 7909–7913.

16. Tannenbaum, G. S., and Ling, N. (1984). The interrelationship of growth hormone (GH)-releasing factor and somatostatin in generation of the ultradian rhythm of GH secretion. *Endocrinology (Baltimore)* **115**, 1952–1957.

17. Plotsky, P. M., and Vale, W. (1985). Patterns of growth hormone-releasing factor and somatostatin secretion in the hypophysial-portal circulation of the rat. *Science* **230**, 461–463.

18. Kraicer, J., Cowan, J. S., Sheppard, M. S., Lussier, B., and Moor, B. S. (1986). Effect of somatostatin withdrawal and growth hormone (GH)-releasing factor on GH release *in vitro*: Amount available for release after disinhibition. *Endocrinology (Baltimore)* **119**, 2047–2051.

19. Spiess, J., Rivier, J., and Vale, W. (1983). Characterization of rat hypothalamic growth hormone-releasing factor. *Nature (London)* **303**, 532–535.

20. Barinaga, M., Yamamoto, G., Rivier, C., Vale, W., Evans, R., and Rosenfeld, M. G. (1983). Transcriptional regulation of growth hormone gene expression by growth hormone-releasing factor. *Nature (London)* **306**, 84–85.

21. Clayton, R. N., Bailey, L. C., Abbot, S. D., Detta, A., and Docherty, K. (1986). Cyclic adenosine nucleotides and growth hormone-releasing factor increase cytosolic growth hormone messenger RNA levels in cultured rat pituitary cells. *J. Endocrinol.* **110**, 51–57.

22. Schwander, J. C., Hauri, C., Zapf, J., and Froesch, E. R. (1983). Synthesis and secretion of insulin-like growth factor and its binding protein by the perfused rat liver: Dependence on growth hormone status. *Endocrinology (Baltimore)* **113**, 297–305.

23. d'Ercole, A. J., Stiles, A. D., and Underwood, L. E. (1984). Tissue concentrations of somatomedin C: Further evidence for multiple sites of synthesis and paracrine or autocrine mechanisms of action. *Proc. Natl. Acad. Sci. U.S.A.* **81**, 935–939.

24. Hynes, M. A., van Wyk, J. J., Brooks, P. J., d'Ercole, A. J., Jansen, M., and

Lund, P. K. (1987). Growth hormone dependence of somatomedin-C/insulin-like growth factor-I and insulin-like growth factor-II messenger ribonucleic acids. *Mol. Endocrinol.* **1**, 233–242.

25. Schoenle, E., Zapf, J., Humbel, R. E., and Froesch, E. R. (1982). Insulin-like growth factor I stimulates growth in hypophysectomized rats. *Nature (London)* **296**, 252–253.

26. Berelowitz, M., Szabo, M., Frohman, L. A., Firestone, S., and Chu, L. (1981). Somatomedin-C mediates growth hormone negative feedback by effects on both the hypothalamus and the pituitary. *Science* **212**, 1279–1281.

27. Voogt, J. L., Clemens, J. A., Negro-Vilar, A., Welsch, C., and Meites, J. (1971). Pituitary GH and hypothalamic GHRF after median eminence implantation of ovine or human GH. *Endocrinology (Baltimore)* **88**, 1363–1367.

28. Tannenbaum, G. S., Guyda, H. J., and Posner, B. I. (1983). Insulin-like growth factors: A role in growth hormone negative feedback and body weight regulation via brain. *Science* **220**, 77–79.

29. Abe, H., Molitch, M. E., van Wyk, J. J., and Underwood, L. E. (1983). Human growth hormone and somatomedin C suppress the spontaneous release of growth hormone in unanesthetized rats. *Endocrinology (Baltimore)* **113**, 1319–1324.

30. Berelowitz, M., Firestone, S. L., and Frohman, L. A. (1981). Effects of growth hormone excess and deficiency on hypothalamic somatostatin content and release and on tissue somatostatin distribution. *Endocrinology (Baltimore)* **109**, 714–719.

31. Conway, S., McCann, S. M., and Krulich, L. (1985). On the mechanism of growth hormone autofeedback regulation: Possible role of somatostatin and growth hormone-releasing factor. *Endocrinology (Baltimore)* **117**, 2284–2292.

32. Sato, M., Chihara, K., Kita, T., Kashio, Y., Okimura, Y., Kitajima, N., and Fujita, T. (1989). Physiological role of somatostatin-mediated autofeedback regulation for growth hormone: Importance of growth hormone in triggering somatostatin release during a trough period of pulsatile growth hormone release in conscious male rats. *Neuroendocrinology* **50**, 139–151.

33. Fernandez, G., Cacicedo, L., Lorenzo, M. J., Lopez, J., and Sanchez-Franco, F. (1991). GRF and somatostatin regulation by GH and IGF-I in cultured hypothalamic neurons. *J. Endocrinol. Invest.* **14**(Suppl. 4), 68.

34. Lanzi, R., and Tannenbaum, G. S. (1992). Time course and mechanism of growth hormone's negative feedback effect on its own spontaneous release. *Endocrinology (Baltimore)* **130**, 780–788.

35. Chomczynski, P., Downs, T. R., and Frohman, L. A. (1988). Feedback regulation of growth hormone (GH)-releasing hormone gene expression by GH in rat hypothalamus. *Mol. Endocrinol.* **2**, 236–241.

36. Rogers, K. V., Vician, L., Steiner, R. A., and Clifton, D. K. (1988). The effect of hypophysectomy and growth hormone administration on pre-prosomatostatin messenger ribonucleic acid in the periventricular nucleus of the rat hypothalamus. *Endocrinology (Baltimore)* **122**, 586–591.

37. Bertherat, J., Timsit, J., Bluet-Bajot, M.-T., Mercadier, J.-J., Gourdji, D., Kordon, C., and Epelbaum, J. (1993). Chronic growth hormone (GH) hypersecretion induces reciprocal and reversible changes in mRNA levels from hypothalamic

GH-releasing hormone and somatostatin neurons in the rat. *J. Clin. Invest.* **91**, 1783–1791.

38. Sato, M., and Frohman, L. A. (1993). Differential effects of central and peripheral administration of growth hormone (GH) and insulin-like growth factor on hypothalamic GH-releasing hormone and somatostatin gene expression in GH-deficient dwarf rats. *Endocrinology (Baltimore)* **133**, 793–799.

39. Aguila, M. C., and McCann, S. M. (1985). Stimulation of somatostatin release *in vitro* by synthetic human growth hormone-releasing factor by a nondopaminergic mechanism. *Endocrinology* **117**, 762–765.

40. Katakami, H., Arimura, A., and Frohman, L. A. (1986). Growth hormone (GH)-releasing factor stimulates hypothalamic somatostatin release: An inhibitory feedback effect on GH secretion. *Endocrinology (Baltimore)* **118**, 1872–1877.

41. Katakami, H., Downs, T. R., and Frohman, L. A. (1988). Inhibitory effect of hypothalamic medial preoptic area somatostatin on growth hormone-releasing factor in the rat. *Endocrinology (Baltimore)* **123**, 1103–1109.

42. Miki, N., Ono, M., and Shizume, K. (1988). Withdrawal of endogenous somatostatin induces secretion of growth hormone-releasing factor in rats. *J. Endocrinol.* **117**, 245–252.

43. Mitsugi, N., Arita, J., and Kimura, F. (1990). Effects of intracerebroventricular administration of growth hormone-releasing factor and corticotropin-releasing factor on somatostatin secretion into rat hypophysial portal blood. *Neuroendocrinology* **51**, 93–96.

44. Yamauchi, N., Shibasaki, T., Ling, N., and Demura, H. (1991). *In vitro* release of growth hormone-releasing factor (GRF) from the hypothalamus: Somatostatin inhibits GRF release. *Regul. Pept.* **33**, 71–78.

45. Tannenbaum, G. S., Farhadi-Jou, F., and Beaudet, A. (1993). Ultradian oscillation in somatostatin binding in the arcuate nucleus of adult male rats. *Endocrinology (Baltimore)* **133**, 1029–1034.

46. Epelbaum, J. (1992). Intrahypothalamic neurohormonal interactions in the control of growth hormone secretion. *In* "Functional Anatomy of the Neuroendocrine Hypothalamus," pp. 54–68. Wiley, Chichester.

47. Katakami, H., Hidaka, H., Romanelli, R. J., Mayo, K. E., and Matsukuru, S. (1991). Inhibition by somatostatin of growth hormone-releasing factor gene expression in the hypothalamic arcuate nucleus. *73rd Endocrine Society Annual Meeting*, Abstract 1608.

48. Tannenbaum, G. S., McCarthy, G. F., Zeitler, P., and Beaudet, A. (1990). Cysteamine-induced enhancement of growth hormone-releasing factor (GRF) immunoreactivity in arcuate neurons: Morphological evidence for putative somatostatin/GRF interactions within hypothalamus. *Endocrinology (Baltimore)* **127**, 2551–2560.

49. Bertherat, J., Berod, A., Normand, E., Bloch, B., Rostene, W., Kordon, C., and Epelbaum, J. (1991). Somatostatin depletion by cysteamine increases somatostatin binding and growth hormone-releasing factor messenger ribonucleic acid in the arcuate nucleus. *J. Neuroendocrinol.* **3**, 115–118.

50. Peterfreund, R. A., and Vale, W. (1984). Somatostatin analogs inhibit somato-

statin secretion from cultured hypothalamus cells. *Neuroendocrinology* **39**, 397–402.

51. Lumpkin, M. D., Negro-Vilar, A., and McCann, S. M. (1981). Paradoxical elevation of growth hormone by intraventricular somatostatin: Possible ultrashort-loop feedback. *Science* **211**, 1072–1074.

52. Yamashita, S., and Melmed, S. (1986). Insulin-like growth factor I action on rat anterior pituitary cells: Suppression of growth hormone secretion and messenger ribonucleic acid levels. *Endocrinology (Baltimore)* **118**, 176–182.

53. Baumann, G., Stolar, M. W., Amburn, K., Barsano, C. P., and deVries, B. C. (1986). A specific growth hormone-binding protein in human plasma: Initial characterization. *J. Clin. Endocrinol. Metab.* **62**, 134–141.

54. Herington, A. C., Ymer, S., and Stevenson, J. (1986). Identification and characterization of specific binding proteins for growth hormone in normal human sera. *J. Clin. Invest.* **77**, 1817–1823.

55. Leung, D. W., Spencer, S. A., Cachianes, G., Hammonds, R. G., Collins, C., Henzel, W. J., Barnard, R., Waters, M. J., and Wood, W. I. (1987). Growth hormone receptor and serum binding protein: Purification, cloning and expression. *Nature (London)* **330**, 537–543.

56. Baumann, G., Amburn, K. D., and Buchanan, T. A. (1987). The effect of circulating growth hormone-binding protein on metabolic clearance, distribution, and degradation of human growth hormone. *J. Clin. Endocrinol. Metab.* **64**, 657–660.

57. Baumann, G., and Shaw, M. A. (1990). A second, lower affinity growth hormone-binding protein in human plasma. *J. Clin. Endocrinol. Metab.* **70**, 680–686.

58. Baxter, R. C. (1993). Circulating binding proteins for the insulinlike growth factors. *Trends Endocrinol. Metab.* **4**, 91–96.

59. Shimasaki, S., and Ling, N. (1991). Identification and molecular characterization of insulin-like growth factor binding proteins (IGFBP-1, -2, -3, -4, -5, and -6). *Prog. Growth Factor Res.* **3**, 243–266.

60. Wilson, J. D., and Foster, D. W. (eds.) (1992). "Textbook of Endocrinology," 8th Ed., Chap. 6. Saunders, Philadelphia, Pennsylvania.

61. Berne, R. M., and Levy, M. N. (eds.) (1990). "Principles of Physiology." Mosby, St. Louis, Missouri.

62. Refetoff, S., and Sonksen, P. H. (1970). Disappearance rate of endogenous and exogenous human growth hormone in man. *J. Clin. Endocrinol.* **30**, 386–392.

63. Frohman, L. A., Downs, T. R., Williams, T. C., Heimer, E. P., Pan, Y.-C. E., and Felix, A. M. (1986). Rapid enzymatic degradation of growth hormone-releasing hormone by plasma *in vitro* and *in vivo* to a biologically inactive product cleaved at the NH_2 terminus. *J. Clin. Invest.* **78**, 906–913.

64. Sheppard, M., Shapiro, B., Pimstone, B., Kronheim, S., Berelowitz, M., and Gregory, M. (1979). Metabolic clearance and plasma half-disappearance time of exogenous somatostatin in man. *J. Clin. Endocrinol. Metab.* **48**, 50–53.

65. Veldhuis, J. D., Carlson, M. L., and Johnson, M. L. (1987). The pituitary gland secretes in bursts: Appraising the nature of glandular secretory impulses by simultaneous multiple-parameter deconvolution of plasma hormone concentrations. *Proc. Natl. Acad. Sci. U.S.A.* **84**, 7686–7690.

66. Pincus, S. M., Gladstone, I. M., and Ehrenkranz, R. A. (1991). A regularity statistic for medical data analysis. *J. Clin. Monit.* **7**, 335–345.

67. Pincus, S. M., and Goldberger, A. L. (1994). Physiological time-series analysis: What does regularity quantify? *Am. J. Physiol.* **266,** H1643–1656.
68. Pincus, S. M. (1994). Quantification of evolution from order to randomness in practical time series analysis. *In* "Methods in Enzymology" (M. L. Johnson and L. Brand, eds.), Vol. 240, pp. 68–89. Academic Press, San Diego.
69. Pincus, S. M. (1995). Quantifying complexity and regularity of neurobiological systems. *Methods Neurosci.* (this volume).
70. Pritchard, W. S., and Duke, D. W. (1992). Measuring chaos in the brain: A tutorial review of nonlinear dynamical EEG analysis. *Int. J. Neurosci.* **67,** 31–80.
71. Yates, F. E. (1992). Fractal applications in biology: Scaling time in biochemical networks. *In* "Methods in Enzymology" (L. Brand and M. L. Johnson, eds.), Vol. 210, pp. 636–675. Academic Press, San Diego.
72. Garfinkel, A., Spano, M. L., Ditto, W. L., and Weiss, J. N. (1992). Controlling cardiac chaos. *Science* **257,** 1230–1235.
73. Goldberger, A. L., and Rigney, D. R. (1991). Nonlinear dynamics at the bedside, *in* "Theory of Heart: Biomechanics, Biophysics, and Nonlinear Dynamics of Cardiac Function" (L. Glass, P. Hunter, and A. McCulloch, eds.), pp. 583–605. Springer-Verlag, New York.
74. Friesen, W. O., and Block, G. D. (1984). What is a biological oscillator? *Am. J. Physiol.* **246** (*Regul. Integrative Comp. Physiol.* **15**), R847–R851.
75. Friesen, W. O., Block, G. D., and Hocker, C. G. (1993). Formal approaches to understanding biological oscillators. *Annu. Rev. Physiol.* **55,** 661–681.
76. Hill, A. V. (1910). The possible effects of the aggregation of the molecules of hemoglobin on its dissociation curve. *J. Physiol. (London)* **40,** 4–7.
77. Thomas, R., and d'Ari, R. (1990). "Biological Feedback." CRC Press, Boca Raton, Florida.
78. Mathews, C. K., and van Holde, K. E. (1990). "Biochemistry." Benjamin/Cummings, Redwood City, California.
79. Jansson, J.-O., Ishikawa, K., Katakami, H., and Frohman, L. A. (1987). Pre- and postnatal developmental changes in hypothalamic content of rat growth hormone-releasing factor. *Endocrinology (Baltimore)* **120,** 525–530.
80. O'Sullivan, D., Millard, W. J., Badger, T. M., Martin, J. B., and Martin, R. J. (1986). Growth hormone secretion in genetic large (LL) and small (SS) rats. *Endocrinology (Baltimore)* **119,** 1948–1953.
81. Forsythe, G. E., Malcolm, M. A., and Moler, C. B. (1977). "Computer Methods for Mathematical Computations." Prentice-Hall, Englewood Cliffs, New Jersey.
82. Gear, C. W. (1971). "Numerical Initial Value Problems in Ordinary Differential Equations." Prentice-Hall, Englewood Cliffs, New Jersey.
83. Clark, R. G., Carlsson, L. M. S., Rafferty, B., and Robinson, I. C. A. F. (1988). The rebound release of growth hormone (GH) following somatostatin infusion in rats involves hypothalamic GH-releasing factor release. *J. Endocrinol.* **119,** 397–404.
84. van Cauter, E., Kerkhofs, M., Caufriez, A., van Onderbergen, A., Thorner, M. O., and Copinschi, G. (1992). A quantitative estimation of growth hormone secretion in normal man: Reproducibility and relation to sleep and time of day. *J. Clin. Endocrinol. Metab.* **74,** 1441–1450.

85. van Cauter, E., Caufriez, A., Kerkhofs, M., van Onderbergen, A., Thorner, M. O., and Copinschi, G. (1992). Sleep, awakenings, and insulin-like growth factor-I modulate the growth hormone (GH) secretory response to GH-releasing hormone. *J. Clin. Endocrinol. Metab.* **74,** 1451–1459.

86. Thorner, M. O., Vance, M. L., Hartman, M. L., Holl, R. W. Evans, W. S., Veldhuis, J. D., van Cauter, E., Copinschi, G., and Bowers, C. Y. (1990). Physiological role of somatostatin on growth hormone regulation in humans. *Metabolism* **39**(No. 9, Suppl. 2), 40–42.

87. Hanson, S. (1990). What connectionist models learn: Learning and representation in connectionist neural networks. *Brain Behav. Sci.* **13,** 471–511.

88. Sturis, J., Polonsky, K. S., Mosekilde, E., and van Cauter, E. (1991). Computer model for mechanisms underlying ultradian oscillations of insulin and glucose. *Am. J. Physiol.* **260** (*Endocrinol. Metab.* **23**), E801–E809.

[12] Implementation of a Stochastic Model of Hormonal Secretion

Daniel Keenan

Introduction

Neuroendocrine ensembles typically communicate via intermittent signals encoded via amplitude and frequency variations (1–6). Episodic neurohormone release is subject further to physiological regulation and pathological disturbances in various disease states (2–17). However, mechanisms controlling the apparently random timing of neurohormone secretory bursts have been difficult to investigate in part because of the absence of a well-defined and tractable mathematical formulation, or biophysical model, which embraces the stochastic nature of so-called (neuroendocrine) pulse-generator systems (5). Recent studies suggest that such a model should also include slow deterministic trends in the data (e.g., 24-hr variability in secretion).

This chapter offers a stochastic formulation of the problem as the basis for the incremental amount of secretion (z_i) between times t_{i-1} and t_i, $i = 1, \ldots, n$ during some period $\lfloor t_0, t_n \rfloor$. We formulate a framework in which one can define and estimate such quantities as expected number of bursts, total basal secretion, pulsatile secretion, secretion within a pulse, and maximum of a pulse. We present a model for partial frequency and amplitude control of the pulse generator by an underlying oscillator; for simplicity, we consider one hormone, for example, gonadotropin-releasing hormone (GnRH) which constitutes the pulse generator driving intermittent activity of the gonadotropin axis. Understanding the pulse generator is crucial as a prerequisite to a later comprehension of positive feedback control of the reproductive axis. Here we examine the application of this formulation to synthetic (computer simulated) data so as to estimate physiological measures of secretion.

Overview and General Remarks

There are several basic issues, from a modeling perspective, which need to be addressed. One question is: What is the appropriate level at which to formulate the model? This depends greatly on what one is able to observe. We are assuming that for a given subject the observations are made at times t_i, $i = 1, \ldots, n$ during the period $[L_1, L_2]$ with $t_0 = L_1$, $t_n = L_2$ and that what is observed is, z_i, the incremental amount secreted during the time $[t_{i-1}, t_i]$; the sampling interval $\Delta t = t_i - t_{i-1}$ is assumed to be constant. We have chosen to model the rate of secretion and to do so in continuous time. Using the notation of the next section, dZ_t/dt is the rate of secretion where Z_t is the cumulative amount secreted up to time t and z_i is, therefore, the integral of the rate of secretion between times t_{i-1} and t_i. If one did not model at this level, then it would not be possible to compare data observed under different sampling schemes.

We will model the rate of secretion as arising in two stages. The first stage concerns the mechanism which governs the time occurrence of bursts (or pulses); the second stage concerns the resulting shapes of the bursts at the various burst times.

A standard model for the time occurrences of bursts is via a renewal process. A renewal process is given by the partial sums (S_m, $n \geq 1$) of independent and identically distributed (IID) positive random variables (X_i, $i = 1, 2, ...$):

$$S_n = \sum_{i=1}^{n} X_i.$$

The nth burst time is then given as the value of the nth partial sum, S_n. The X_is are the interburst lengths in this setting. For a renewal process the interburst lengths are, by construction, IID. If the $\{X_i\}$ are from an exponential distribution with parameter λ (a positive real number), then the number of bursts is described by the time-homogeneous (or stationary) Poisson process.

The time-homogeneous Poisson process has special status within the family of renewal processes. Ordinarily the superposition of two independent renewal processes is not again a renewal process; that is, the union of the burst times from the two will not, ordinarily, correspond to those of a renewal process. If the superposition is in fact a renewal process, then all three must have been time-homogeneous Poisson processes. Consequently, if these are justified scientific reasons for viewing the burst times as the superposition of a large number of sparse arrival processes, then the stationary Poisson model is appropriate. It is defined as homogeneous or stationary because the probabilistic description of the number of bursts which will occur in any given interval depends only on the length of the interval (and, of course, on the value of λ). The probability distribution of the time of the $(n + 1)^{st}$ burst, T_{n+1}, given the times of the previous bursts, $T_1, ..., T_n$, depends only on the time of the last burst, T_n. Consequently, the times form a Markov chain; in the case of the homogeneous Poisson process, this distribution, in fact, does not depend on n and the burst times form a homogeneous Markov chain. One consequence of all of the above is that one cannot, using a homogeneous Poisson process, incorporate into the model any tendency for patterns of highs and lows in the number of pulses (or bursts). If one expects more pulses to occur during a certain period of the day than during another period, the homogeneous Poisson process cannot accommodate this.

If one replaces the positive value λ with a positive deterministic function $\{\lambda(s), s \geq 0\}$, thus allowing λ to change over time, one can define the time-inhomogeneous (or nonstationary) Poisson process. It no longer has independent interburst lengths; it should not, because, when $\lambda(\cdot)$ is large, there tend to be more bursts than when $\lambda(\cdot)$ is small. One can, therefore, build in the cyclical nature of secretion variation via the function $\lambda(\cdot)$. The interburst lengths do have the property that the conditional distribution of the $(n + 1)^{st}$ interburst length, $T_{n+1} - T_n$, given the preceding lengths, has the exponential distribution, now however depending on T_n. The burst times $\{T_n\}$ again form a Markov chain, however, now the probability distribution T_{n+1}, given T_1, ... T_n, depends not only on T_n but also on n.

In Keenan and Veldhuis (18) and in this chapter, the burst times are modeled by a time-inhomogeneous Poisson process. The function $\lambda(\cdot)$ is taken to be periodic, describing the possible cyclical variability in secretion (e.g., circadian behavior). The nonstationary Poisson process can be viewed as the standard (i.e., $\lambda \equiv 1$) stationary Poisson process where there has been a deterministic transformation of the time axis; this fact is of great practical importance.

One major difficulty with the use of the stationary and nonstationary Poisson processes to model burst times is that the mean and variance of a Poisson random variable (e.g., the number of bursts in a certain fixed interval of time) are constrained to be equal. One would like to have more flexibility, for example, allowing for the expected number of bursts to be large and the variance to be small.

One possibility is to replace the above deterministic $\lambda(\cdot)$ function, the so-called intensity function, by a stochastic intensity. Such models have also been relatively well studied. One can allow for the intensity to thus be self-adapting based on the feedback, possibly time delayed, from the concentration levels of various associated hormones. In fact, just as the nonstationary Poisson process can be viewed as the stationary Poisson process resulting from a deterministic transformation of the time axis, the general stochastic intensity models can be viewed as random transformations of the time axis (see Ref. 19 for a complete discussion). A second possibility is to replace the rather restrictive Markov chain models for the burst times $\{T_n\}$ which resulted from the stationary and nonstationary Poisson assumption with more general Markov chain models. Both of these two possibilities allow for more flexibility in the relationship between the mean and variance of the number of bursts. At present, work (20) is being done, using the above, on modeling the gonadotropin axis allowing for beedback and time delayed interactions between the various hormones.

Another issue in the modeling of the rate of secretion concerns the shape (or shapes) of the bursts. In the next section we discuss such modeling, and in the section on Estimation we discuss estimation of the parameters associated with such models. An important concern is that parameters be chosen which appropriately capture the relevant structure and that the parameterization be identifiable. That is, different parameter choices need to produce probabilistically different models; this is the most basic requirement from a modeling perspective. In a dissertation by Yang (21) this issue is discussed along with the development of estimators of the parameters. We are at present (20, 21) modeling both hormonal secretions and concentrations of a system (e.g., gonadotropin axis), allowing for time-delayed feedback between the concentration and secretion.

Model Description

We will view the model as arising in two stages. The first stage concerns the mechanism which governs the time occurrences of the bursts; the second concerns the resulting shapes of the bursts (at the various burst times). In the first stage we assume there is a function $\{\lambda(t), t \in \mathbf{R}\}$ which describes the relationship between the pulse generator and the underlying oscillator. The unknown function $\lambda(\cdot)$ describes the tendency for patterns of bursts to occur over time. We assume that $\lambda(\cdot)$ is a strictly positive, periodic function of a known period Γ:

$$\lambda(t) = \lambda(\theta, t) = A_0 + \sum_{j=1}^{k} \left[A_j \cos \left(\frac{j2\pi t}{\Gamma} \right) + B_j \sin \left(\frac{j2\pi t}{\Gamma} \right) \right] \tag{1}$$

involving a finite number of unknown terms (k) with $\theta = (A_0, A_1, ..., A_k, B_1, ..., B_k)$. Define $\Theta \subset \mathbf{R}^{2k+1}$ as

$$\Theta = \{ \theta \in \mathbf{R}^{2k+1} | A_0 > 0, \lambda(\theta, t) > 0, \forall t \in \mathbf{R} \}. \tag{2}$$

In the simulations included in this paper, we used $k = 2$, $\Gamma = 24$ hr, $A_0 = 1.04$, $A_1 = 0.4167$, $A_2 = 0.1042$, $B_1 = 0.4167$, and $B_2 = 0.1042$.

We assume that for the M subjects, their individual clocks differ from $\lambda(\cdot)$ by an unknown phase shift: $\lambda^{(r)}(s) = \lambda(\theta, s + \phi^{(r)})$, $s \in \mathbf{R}$, and define $a^{(r)}(\cdot)$ as,

$$a^{(r)}(t) = \int_{t_0}^{t} \lambda^{(r)}(s) \, ds. \tag{3}$$

The expected number of occurrences of bursts in the interval $[t_0, t_n]$ for the rth subject will be $a^{(r)}(t_n) - a^{(r)}(t_0)$. The periodic nature of $\lambda(\cdot)$ allows for the tendency for the number of occurrences to show patterns of highs and lows over time.

Let $\{N_t^{(r)}, t \geq 0, r = 1, ..., M$, be M independent, inhomogeneous Poisson processes with rate functions $\{a^{(r)}(t), t \geq 0\}$, $r = 1, ..., M$, and let

$$0 = T_0^{(r)} < T_1^{(r)} < T_2^{(r)} < \cdots \tag{4}$$

be the occurrence times of the rth Poisson process; these are the times that the gland of the rth subject is sent signals to activate the production of granules, the transport of the granules and individual molecules to the cell membrane, and so on [see Cinlar (17) as a general reference on Poisson processes]. Figure 1a is a simulation of the Poisson process for one of the M subjects. A random phase shift produces $\lambda^{(1)}(\cdot)$, shown in Fig. 1a. The marks on the time axes are the corresponding random occurrence times. Figure 1b displays the function $a^{(1)}(\cdot)$; the expected number of occurrences was taken to be 25.

Based on the results of the first stage, the second stage then models the rate of secretion. We will assume that, on receiving a signal, the rate of secretion increases abruptly and then decreases less rapidly. The shape of this burst can take various forms under different situations. Models of such bursts should allow for both symmetric, Gaussian-like shapes as well as skewed (to the right) shapes. We provisionally model this increase and decrease, for the rth-subject, by a Gamma family of functions:

$$\nu^{(r)}(s) = s^{\beta_2^{(r)}-1} e^{-s/\beta_3^{(r)}}, \tag{5}$$

where $\beta_2^{(r)} > 1$ and $\beta_3^{(r)} > 0$ are two parameters that model the secretion kinetics. Appropriate choices of β allow a nearly Gaussian secretory burst, as suggested in a recent deconvolution model, and a host of variably skewed representations of putative secretory bursts. The motivation for using the Gamma family is due to the broad range of unimodal shapes which are allowed. The maximum of $\nu^{(r)}(\cdot)$ occurs at $\xi^{(r)} = (\beta_2^{(r)} - 1) \times \beta_3^{(r)}$, and, following the occurrence of the maximum, $\nu^{(r)}(\cdot)$ will be at two-thirds of its maximum amplitude at the value $\delta^{(r)}$. We have nominally (or arbitrarily) taken the refractory length to be $\delta^{(r)}$; later this will be allowed to be random, since this value would likely differ for different conditions, etc.

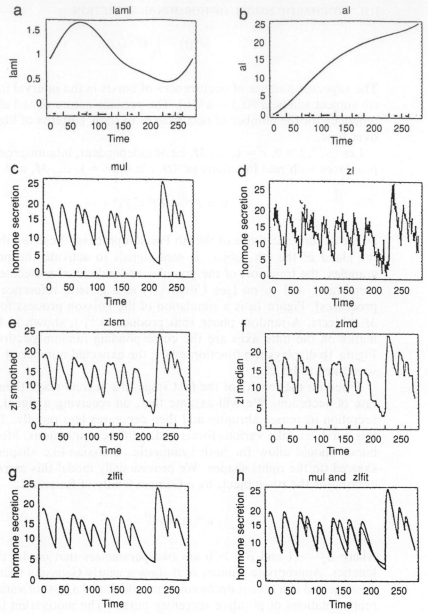

FIG. 1 (a) *lam*1 is a phase-shifted $\lambda(\cdot)$ function, describing the relationship of the pulse generator to the oscillator; (b) *a*1, the integral of *lam*1, describes the expected number of bursts; (c) *mu*1 is the noiseless incremention secretions, and the marks on the time axis are the burst occurrences; (d) *z*1 is the observed incremental secretions: *mu*1 plus noise; (e) *z*1*sm* is *z*1 smoothed by a filter; (f) *z*1*md* is the result of a median filter applied to *z*1; (g) *z*1*fit* is the maximum likelihood estimation of *mu*1; (h) *mu*1 and *z*1*fit* are displayed together.

Let $\tau_0^{(r)} = 0$ for $r = 1, \ldots, M$, and define recursively

$$\tau_k^{(r)} = \underset{j}{\text{Min}} \{T_j^{(r)} \mid T_j^{(r)} \geq \tau_{k-1}^{(r)} + \delta^{(r)}\}. \tag{6}$$

Thus, $\tau_k^{(r)}$ is the kth signal which is received; signals occurring between $\tau_{k-1}^{(r)}$ and $\tau_{k-1}^{(r)} + \delta^{(r)}$ are disregarded. In Fig. 1a the vertical marks are the received signals (τ values); the asterisks are the signals disregarded due to the refractory period. The longer the preceding interburst interval the more granules are released at time $\tau_k^{(r)}$, suggesting a model which allows for the magnitude of a burst to depend on the length of the previous interburst interval: $(\tau_{k-1}^{(r)} + \xi^{(r)}, \tau_k^{(r)})$. Our model of hormonal secretion in the second stage is that the rate of secretion at time s, for subject r, is given by the stochastic differential equation [see Protter (22) as a general reference]:

$$dZ_s^{(r)} = \left[\beta_0^{(r)} + \sum_{\tau_k \leq s} \{\eta_0^{(r)} + \eta_1^{(r)}[\tau_k^{(r)} - (\tau_{k-1}^{(r)} + \xi^{(r)})]\} \times (s - \tau_k^{(r)})_+^{\beta_2^{(r)}-1} \right.$$
$$\left. \times e^{-(s-\tau_k^{(r)})+/\beta_3^{(r)}} \right] ds + c^{(r)} dW_s^{(r)}, \tag{7}$$

where

$$(t)_+ = \begin{cases} t, & \text{if } t > 0 \\ 0, & \text{if } t \leq 0 \end{cases},$$

and where $\eta_0^{(r)}$ and $\eta_1^{(r)}$ are the intercept and slope of the linear function of the previous interburst length which governs the pulse magnitude and $\{W_s^{(r)}, s \geq 0\}$, $r = 1, \ldots, M$ are M independent standard Brownian motions with $c^{(r)} \geq 0$ unknown. The Brownian motions are assumed to be independent of the Poisson processes. The stochastic differential $dW_s^{(r)}$ is the formulation of noise. One could formulate the above so that the rate of secretions is always nonnegative. From a practical point of view this is not so important. The parameters of the second stage are $(\beta_0^{(r)}, \eta_0^{(r)}, \eta_1^{(r)}, \beta_2^{(r)}, \beta_3^{(r)})$ and $c^{(r)}$, $r = 1, \ldots, M$. The parameter $\beta_0^{(r)}$ is the rate of baseline secretion which is here assumed to be unaffected by the signals. The summation is a sum of Gamma functions, positioned at each of the signal times, with each linearly scaled in amplitude by the previous interburst length. This is illustrated in Fig. 2. One consequence of the above model, which has the realized pulses being the result of a basic pulse shape multiplied by an amplitude, is that the half-width of the pulses is constant; this is built into this model.

Therefore, in our model, the amount of hormonal secretion between t_{i-1} and t_i: $z_i^{(r)}$, is given by

$$z_i^{(r)} = \int_{t_{i-1}}^{t_i} dZ_s^{(r)} = \beta_0^{(r)} + \Delta t_i + \left(\int_{t_{i-1}}^{t_i} \sum_k \{\eta_0^{(r)} + \eta_1^{(r)}[\tau_k^{(r)} - (\tau_{k-1}^{(r)} + \xi^{(r)})]\} \right.$$
$$\left. \times (s - \tau_k^{(r)})_+^{\beta_2^{(r)}-1} \times e^{-(s-\tau_k^{(r)})+/\beta_3^{(r)}} ds \right) + \varepsilon_i^{(r)} \stackrel{\text{def}}{=} \mu_i^{(r)} + \varepsilon_i^{(r)} \quad (8)$$

for $i = 1, \ldots n$; $r = 1, \ldots, M$, where $\{\varepsilon_i^{(r)}\}$ are I.I.D. $N(0, \sigma_r^2)$, $\sigma_r^2 = c^{(r)}\Delta t_i$. For small Δt the integration is basically just a multiplication by Δt [i.e., a rescaling of $(\beta_0, \eta_0, \eta_1)$]; one could just work with this rescaled model, forgetting about the integration. However, the modeling is most appropriately done at the level of the secretion rate rather than at the level of the observation, for only then can one account for the affect of a given sampling interval

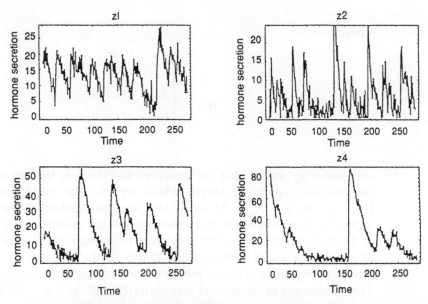

FIG. 2 $z1$–$z4$ are four simulations of incremental secretions under four different choices of the secretion parameters. The development of $z1$ is that described in Fig. 1.

length (Δt). The times of the two most recent signals prior to the time t_0, denoted by $\tilde{\tau}_{-1}^{(r)}$ and $\tilde{\tau}_{0}^{(r)}$, influence the observations from t_0 until the first observed signal. Let $\tilde{\underline{\tau}}^{(r)}$ be the times of the observed signals, in addition to $\tilde{\tau}_{-1}^{(r)}$ and $\tilde{\tau}_{0}^{(r)}$, for the rth subject. Define the parameter of the second stage, ψ, as

$$\psi = (\psi^{(1)}, \psi^{(2)}, ..., \psi^{(m)}),$$

where

$$\psi^{(r)} = (\beta_0^{(r)}, \eta_0^{(r)}, \eta_1^{(r)}, \beta_2^{(r)}, \beta_3^{(r)}, \sigma_r^2, \tilde{\underline{\tau}}^{(r)}), \qquad r = 1, 2, ..., M, \qquad (9)$$

Let Ψ be the parameter space of ψ characterized by the various constraints on the components: $\beta_0^{(r)} > 0$, $\beta_2^{(r)} > 0$, $\beta_3^{(r)} > 0$, elements of $\tilde{\underline{\tau}}^{(r)}$ increasing, and so on.

From expression (8) it follows that the log likelihood in the second stage is approximately

$$l_2(\psi \,|\, \{z_i^{(r)}\}_{i=1}^n, r = 1, ..., M) = -\frac{1}{2} \sum_{r=1}^{M} \frac{1}{\sigma_r^2} \sum_{i=1}^{n} [z_i^{(r)} - \mu_i^{(r)}(\psi^{(r)})]^2, \qquad (10)$$

where $\mu_i^{(r)}$ is the function of $\psi^{(r)}$ defined within expression. It is not the exact likelihood because of the nature of the functions $\nu^{(r)}(\cdot)$ defined in expression (7), however, the difference is negligible; the functions $\nu^r(\cdot)$ die out exponentially but are nevertheless always positive. Consequently, strictly speaking, all of the previous pulses should be included in the definition of ψ rather than just the two most recent prior to the time t_0.

The above is a description of the forward equation describing how the data came about. Our goal is to solve the inverse problem, having observed the data (given in Fig. 2) to first estimate the $\psi^{(r)}$, $r = 1, ..., M$ and then, using the estimates of the $\underline{\tau}^{(r)}$, $r = 1, ..., M$, estimate the function $\lambda(\cdot)$. If one views the $\underline{\tau}^{(r)}$, $r = 1, ..., M$ as missing data, one could alternatively use the Expectation-Maximization (EM) algorithm for the estimation. [See Dempster et al. (23) for general reference.] In the next section we discuss the estimation.

Estimation

Initial Estimates

Having observed the incremental hormonal secretions of the rth subject: $\{z_i^{(r)}\}_{i=1}^n$, we first get a variety of initial estimates of $\psi^{(r)}$ [expression (9)], which will be used as starting points for the maximum likelihood estimation of $\psi^{(r)}$,

$r = 1, 2, \ldots, M$. For each of the series $\{z_i^{(r)}\}_{i=1}^n$, $r = 1, \ldots, M$, we first (A) smooth the data with a moving average filter $(a_{-q}, a_{-q+1}, \ldots, a_0, a_1, \ldots, a_q)$; this is repeated $n$smooth times. For each of the smoothed series the Min, Max, and Range = Max−Min are calculated and (B) the Min of each series is used to estimate the baseline ($\beta_0^{(r)} \times \Delta t$). The variance σ_r^2 is estimated from the deviations of the data about the smoothed data. For each of the smoothed series (C) the locations where the smoothed process crosses the level of Min $+(QA \times$ Range) are determined; at each of these locations the sign of the 1st difference is determined and these locations are used to calculate estimates of the $\bar{\tau}$s and ξs. The value QA is to be chosen. Another estimate (D) is constructed by using these estimates to break up $[t_0, t_n]$ into segments with step C repeated on each segment, producing new estimates of the τs and ξs; this repeated step produces a good starting estimate in the case where there are large differences in the magnitudes of the bursts, since the first application of step C would only identify the large bursts in that case. The steps C and D will not, by their construction, detect small bursts. To do this (E), we calculate from the smoothed data, the locations of where the 1st differences change sign, and thus are used as estimates of the τs and ξs. The values of $q = 2$, $QA = 0.5$, were used in the simulations, where $n$smooth = 8 was used in steps A–D and $n$smooth = 3 in step E. The results of the smoothing are displayed as $z1sm$ in Fig. 1e for $z1$. The estimates of the τs are displayed as vertical marks on the time axis.

One difficulty with the above approach is that smoothing tends to make asymmetrical curves less asymmetrical. Because our estimation of the $\tau^{(r)}$s and the other components of $\psi^{(r)}$ is based on the location and shape of the rapid changes due to $\nu^{(r)}(\cdot)$ being positioned at the $\tau^{(r)}$s, smoothing greatly affects the estimation. Because of this we have also performed the above step with a median filter replacing the moving average filter; the results are usually better. The results of the median filter are shown as $z1md$ in Fig. 1f for $z1$. The estimates of the τs are displayed as vertical marks on time axis. Initial estimates of $(\eta_0, \eta_1, \beta_2, \beta_3)$ are constructed from step C using properties of Gamma densities: the location of the maximum and the maximum value. Because there will be many local maxima for $l_2(\cdot)$, a variety of starting points will aid in finding the global maximum.

Maximum Likelihood Estimation of ψ

The log likelihood, l_2, of ψ given the data $\{\{z_i^{(r)}\}_{i=1}^n, r = 1, \ldots, M\}$ is given by expression (10). The maximization procedure that we have used to find the maximum likelihood estimates is the Levenberg–Marquardt algorithm. Because our approach is model based, the (minus) inverse of the Hessian of l_2, evaluated at the maximum likelihood estimate, $\hat{\psi}$, is an estimate of the covariance matrice of $\hat{\psi}$. The Hessian is an integral part of the Levenberg–Marquardt algorithm and is thus produced by the procedure.

As discussed in the Initial Estimates section above we implement the algorithm for l_2 at a variety of starting values. It does not seem unreasonable to have on the order of 100 different starting locations, although in the simulations we have only used 25.

For the purpose of exposition consider the case $M = 1$ with $z1$ being the observed data. In Fig. 1g, $z1\mathit{fit}$ is the best fit of the 25 starting points; that is, $z1\mathit{fit}$ is the function $\mu1$ calculated from the $\hat{\psi}^{(1)}$ rather than the true value $\psi^{(1)}$. As part of the algorithm, the observed information [i.e., (minus) the second derivative of $l_2(\cdot)$ evaluated at $\hat{\psi}^{(1)}$] is calculated, from which standard errors of our estimates can be obtained. The true and estimated values of the parameters and the standard errors of the estimates are as follows:

True	2.17	4.67	5.92	7.75	10.0	11.25	12.92	14.92	18.92	21.0	22.25
Parameter	τ_1	τ_2	τ_3	τ_4	τ_5	τ_6	τ_7	τ_8	τ_9	τ_{10}	τ_{11}
Estimate	2.18	4.57	6.01	7.76	10.08	12.68	12.91	14.67	18.80	21.01	22.29
Standard error	0.061	0.038	0.052	0.046	0.34	0.043	0.179	0.062	0.035	0.046	0.059

True	4	20	1.5	1.0	50	175	-3.25	-0.25
Parameter	σ^2	β_0	β_2	β_3	η_0	η_1	τ_{-1}	τ_0
Estimate	4.01	43.14	1.61	0.92	17.62	186.08	-3.77	-0.33
Standard error		8.568	0.133	0.125	27.29	25.54	0.602	0.103.

In the second stage σ^2 is estimated and the MLE procedure then uses that estimated value of σ^2 so that no standard error of our estimated variance is given by Levenberg–Marquardt; it could be modified to do so.

In the present model, the pulse amplitudes are not each described by a separate parameter but rather each amplitude is assumed to be a function (involving two parameters η_0 and η_1) of the preceding interpulse length. The estimates of β_0 (basal secretion rate) and η_0 and η_1 are highly correlated; this is a phenomenon common to nonlinear models and when extreme, can be an indication that the model is overparameterized. Because of the correlation one may wish to display simultaneous confidence regions for two or more of the parameters that are part of the analysis.

Summary

We have developed and implemented a stochastic model for a neurohormone pulse generator that is under frequency and amplitude control by an underlying oscillator. The diversity of output by this model is illustrated in Fig. 2, typifying the range of neurohormone secretion recognized in health and disease (1–8). The actual incremental secretions are viewed as arising in two stages, in which a Poisson process contributes the random occurrence of release events over time, as modified by an underlying deterministic trend imposed by an oscillator. Secretory burst amplitudes are described by stochastic differential equations that assume Brownian-like motions independent of the Poisson (frequency) process. Thus there are two stochastic components in our model of hormonal secretion: the burst times described by a nonhomogeneous Poisson process and a noise term which allows for the randomness in the secretory burst amplitude. The stochastic rate of secretion is given by the stochastic differential equation [expression (7)]. Estimation of the parameters for the timing of secretory events (τ values) and the shape of the secretory events (β values) was discussed and applied to synthesized data. In forthcoming work we will discuss how the temporal structure of the underlying deterministic oscillations [described by $\lambda(\cdot)$] can be estimated given observations on M subjects. Consequently, this biomathematical representation of the intermittent secretory behavior of a neuroendocrine apparatus provides a new framework for investigating quantitative aspects of episodic glandular signaling.

Acknowledgment

Supported by Office of Naval Research Contract No. 00014-90-J-1007.

References

1. Yates, F. E. (1981). Analysis of endocrine signals: The engineering and physics of biochemical communication systems. *Biol. Reprod.* **24**, 73.
2. Desjardins, C. (1981). Endocrine signaling and male reproduction. *Biol. Reprod.* **24**, 1–21.
3. Veldhuis, J. D., Iranmanesh, A., Johnson, M. L., and Lizarralde, G., (1990). Twenty-four hour rhythms in plasma concentrations of adenohypophyseal hormones are generated by distinct amplitude and/or frequency modulation of underlying pituitary secretory burst. *J. Clin. Endocrinol. Metab.* **71**, 1616–1623.
4. Plotsky, P. M., and Vale, W. (1985). Patterns of growth hormone-releasing factor and somatostatin secretion in the hypophysial-portal circulation of the rat. *Science* **230**, 461–465.
5. Weitzman, E. E., Fukushima, D., Nogeire, C., Roffwarg, H., Gallagher, T. F., and Hellman, L. (1971). Twenty-four hour pattern of the episodic secretion of cortisol in normal subjects. *J. Clin. Endocrinol.* **33**, 14–22.
6. Hartman, M. L., Faria, A. Cs., Vance, M. L., Johnson, M. L., Thorner, M. O.,

and Veldhuis, J. D. (1991). Temporal structure of *in vivo* growth hormone secretory events in man. *Am. J. Physiol.* **260**, E101–E110.

7. Walter-Van Cauter, E., Virasoro, E., Leclerq, R., and Copinschi, G. (1981). Seasonal, circadian, and episodic variations of human immunoreactive beta-MSH, ACTH and cortisol. *Int. J. Pept. Protein Res.* **7**, 3–13.

8. Weitzman, E. D., Boyar, R. M., Kapen, S., and Hellman, L. (1975). The relationship of sleep and sleep stages to neuroendocrine secretion and biological rhythms in man. *Prog. Horm. Res.* **31**, 399–446.

9. Winer, L. M., Shaw, M. A., and Baumann, G. (1990). Basal plasma growth hormone levels in man: New evidence for rhythmicity of growth hormone secretion. *J. Clin. Endocrinol. Metab.* **70**, 1678–1686.

10. Marshall, J. C., and Kelch, R. P. (1986). Gonadotropin-releasing hormone: Role of pulsatile secretion in the regulation of reproduction. *N. Engl. J. Med.* **315**, 1459–1467.

11. Naylor, M. R., Krishman, K. R. R., Manepalh, A. N., Ritchie, J. C. Wilson, W. H., and Carroll, B. J. (1988). Circadian rhythm of adrenergic regulation of adrenocorticotropin and cortisol secretion in man. *J. Clin. Endocrinol. Metab.* **67**, 404–406.

12. Stewart, J. K., Clifton, D. K., Koerker, D. J., Rogol, A. D., Jaffe, T., and Goodner, C. J. (1985). Pulsatile release of growth hormone and prolactin from the primate pituitary *in vitro*. *Endocrinology (Baltimore)* **116**, 1–5.

13. Krieger, D. T. (1979). Rhythms in CRF, ACTH and corticosteroids. *in* "Endocrine Rhythms" (D. T. Krieger, ed.), pp. 123–142. Raven, New York.

14. Butler, J. P., Spratt, D. I., O'Dea, L. S., and Crowley, W. F. (1986). Interpulse interval sequence of LH in normal men essentially constitutes a renewal process. *Am. J. Physiol.* **250**, E338–E340.

15. Knobil, E. (1980). Neuroendocrine control of the menstrual cycle. *in* "Recent Progress in Hormone Research" (R. O. Greep, ed.), pp. 53–88. Academic Press, New York.

16. Lincoln, D. W., Fraser, H. M., Lincoln, G. A., Martin, G. B., and McNeilly, A. S. (1985). Hypothalamic pulse generators. *Recent Prog. Horm. Res.* **41**, 369.

17. Cinlar, E. (1975). "Introduction to Stochastic Processes." Prentice-Hall, Englewood Cliffs, New Jersey.

18. Keenan, D. M., and Veldhuis, J. D. (1994). A stochastic model of pulsatile hormonal secretion. submitted for publication.

19. Brémund, P. (1981). "Point Processes and Queues." Springer-Verlag, New York.

20. Sun, W. (1994). Ph.D. Dissertation, Division of Statistics, Dept. of Mathematics, University of Virginia, Charlottesville.

21. Yang, R. (1994). Ph.D. Dissertation, Division of Statistics, Dept. of Mathematics, University of Virginia, Charlottesville.

22. Protter, P. (1990). "Stochastic Integration and Differential Equations." Springer-Verlag, New York.

23. Dempster, A. P., Laird, N., and Rubin, D. B. (1977). Maximum likelihood from incomplete data via the EM algorithm (with Discussion), *Journal of the Royal Statistical Society, Series B* **39**, 1–38.

[13] Modeling the Impact of Neuroendocrine Secretogogue Pulse Trains on Hormone Secretion

James E. A. McIntosh and Rosalind P. Murray-McIntosh

Introduction

Pulses and rhythms from milliseconds to decades are ubiquitous in biological systems. They are a basis for exerting control, for communication, and for organization of function by both external and internal factors. The correct interplay of time-varying rhythmic components or concentrations controls the initiation of many events. For example, to ensure prolonged responsiveness of gonadotropic cells of the pituitary gland to the hypothalamic gonadotropin-releasing hormone (GnRH), this stimulant must be delivered in pulses. In contrast, continuous stimulation causes desensitization of the target cells and secretion ceases (1–5).

We have used sheep anterior pituitary cells in perfusion to explore *in vitro* the dynamic characteristics of this process and applied mathematical models to analyze our observations (5–11). We have also investigated the dynamics of this endocrine signaling system *in vivo* (12, 13). Patterns of hypothalamic stimulation causing pituitary hormone release cannot be studied directly in humans. Instead, it is possible to infer mechanism from the nature of the response of the target organ as revealed by patterns of pituitary hormones in blood. We have used replicated, precise assays of luteinizing hormone (LH) in frequently sampled blood of women at different stages of the menstrual cycle to demonstrate that secretion of LH is compatible with a model of discrete, instantaneous episodes of hormone output, each of which is assumed to be stimulated by isolated bursts of increased stimulatory hypothalamic GnRH, followed by a relatively slow decline which is related to clearance of the hormone.

Frequent episodes of hormone secretion cause permanently raised levels in the blood, which have led to the suggestion that there are two forms of LH secretion: a tonic one represented by the baseline plus an imposed, pulsatile component. Our experiments, however, which give weight to a hypothesis in which a model of LH secretion describing a single mode of episodic release which is based on theory, is physiologically plausible and is sufficient to explain the observed behavior. Our data have shown that this

Methods in Neurosciences, Volume 28

model can provide an adequate and useful description of pituitary function as expressed by LH levels in blood and that there is neither a need to invoke a tonic release of LH nor evidence to support its existence. This model has been applied in comparing magnitudes of blood LH peaks at different stages of human menstrual cycles when the frequency of pulsatile LH release changes, and to illustrate the variations in frequency and magnitude of LH pulses that can occur between cycles of a single individual (12). Analogous measurements of the dynamics of changes in growth hormone (GH) concentrations in blood of women were used to show that similar modes of hypothalamic stimulation and pituitary secretion could not describe the data; another model was required (13).

This chapter concentrates on aspects of quantitative analysis and mathematical modeling used in our research. Also discussed are some aspects of how the assumptions about the physiological, sampling or experimental processes generating the data affect comparisons, usefulness, or physiological relevance of the conclusions.

Impact of Neuroendocrine Secretogogue Pulse Trains on Hormone Stimulation by Cells in Perfusion

We have previously described the development of our methods for preparing and culturing dispersed sheep pituitary cells, and for perfusing them with medium containing stimulatory pulse trains of hormones (5–11, 14–17). All of this methodology has been reviewed (18). The quantitation of hormone secretion from the cells is readily susceptible to distortion by spurious experimental factors unless careful validation of the technique is carried out, and replicates and controls are included with every experiment, as outlined in the preceding references.

Effects of Stimulatory Pulse Length, Interval, and Concentration on Cell Secretion in Perfusion

Many of our experiments have aimed to characterize the influence of pulse shape and interval on the release of secreted hormones from dispersed sheep pituitary cells. Using a microcomputer-controlled device, we generated a wide range of pulse patterns of releasing hormone and applied them to sets of identical samples of dispersed cells suspended and perfused in microcolumns, and measured the dynamics of hormone release into effluent fractions for up to 10 hr. The results obtained have been reported elsewhere (5, 6, 8–10, 14, 15) and reviewed (18).

Stimulatory GnRH, corticotropin-releasing factor (CRF), or arginine vasopressin (AVP) were applied continuously or in pulses varying in width and interval from 2 min every 5 min to 16 min every 128 min, to up to 15 equal portions of single preparations of sheep pituitary cells (5, 6, 10, 17). Effluent was collected in usually 4-min fractions and measured by radioimmunoassay (RIA) for content of LH, follicle-stimulating hormone (FSH), adrenocorticotropic hormone (ACTH), β-endorphin, or β-lipotropin. Analysis of response to early pulses was carried out to avoid effects from depletion of secretable hormone from the cells. Baseline levels of GH in the perfusate were always low or below the sensitivity of the assay.

The variation of the molar ratio of secretory response hormone to stimulating hormone with pulse interval was compared for different widths of pulses and with inverse of pulse width for different pulse intervals. An example of a three-dimensional plot summarizing results from experiments in which cells were stimulated with CRF is shown in Fig. 1.

Similar experiments studied the effect of amplitude of the stimulatory pulse, and for GnRH stimulation demonstrated a dose response curve reaching a maximum total response output which did not decline despite the anticipated increased receptor occupancy; nor did the response pulses desensitize faster. A simple mathematical model based on the logit transformation was used to describe the interaction between stimulating hormone and secreted response hormones. This model was expressed in terms of a FORTRAN subroutine (LOGIT) compatible with our general purpose computer program MODFIT (19), which is suitable for fitting almost any mathematical model, linear or nonlinear in its parameters, to experimental data. MODFIT can be adapted to fit models of great variety simply by selecting a single subprogram that provides a mathematical description of the model. Our current implementation runs on a Macintosh microcomputer. We used MODFIT to estimate values of ED_{50}, the concentration of stimulant causing half-maximal secretory release, as a measure of dose–response relationships or sensitivity of the cells to stimulation.

These experiments revealed several interesting features. (1) Shorter stimulatory pulses with longer intervals between them produce proportionately greater and more sustainable responsiveness from the cells. (2) The results illustrate the enhanced sensitivity of pituitary cells to the leading edge of a neuroendocrine signal. (3) A correlation between secreted pulse amplitude and interval since the preceding pulse suggested the presence of a recovery time after stimulation which controls sensitivity of the pituitary. (4) It is noteworthy that equivalent dynamics of release were observed with stimu-

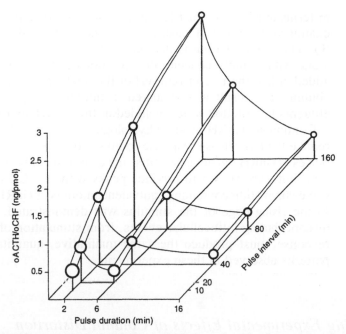

FIG. 1   Specific response of ACTH to CRF from freshly dispersed sheep anterior pituitary cells to the initial pulses of CRF before desensitization (second, third, and fourth pulses) as functions of both pulse interval and duration. The smooth curves emphasize the trends. Similar graphs can be drawn for responses of ACTH to AVP and for responses of LH and FSH to GnRH. The relative effectiveness of short pulses of stimulant at longer intervals in producing a response is evident.

lants binding to receptors which activate quite different intracellular transduction mechanisms. (5) Potentiation of release of ACTH to combined AVP and CRF stimulation compared with the sum of their individual responses was also quantitated. (6) In addition, a difference was observed in rates of decline of secretion after cessation of stimulation by AVP and CRF (17).

Effects of Low Continuous Stimulation and Slowly Rising Pulses on Cell Secretion in Perfusion

To further test the sensitivity of pituitary cells to the leading edge of a neuroendocrine signal, we extended these studies to include the influence on pulses of GnRH of continuous low levels of the same stimulant, measured in terms of efficiency of LH output from dispersed pituitary cells. We also examined the effect of raising slowly the leading edge of the GnRH pulse (9). The shape of each GnRH gradient was generated by mixing medium alone with suitably concentrated solutions of GnRH in medium, the latter added at half the rate of removal of the mixture as it was pumped onto the column. The result was validated quantitatively by analyzing the column effluents for tritiated water included in the GnRH solutions.

Continuous GnRH at low physiological levels (5–60 pM) were shown to reduce the LH release response to pulses of GnRH (850 pM). Slowly rising pulses of GnRH did not desensitise the cells as quickly as continuous stimulation, but were shown to produce peaks with less LH output than square-wave stimulation containing equivalent amounts of GnRH. Again, sensitivity to the rising edge of the stimulus was demonstrated. The summed total of contributing intracellular events, from stimulation of the cell to cellular response, must produce the same qualitative, quantitative, and dynamic patterns observed in these experiments.

Modeling Experimental Effects of Column Distortion

From the preceding experiments it was clear that the actual experimental shape of the stimulatory pulse applied to the perfused cells (particularly the leading edge) was very important to the outcome. The physical spatial characteristics of the perfusion apparatus may modify the expected shape of the stimulatory signal. It was also necessary to ascertain how passage of the released hormone through the column might reshape the signal that was sampled for analysis. In other words, it was important to know what influence the physical conditions of column transit might have on the pattern of released hormone in order to ensure physiological relevance of the results. To do this we derived mathematical models to predict the elution concentration in perfusion columns as a function of time when the concentration–time data of the hormone were specified at the column entrance. Of particular interest was the behavior of the system when a square-wave slug of stimulant was applied to the entrance of the columns (7).

The model revealed that, for columns such as these in which combined convective flow and dispersion occur (which also includes elution chromatography, adsorption columns, and studies of ground water flow), the column distorts the incoming signal at the column inlet, and that this effect increases with decreased flow rate of the perfusion medium. However, we showed that at the flow rates we used (measured by timed collection of radioactivity in the effluent), and with the experimental precision of interest in our experiments, there was no need to correct our observed output signals (7, 9).

Modeling Neuroendocrine Stimulation through Measurement of Hormonal Rhythms in Human Blood

Blood Sampling and Assaying

Our techniques for studying the time course of hormone secretion in women at several stages of the menstrual cycle have been described (12, 13).

It is noted that blanket application of complex methods of quantitative analysis of rhythms to laboratory measurements of hormones do not always lead to conclusions that are physiologically relevant. Some of the many potential sampling and assay considerations related to physiological aspects that we have found need considering, include the following. The neuroendocrine response to the stress of the experiment itself and of events which may occur during the experimental period may influence hormonal secretions or their processing in blood. For example, measurements of circadian rhythms may need to continue beyond 24 hr to ensure that values coincide at the same time next day, and to allow initial disturbances to stabilize. We have found that, in general, protein concentration in blood decreases by up to 14% on lying down (20). Clearly, with a marked periodical change, the time of a cycle at which a single sample of blood is taken for clinical analysis will influence interpretation of the result; either the range will encompass the full cycle and information will be lost on potentially pathological variation in cyclicity, or a peak measurement may incorrectly be deemed to fall outside the accepted range.

Many analyses of rhythms use a sine wave model; for a number of purposes we have found this smoothly varying symmetrical mathematical description to be inappropriate when applied to biological events. The intervals between samples determines the frequency of the episodes that can be detected, and aliasing can occur if insufficient samples are taken in a given time period (19). Replication of measurements, randomization of samples in a time series, and repeat assays of time series data are essential to confirm subtle rhythms; reproducibility of the assay for each sample needs to be known in such cases because the size of the peak that can be detected is totally dependent on the particular local variance. The nature of the antisera used can distort interpretation by reacting disproportionately to the presence of less physiologically relevant forms of hormone or even long-lived degradation products of multiform hormones such as LH, and may not reflect changes in bioactive forms related to different physiological or pathological states (12).

For our experiments investigating patterns of LH and GH secretion in humans, blood was sampled from an antecubital vein at regular intervals 5–15 min apart for 4–6 hr from women with normal menstrual cycles at luteal and follicular phases, and during the mid-cycle LH surge. Details of sample preparation and storage have been reported (12, 13).

Plasma samples were analyzed for hormone content by RIA, precautions being taken to eliminate inaccuracy and maximize precision; details of the procedures are described elsewhere (6, 12–14, 19, 21). Assay results were calculated with a computer program based on a four parameter logistic model incorporated into a tailored version of MODFIT, which estimated 95% confidence limits on all results (21).

Mathematical and Statistical Analyses

We assumed that releasing hormones stimulate the pituitary to release gonadotropins or other hormones in discrete episodes approximately consistent in shape from one episode to another within each sampling session. When observed in blood the shape of the secretory episode from the pituitary is distorted by mixing and clearance from the circulation. At any time the level of hormone in blood represents the sum of all previous episodes, each diminished by clearance by an amount dependent on time since its secretion. If these episodes are infrequent, clearance ensures that concentrations decline to levels below assay sensitivity and estimates of peak maxima, interpeak interval and decay rates are simple to make, but frequent episodes of release are more difficult to analyze.

Estimates of the rates of clearance of LH and GH from blood were made in two ways. The direct approach was to estimate a mean clearance coefficient by fitting a single exponential decay process to measurements on the descending sides of peaks which contained at least four blood samples (12 data points; range 4–16 samples) in which hormone levels decayed to less than one-third maximum height during the time of blood sampling. This was possible for LH only with samples collected frequently during the late luteal stages of cycles when periods of about 150 min between episodes ensured the almost complete decay of each, or with peaks of GH which occur less frequently. Fitting was done using MODFIT together with subroutine RESERVR (see below; also Ref. 19). The data were neither precise nor extensive enough to justify attempting to describe the decay in terms of more than one process, although such an underlying model is theoretically probable.

We also used an indirect approach which was to introduce the clearance coefficient as an adjustable parameter into the model to be described below so as to find a value which minimized the variance of the model about the experimental data. This had the advantage of being applicable to all data, rather than only to that in which episodes were widely spaced. The model for analyzing secretory episodes based on summing exponential decays is summarized by the following equation:

$$[\text{hormone}]_t = \sum_{i=1}^{n} [A_i \, e^{-\lambda(t-t_i)}],$$

where $[\text{hormone}]_t$ is the concentration of hormone in blood sample at time t since the beginning of sampling, n the number of episodes analyzed, A_i the maximum amplitude of the ith episode in the blood occurring at time t_i, and λ the mean exponential decay coefficient for the sampling session, estimated as described above from the declining sides of widely spaced peaks or by inclusion as an adjustable parameter. This analysis allowed estimation of the relative amplitudes of secretory episodes of LH as observed in blood, free of interference from the slow clearance of previous episodes during times of rapid frequency of release, as in the early follicular stage.

The secretory episode model was defined in the form of subroutine UNCOV suitable for use with MODFIT. Inputs required were blood levels of hormone measured at frequent (but not necessarily equal) intervals, the clearance coefficient (either estimated separately or included as a parameter to be adjusted), starting estimates of the parameters (the concentration of hormone in the blood at the beginning of the sampling interval, as well as the initial amplitudes of the secretory episodes), and the times since the beginning of blood sampling of the secretory episodes found by prior inspection of the data. Output was the best-fitting estimates of the initial amplitude of each episode as observed in blood free of the effects of the others.

One or two points on the rising side of each peak were omitted from the analysis when a maximal value was not reached in a single sampling interval following a nadir. These few omitted values were assumed to result from the time required for mixing of the peak in the blood volume (probably a constant for all peaks) and might also be expected to incorporate any effects of noninstantaneous stimulation and response. Peaks of GH were slower to reach a maximum than those of LH.

Because it is unlikely that the particular times of sampling used coincided with the maximum level of hormone reached in every pulse, the amplitudes of secretory episodes were inevitably underestimated. The mean level of underestimation would be expected to increase with sampling interval. A mean multiplication factor for the amplitude of episodes was used to approximately correct all peaks using the following formula:

$$\text{Multiplication factor} = 1/[e^{-\lambda(\delta/2)}],$$

where λ is the clearance coefficient (per min) and δ is the sampling interval (min). This correction amounted to between 3 and 10% depending on the sampling intervals and clearance coefficients of our data.

Rise times of peaks were estimated between a minimum and maximum hormone level defined as follows. Best fitting straight lines were drawn by eye through the rising and falling sides of each peak, the intersections of which defined the maximum. Minima were assumed to occur at the point where the rising line intersected an extrapolation of the tail of the preceding peak or baseline.

Nonparametric statistical tests were used so as to avoid assumptions of normality and homoscedasticity in the data (22). All statistical procedures described here can be carried out conveniently with the SYSTAT (23) or STATVIEW (24) packages on the Macintosh.

Results and Discussion

We have assumed that episodic release of hormones from the pituitary causes fluctuations in their blood levels, and that these fluctuations give information about both the intervals between pulses of releasing hormone stimulating each peak and the sensitivity of the pituitary. It has been assumed that the observed levels of LH in the blood represented the combination of a tonic, baseline secretion and an imposed, pulsatile component. We have tested an alternative hypothesis, namely that discrete episodes of secretion alone are sufficient to produce the observed blood levels without the need to postulate a tonic release. We have expressed this alternative model mathematically and applied it to successfully analyze the secretion of LH during various stages of the menstrual cycle (12).

Detailed measurements of the dynamic secretion patterns of LH and GH in women revealed much slower rates of increase of GH levels (median time to maximum concentration 38 min) compared with LH (13 min) assayed in the same blood samples (13). Thus, estimates of the heights of peaks of GH in the blood are likely to be more accurate than those of LH because with slower rise rates the true peak height is more likely to be measured (for the reason given above).

Figure 2 shows changes with time in the levels of LH and GH measured in the same blood samples in a woman 10 days after the onset of menstruation, a time at which the frequency of secretory episodes of LH is maximal (12). The clearance rate of GH estimated from the data was greater (-0.038/min) than that of LH (-0.0089/min) and this, together with the less frequent

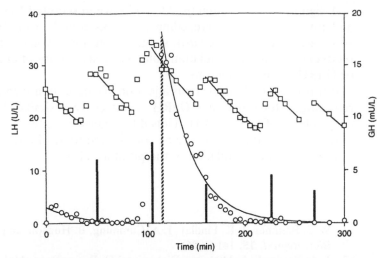

FIG. 2 Changes with time in the levels of LH (□) and GH (○) measured in the same blood samples in a woman 10 days after the onset of menstruation. The smooth curves represent individual fitted decay processes estimated using subroutine UNCOV together with the computer program MODFIT. Vertical bars represent the positions and amplitudes of individual secretory episodes of LH (solid bars) and GH (hatched bar) also estimated using subroutine UNCOV. (The GH peak is shown for comparative purposes. Most peaks of GH did not lend themselves to the approximation of an instantaneous release, as shown here, because of longer rise times and discontinuities.)

release of GH, ensured that the concentration of GH declined between peaks to levels less than the sensitivity of the assay. The smooth curves show the results of fitting the model described above to the decay in LH concentration with time of the combined secretory episodes, and the solid bars are estimates of peak height of each episode.

It is not, however, appropriate to approximate as instantaneous, release of hormone from the pituitary when analyzing most peaks of GH because in our experience the rise rates of GH in the blood contain discontinuities, and are slower than those of LH (13). However, for comparative purposes, we have included an estimate of peak height from application of the model (hatched bar).

From the analysis of numerous secretory episodes of LH in this way we conclude that the model of very brief episodes of stimulation by the hypothalamus followed by rapid responses from the pituitary, gives an excellent description of the data and that there is no need to invoke a constant, tonic stimulation by the hypothalamus nor continuous release of LH (12).

A significant correlation ($p < 0.01$) between secreted LH pulse amplitude and interval since the preceding pulse, as found in perfusion of GnRH-stimulated sheep cells *in vitro*, suggested in these women the presence of a recovery time after stimulation controlling sensitivity of either the pituitary or GnRH-secreting neurons.

It was noted that intravenous injection of a bolus of mixed growth hormone releasing factor (GRF) and GnRH produced very similar dynamics of pituitary release of GH and LH. Thus differences in patterns of natural release of the two hormones appear to be contributed to by differences in the modes of hypothalamic stimulation of the pituitary (13).

References

1. W. J. Bremner, J. K. Findlay, I. A. Cumming, B. Hudson, and D. M. de Kretser, *Biol. Reprod.* **15,** 141 (1976).
2. J. de Konig, J. A. M. J. van Dieten, and G. P. van Rees, *Mol. Cell. Endocrinol.* **5,** 151 (1976).
3. C. R. Hopkins, *J. Cell Biol.* **73,** 685 (1977).
4. D. Rabin and L. W. McNeil, *J. Clin. Endocrinol. Metab.* **51,** 873 (1980).
5. R. P. McIntosh and J. E. A. McIntosh, *J. Endocrinol.* **98,** 411 (1983).
6. J. E. A. McIntosh and R. P. McIntosh, *J. Endocrinol.* **109,** 155 (1986).
7. W. R. Smith, G. C. Wake, J. E. A. McIntosh, R. P. McIntosh, M. Pettigrew, and R. Kao, *Am. J. Physiol.* **261,** R247 (1991).
8. J. E. A. McIntosh, R. P. McIntosh, and R. J. Kean, *Med. Biol. Eng. Comput.* **22,** 259 (1984).
9. R. P. McIntosh and J. E. A. McIntosh, *Endocrinology (Baltimore)* **117,** 169 (1985).
10. M. J. Evans, J. T. Brett, R. P. McIntosh, J. E. A. McIntosh, H. K. Roud, J. H. Livesey, and R. A. Donald, *Endocrinology (Baltimore)* **117,** 893 (1985).
11. L. Starling, R. P. McIntosh, and J. E. A. McIntosh, *J. Endocrinol.* **111,** 167 (1986).
12. R. P. McIntosh and J. E. A. McIntosh, *J. Endocrinol.* **107,** 231 (1985).
13. R. P. McIntosh, J. E. A. McIntosh, and L. A. Lazarus, *J. Endocrinol.* **118,** 339 (1988).
14. R. P. McIntosh and J. E. A. McIntosh, *Acta Endocrinol.* **102,** 42 (1983).
15. R. P. McIntosh, *Trends Pharmacol. Sci.* **5,** 429 (1984).
16. R. P. McIntosh, J. E. A. McIntosh, and L. Starling, *J. Endocrinol.* **112,** 289 (1987).
17. M. J. Evans, J. T. Brett, R. P. McIntosh, J. E. A. McIntosh, J. L. McLay, J. H. Livesey, and R. A. Donald, *J. Endocrinol.* **117,** 387 (1988).
18. R. P. McIntosh and J. E. A. McIntosh, *Methods Neurosci.* **20,** 423 (1994).
19. J. E. A. McIntosh and R. P. McIntosh, "Mathematical Modelling and Computers in Endocrinology." Springer-Verlag, New York, 1980.
20. R. R. Cooke, J. E. A. McIntosh, and R. P. McIntosh, *Clin. Endocrinol.* **39,** 163 (1993).
21. J. E. A. McIntosh, A program for analyzing the results of protein binding assays on the Macintosh personal computer. Unpublished, 1989.

22. D. Colquhoun, "Lectures on Biostatistics," p. 103. Oxford Univ. Press (Clarendon), Oxford, 1971.
23. SYSTAT version 5.2, SYSTAT, Inc., 1800 Sherman Ave, Evanston, IL 60201-3793 (1992).
24. STATVIEW version 4.0, Abacus Concepts Inc., 1984 Bonita Ave, Berkeley CA 94704-1038 (1992).

[14]   Quantifying Complexity and Regularity of Neurobiological Systems

Steven M. Pincus

Introduction

Time series are encountered frequently in analysis of biological signals. Within endocrinology, hormone concentration time series that are based on frequent, fixed-increment samples have been the subject of intensive study (1); heart rate and the EEG (electroencephalogram) are two further examples of physiological time series. The biologist is often interested in time series for either of two important purposes: (i) to distinguish (discriminate) systems, on the basis of statistical characteristics; (ii) to model systems mathematically. In both cases, effective statistics and models need to account for the sequential interrelationships among the data: the study of autocorrelation and of power spectra are motivated by this recognition. Below, we focus on the first of these purposes, statistical discrimination, via a quantification of regularity of a time series. This approach also calibrates the extent of sequential interrelationships, from a relatively new perspective, based on quantifying a notion of orderliness (as opposed to randomness) of the data.

Before presenting a detailed discussion of regularity, we consider three sets of time series (Figs. 1–3) to illustrate what we are trying to measure. In Fig. 1, the data represent growth hormone (GH) levels from three subjects, taken at 5-min intervals during a fed state (2): Fig. 1A,B shows data from normal subjects, and Fig. 1C data are from an acromegalic (giant). The mean levels associated with the three subjects are similar, yet the series in Fig. 1A,B appear to be regular, less random than that in Fig. 1C. In Fig. 2, the data represent the beat-to-beat heart rate, in beats per minute, at equally spaced time intervals. Figure 2A is from an infant who had an aborted SIDS (sudden infant death syndrome) episode 1 week prior to the recording, and Fig. 2B is from a healthy infant (3). The standard deviations (SD) of these two tracings are approximately equal, and although the SIDS infant has a somewhat higher mean heart rate the data from both subjects both are well within the normal range. Yet the tracing in Fig. 2A appears to be more regular than the tracing in Fig. 2B. Figure 3, taken from the mathematical MIX(p) process discussed below, shows a sequence of four time series, each of which have mean 0 and SD 1; yet it appears that the time series are

Methods in Neurosciences, Volume 28

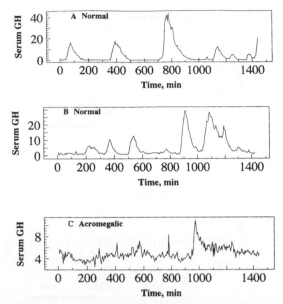

FIG. 1 Growth hormone (GH) serum concentrations, in milliunits/ml, measured at 5-min intervals for 24 hr (fed state). (A) Normal subject, mean concentration = 4.743, ApEn(2, 0.20SD, 288) = 0.318. (B) Normal subject, mean concentration = 5.617, ApEn(2, 0.20SD, 288) = 0.341. (C) Acromegalic subject, mean concentration = 4.981, ApEn(2, 0.20SD, 288) = 1.081.

becoming increasingly irregular as we proceed from Fig. 3A to Fig. 3D).[1] In each of these instances, we ask the following questions. (i) How do we quantify this apparent difference in regularity? (ii) Do the regularity values significantly distinguish the data sets? (iii) How do inherent limitations posed by moderate length time series, with noise and measurement inaccuracy present as in Figs. 1 and 2, affect statistical analyses? (iv) Is there some general mechanistic hypothesis, applicable to diverse contexts, that might explain such regularity differences?

A new mathematical approach and formula, Approximate Entropy (ApEn), has been introduced as a quantification of regularity in data, motivated by the questions above (4). Mathematically, ApEn is part of a general theoretical

---

[1] A useful (regularity) statistic should (i) agree with intuition, that is, confirm differences that are visually "obviously distinct", as in the comparisons in Figs. 1 and 2, and in Fig. 3A and 3D; and (ii) provide information in more subtle comparisons, that is, distinguish less obviously different time series, such as those in Fig. 3C and 3D.

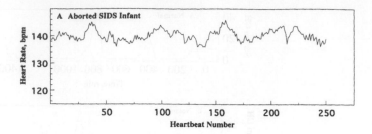

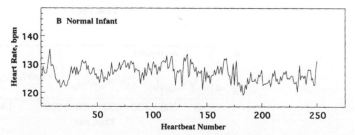

FIG. 2 Two infant quiet sleep heart rate tracings with similar variability, *SD*. (A) Aborted SIDS episode infant, *SD* = 2.49 beats per minute (bpm), ApEn(2, 015*SD*, 1000) = 0.826. (B) Normal infant, *SD* = 2.61 bpm, ApEn(2, 0.15*SD*, 1000) = 1.463.

development, as the rate of entropy for an approximating Markov chain to a process (5). In applications to a range of medical settings, findings have discriminated groups of subjects via ApEn, applied to heart rate time series, in instances where classic statistics (mean, *SD*) did not show clear group distinctions (3, 6–9).

The development of ApEn evolved as follows. To quantify time-series regularity (and randomness), we initially applied the Kolmogorov–Sinai (K-S) entropy (10) to clinically derived data sets. The application of a formula for K-S entropy (11, 12) yielded intuitively incorrect results. Closer inspection of the formula showed that the low-magnitude noise present in the data greatly affected the calculation results. It also became apparent that to attempt to achieve convergence of this entropy measure, extremely long time series would be required (often 100,000 or more points), which even if available, would then place extraordinary time demands on the computational resources. The challenge was to determine a suitable formula to quantify the concept of regularity in moderate length, somewhat noisy data sets, in a manner thematically similar to the approach given by the K-S entropy.

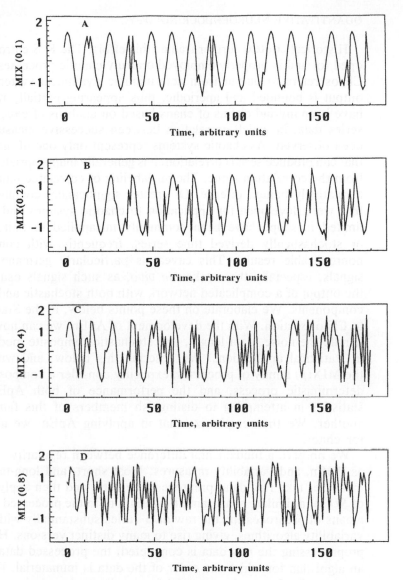

Fig. 3 MIX(p) model time-series output for four parameter values: (A) $p = 0.1$; (B) $p = 0.2$; (C) $p = 0.4$; (D) $p = 0.8$. MIX(p) is a family of processes that samples a sine wave for $p = 0$, samples independent, identically distributed (IID) uniform random variables for $p = 1$, and intuitively becomes more irregular as p increases. ApEn quantifies the increasing irregularity and complexity with increasing p: for $m = 2$, r $= 0.18$, and $N = 1000$, ApEn[MIX(0.1)] $= 0.436$, ApEn[MIX(0.2)] $= 0.782$, ApEn[MIX(0.4)] $= 1.455$, and ApEn[MIX(0.8)] $= 1.801$. In contrast, the correlation dimension of MIX(p) $= 0$ for all $p < 1$, and the K-S entropy of MIX(p) $= \infty$ for all $p > 0$. Thus even given no noise and an infinite amount of data, these latter two measures do not discriminate the MIX(p) family.

Historical context further frames this effort. The K-S entropy was developed for and is properly employed on truly chaotic processes (time series). Chaos refers to output from deterministic dynamical systems, where the output is bounded and aperiodic, thus appearing partially random. There have been myriad claims of chaos based on analysis of experimental time-series data, in which correlation between successive measurements have been observed. As chaotic systems represent only one of many paradigms that can produce serial correlation, it is generally inappropriate to infer chaos from the correlation alone. The mislabeling of correlated data as chaotic is a relatively benign offense. Of greater significance, complexity statistics that were developed for application to chaotic systems and are relatively limited in scope have been commonly misapplied to finite, noisy, and/or stochastically derived time series, frequently with confounding and nonreplicable results. This caveat is particularly germane to biological signals, especially those taken *in vivo*, as such signals usually represent the output of a complicated network with both stochastic and deterministic components. We elaborate on these points below, in the Statistics Related to Chaos section. With the development of ApEn, we can now successfully handle the noise, data length, and stochastic/composite model constraints in statistical applications. We describe this below, and with analysis of the MIX(p) family of processes, explicitly consider a composite stochastic/deterministic process, and the performance of both ApEn and chaos statistics in attempting to distinguish members of this family from one another. We thus emphasize that in applying ApEn, we are not testing for chaos.

We observe a fundamental difference between regularity statistics, such as ApEn, and variability measures: Most short- and long-term variability measures take raw data, preprocess the data, and then apply a calculation of *SD* (or a similar, nonparametric variation) to the processed data (13). The means of preprocessing the raw data varies substantially with the different variability algorithms, giving rise to many distinct versions. However, once preprocessing the raw data is completed, the processed data are input for an algorithm for which the order of the data is immaterial. For ApEn, the order of the data is the essential factor; discerning changes in order from apparently random to very regular is the primary focus of this statistic.

Finally, an absolutely primary concern in any practical time-series analysis is the presence of either artifacts or nonstationarities, particularly clear trending. If a time series is nonstationary or is riddled with artifacts, little can be inferred from moment, ApEn, or power spectral calculations, as these effects tend to dominate all other features. In practice, data with trends suggest a collection of heterogeneous epochs, as opposed to a single homogeneous state. From the statistical perspective, it is imperative that artifacts and trends first be removed before meaningful interpretation can be made from further statistical calculations.

Quantification of Regularity

Definition of Approximate Entropy

Approximate entropy measures the logarithmic likelihood that runs of patterns that are close for m observations remain close on the next incremental comparisons. Greater likelihood of remaining close, regularity, produces smaller ApEn values, and conversely. From the perspective of a statistician, ApEn can often be regarded as an ensemble parameter of process autocorrelation: smaller ApEn values correspond to greater positive autocorrelation, larger ApEn values indicate greater independence. The opposing extremes are perfectly regular sequences (e.g., sinusoidal behavior, very low ApEn) and independent sequential processes (very large ApEn).

Formally, given N data points $u(1)$, $u(2)$, ..., $u(N)$, two input parameters, m and r, must be fixed to compute ApEn [denoted precisely by ApEn(m,r,N)]. The parameter m is the length of compared runs, and r is effectively a filter. Next, form vector-sequences $x(1)$ through $x(N - m + 1)$ from the $\{u(i)\}$, defined by $x(i) = [u(i), ..., u(i + m - 1)]$. These vectors represent m consecutive u values, commencing with the ith point. Define the distance $d[x(i), x(j)]$ between vectors $x(i)$ and $x(j)$ as the maximum difference in their respective scalar components. Use the sequence $x(1)$, $x(2)$, ..., $x(N - m + 1)$ to construct, for each $i \leq N - m + 1$,

$$C_i^m(r) = \{\text{number of } x(j) \text{ such that } d[x(i), x(j)] \leq r\}/(N - m + 1) \quad (1)$$

The $C_i^m(r)$ terms measure within a tolerance r the regularity, or frequency, of patterns similar to a given pattern of window length m. Next, define

$$\Phi^m(r) = (N - m + 1)^{-1} \sum_{i=1}^{N-m+1} \ln C_i^m(r), \quad (2)$$

where ln is the natural logarithm. We define approximate entropy by

$$\text{ApEn}(m,r,N) = \Phi^m(r) - \Phi^{m+1}(r). \quad (3)$$

Via some simple arithmetic manipulation, we deduce the essential observation that

$$-\text{ApEn} = \Phi^{m+1}(r) - \Phi^m(r)$$
$$= \text{average over } i \text{ of } \ln[\text{conditional probability that}$$
$$|u(j + m) - u(i + m)| \leq r, \text{ given that}$$
$$|u(j + k) - u(i + k)| \leq r \text{ for } k = 0,1, \ldots, m-1]. \quad (4)$$

When $m = 2$, as is often employed, we interpret ApEn as a measure of the difference between the probability that runs of value of length 2 will recur within tolerance r, and the probability that runs of length 3 will recur to the same tolerance. A high degree of regularity in the data would imply that a given run of length 2 would often continue with nearly the same third value, producing a low value of ApEn. To develop a more intuitive, physiological understanding of this definition, a multistep description of the algorithm with accompanying figures is developed in Pincus and Goldberger (14).

Implementation and Interpretation

Choice of m, r, and N

The value of N, the number of input data points for ApEn computations is typically between 100 and 5000. This constraint is usually imposed by experimental considerations, not algorithmic limitations, to insure a single homogeneous epoch. Based on calculations that included both theoretical analysis (4, 15, 16) and clinical applications (3, 6–9) we have concluded that for $m = 2$ and $N = 1000$, values of r between 0.1 and 0.25 SD of the $u(i)$ data produce good statistical validity of ApEn(m,r,N). For such r values, we demonstrated (4, 15, 16) the theoretical utility of ApEn$(2,r,N)$ to distinguish data on the basis of regularity for both deterministic and random processes, and the clinical utility in the aforementioned applications. These choices of m and r are made to ensure that the conditional probabilities defined in Eq. (4) above are reasonably estimated from the N input data points. Theoretical calculations indicate that reasonable estimates of these probabilities are achieved with N at least 10^m (preferably at least 30^m points), analogous to a result for correlation dimension noted by Wolf *et al.* (17). For smaller r values than those indicated above, one usually achieves poor conditional probability estimates as well, while for larger r values, too much detailed system information is lost. In the clinical studies referenced above, we used $m = 2$ and either $r = 0.15$ SD or $r = 0.2$ SD.

To ensure appropriate comparisons between data sets, it is imperative that N be the same for each data set. This is because ApEn is what statisticians denote a biased statistic; the expected value of ApEn(m,r,N) generally increases asymptotically with N to a well-defined limit parameter denoted ApEn(m,r). Restated, if we had 3000 data points, and chose $m = 2$, $r = 0.2$

SD, we would expect that ApEn applied to the first 1000 points would be smaller than ApEn applied to the entire 3000-point time series. Biased statistics are quite commonly employed, with no loss of validity. As an aside, it can be shown that ApEn is asymptotically unbiased, an important theoretical property, but that is not so germane to everyday usage.

This bias arises from two separate considerations. The first source is the concavity (and nonlinearity) of the logarithm function in the ApEn definition. In general, unbiased estimators are uncommon in nonlinear estimation; the observation also applies to algorithms estimating the correlation dimension and K-S entropy for dynamical systems. More significant is the second source of the bias. In the definition of $C_i^m(r)$ in forming ApEn, the template vector $x(i)$ itself counts in the $C_i^m(r)$ aggregation of vectors close to $x(i)$. This is done to ensure that calculations involving logarithms remain finite, but has the consequence that the conditional probabilities estimated in Eq. (4) are underestimated. These points, and a proposed family of ε-estimators to reduce the bias, are discussed further by Pincus and Goldberger (14). The comparison of each template vector to itself in the definition of the ApEn statistic is in agreement with historical precedent; the algorithms for correlation dimension (18) and the K-S entropy (eq. 5.7 in Ref. 19) also compare the template vector to itself in ensemble comparisons.

Family of Statistics

Most importantly, despite algorithmic similarities, ApEn(m,r,N) is not intended as an approximate value of K-S entropy (4, 8, 15). It is imperative to consider ApEn(m,r,N) as a family of statistics; for a given application, system comparisons are intended with fixed m and r. For a given system, there usually is significant variation in ApEn(m,r,N) over the range of m and r (3, 8, 15).

For fixed m and r, the conditional probabilities given by Eq. (4) are precisely defined probabilistic quantities, marginal probabilities on a coarse partition, and contain a great deal of system information. Furthermore, these terms are finite, and thus allow process discrimination for many classes of processes that have infinite K-S entropy (see below). ApEn aggregates these probabilities, thus requiring relatively modest data input.

Alternative Choice for r, Normalized Regularity

ApEn decrease frequently correlates with *SD* decrease. This is not a problem, as statistics often correlate with one another, but there are times when we desire an index of regularity decorrelated from *SD*. We can realize such an index, by specifying r in ApEn(m,r,N) as a fixed percentage of the sample SD of the individual (not group, or fixed) subject data set, with possibly a different r for each subject (3, 7). We call this normalized regularity. Its utility was demonstrated, for example, in refining the differences between heart rate data from quiet and rapid eye movement (REM) infant sleep, two sleep states with markedly different overall variability (SD), by establishing that REM sleep has significantly greater normalized regularity (smaller normalized ApEn values) than quiet sleep has, which is juxtaposed with the more classical finding of much larger variability in REM sleep than in quiet sleep (3). This thus enables us to determine if two processes are different both in variation (SD) and in normalized regularity.

Choosing r in this manner allows sensible regularity comparisons of processes with substantially different SDs. In the aforementioned comparison, if REM sleep were a scale multiple of quiet sleep (the multiple given by the ratio of the respective overall variabilities), plus some translated mean heart rate, the normalized regularity ApEn values for REM and quiet sleep would be identical. Below, we employ normalized regularity to report highly significant scale-invariant distinctions between growth hormone secretory patterns of normal and acromegalic subjects, two groups with markedly different means and SDs.

Validation of Intuition

The ApEn values given in the legends of Figs. 1–3 indicate that the quantification of regularity given by ApEn calibrates with what the eye sees, as is discussed in the second paragraph of the introduction.

Relative Consistency

Above, we commented that ApEn values for a given system can vary significantly with different m and r values. Indeed, it can be shown that for many processes, ApEn(m,r,N) grows with decreasing r like log($2r$), thus exhibiting infinite variation with r (15). We have also claimed that the utility of ApEn is as a relative measure; for fixed m and r, ApEn can provide useful information. We typically observe that, for a given time series, ApEn(2, 0.1) is quite different from ApEn(4, 0.01), so the question arises which parameter choices (m and r) to use. The guidelines above address this, but the most important requirement is consistency. For noiseless, theoretically defined deterministic dynamical systems, such as those given by the logistic map or the Henon map (two much-studied chaotic systems), we have found that when K-S entropy(A) $\leq$ K-S entropy(B), then ApEn(m,r)(A) $\leq$ ApEn (m,r)(B) and conversely, for a wide range of m and r. Furthermore, for both theoretically described systems and those described by experimental data, we have found that when ApEn(m_1,r_1)(A) $\leq$ ApEn(m_1,r_1)(B), then ApEn(m_2,r_2)(A) $\leq$ ApEn(m_2,r_2)(B), and conversely. This latter property also generally holds for parametrized systems of stochastic (random) processes, in which K-S entropy is infinite. We call this ability of ApEn to preserve order a relative property. It is the key to the general and clinical utility of ApEn. We see no sensible comparisons of ApEn(m,r)(A) and ApEn(n,s)(B) for systems A and B unless m = n and r = s.

As an illustrative and typical example of this relative consistency property, we determined that the association of very low (heart rate) ApEn values with aborted SIDS infants was replicated for different choices of parameter values m and r for ApEn input, even though the ApEn values themselves changed markedly with different m and r choices (3). The degree to which ApEn varied as a function of m and r was substantial: ApEn(1, 0.15 SD) generally produced ApEn values about a factor of 2 larger than ApEn(2, 0.25 SD) values, for most infant heart rate data sets. This relative consistency property of ApEn was also confirmed via analysis of 15 (m,r) pairs in a study of healthy and severely ill neonates (8). These findings thus impart a robustness to ApEn input parameter choice insofar as the identification of time-series with atypical ApEn values.

From a more theoretical mathematical perspective, the interplay between meshes [(m,r) pair specifications] need not be nice, in general, in ascertaining which of (two) processes is more random. In general, we might like to ask the question: given no noise and an infinite amount of data, can we say that process A is more regular than process B? The flip-flop pair of processes (15) implies that the answer to this question is not necessarily: in general, comparison of relative process randomness at a prescribed level is the best one can do. That is, processes may appear more random than processes on many choices of partitions, but not necessarily on all partitions of suitably small diameter (r). The flip-flop pair are two independent and identically distributed (IID) processes A and B with the property that for any integer m and any positive r, there exists s < r such that ApEn(m,s)(A) < ApEn(m,s)(B), and there exists t < s such that ApEn(m,t)(B) < ApEn(m,t)(A). At alternatingly small levels of refinement given by r, the B process appears more random and less regular than the A process, followed by appearing less random and more regular than the A process on a still smaller mesh (smaller r). In this construction, r can be made arbitrarily small, thus establishing the point, that process regularity is a relative [to mesh, or (m,r) choice] notion.

Fortunately, for many processes A and B, we can assert more than relative regularity, even though both A and B will typically have infinite K-S entropy. For such pairs of processes, which have been denoted as a completely consistent pair (15), whenever $ApEn(m,r)(A) < ApEn(m,r)(B)$ for any specific choice of m and r, then it follows that $ApEn(n,s)(A) < ApEn(n,s)(B)$ for all choices of n and s. Any two elements of $\{MIX(p)\}$ (defined below), for example, appear to be completely consistent. The importance of completely consistent pairs is that we can then assert that process B is more irregular (or random) than process A, without needing to indicate m and r. Visually, process B appears more random than A at any level of view. We indicate elsewhere (14) a conjecture that should be relatively straightforward to prove, that would provide a sufficient condition to ensure that A and B are a completely consistent pair, and would indicate the relationship to the autocorrelation function.

Computer Implementation

ApEn is typically calculated via a short computer code; a FORTRAN listing for such a code can be found elsewhere [Appendix B in Pincus *et al.* (8)]. On a MacIntosh Plus Personal Computer, given $N = 300$ input data points, an ApEn calculation consumes approximately 5 sec of run time; for $N = 1000$ input points, the ApEn calculation consumes approximately 60 sec. (Generally, run time increases approximately quadratically with input data length). On commonly available superminicomputers, ApEn run time is typically reduced by a factor of about 50. This should be compared to computational run times for most moments (e.g., mean, SD), which are virtually 0, and those for some of the chaos algorithms (e.g., Lyapunov spectrum), which often require hours on a supercomputer, even when correctly employed. The nonzero computational time for ApEn results from continued comparisons of short runs of data to all other short runs, through nested loops, a computationally demanding procedure, but one core to this approach. Fortunately, computational speed is currently such that implementation of ApEn can be automated with turnaround in nearly real time, an essential feature for an algorithm to have wide utility.

Robustness to Outliers

ApEn provides for artifact and outlier insensitivity, via the probabilistic form of the comparisons. This can be inferred from Eq. (4)—extremely large and small artifacts and/or process values have small affect on the ApEn calculation, if they occur infrequently. This robustness of ApEn to infrequent artifacts is a useful statistical property for practical applications. It is not always shared by moment statistics, for which single very large outliers can nontrivially affect the statistical estimate of the true value. In a fetal monitoring study, we noted that for typical labor records, about 5% of artifactual heart rates were deleted per record (9). We calculated both ApEn and near-term variability both before and after artifact deletion in one such record. The ApEn value decreased by 4% with artifact deletion, whereas variability (*SD*) decreased by 41%. A useful consequence of this robustness is that if outliers seem to occur infrequently, there is no acute need to preprocess the time series to identify and delete them, an often surprisingly difficult task (20), to obtain a first-order estimate of ApEn.

The counterpoint to this is that ApEn will not be very useful in discerning acute changes that are associated with infrequent data measurements. We anticipate that the primary utility of ApEn will be to uncover subtle, insidious abnormalities or alterations in long-term data that are not otherwise apparent.

Implicit Noise Filtering

Many signals are corrupted by inaccuracy or noise, and a sound, practical algorithm must accommodate this fact of life. In some instances, data filtering is performed prior to algorithmic calculations, but this can be a chancy business, especially if the form and/or distribution of the noise is not well understood. ApEn implicitly provides de facto noise filtering, via choice of r. To avoid a significant contribution from noise in the ApEn calculation, one must choose r larger than most of the noise. Although exact guidelines depend on the distribution of the noise and the nature of the underlying system, we have had both theoretical and clinical success with r of at least three times an estimated mean noise amplitude. If the signal-to-noise ratio is very low (if the system is very noisy), this requirement could force a very large value of r, thus compromising the utility of ApEn. Of course, very noisy systems cause great difficulties in system recognition and distinction for many other statistics as well. If r is not chosen much larger than system noise levels, ApEn will typically be spuriously large, in effect primarily evaluating the randomness of the noise level.

Model Independence

The physiological modeling of many complex biological systems is often very difficult; one would expect accurate models of such systems to be complicated composites, with both deterministic and stochastic components, and interconnecting network features. The advantage of a model-independent statistic is that it can distinguish classes of systems for a wide variety of models. The mean, variability, and ApEn are all model-independent statistics in that they can distinguish many classes of systems, and all can be meaningfully applied to $N > 100$ data points. In applying ApEn, therefore, we are not testing for a particular model form, such as deterministic chaos; we are attempting to distinguish data sets on the basis of regularity. Such evolving regularity can be seen in both deterministic and random (stochastic) models (4, 15, 16, 21).

Statistical Validity: Error Bars for General Processes

Ultimately, the utility of any statistic or algorithm is based on its replicability. Specifically, if a fixed physical process generates serial data, we would expect statistics of the time series to be relatively constant over time; otherwise we would have difficulty assuring that two very different statistical values implied two different systems (distinction). Here, we thus want to ascertain ApEn variation for typical processes (models), so we can distinguish data sets with high probability when ApEn values are sufficiently far apart. This is mathematically addressed by *SD* calculations of ApEn, calculated for a variety of representative models; such calculations provide error bars to quantify probability of true distinction. Via extensive Monte Carlo calculations we established that the *SD* of ApEn(2, 0.15*SD*, 1000) < 0.055 for a large class of candidate models (15, 16). It is this small *SD* of ApEn, applied to 1000 points from various models, that provides its utility to practical data analysis of moderate length time series. For instance, applying this analysis, we deduce that ApEn values that are 0.15 apart represent nearly three ApEn *SD*s, indicating true distinction with error probability nearly $p = 0.001$.

Applicability to Endocrine Hormone Secretion Data

In Pincus and Keefe (16), the potential applicability of ApEn to endocrinology was examined, to discern abnormal pulsatility in hormone secretion time-series data. We concluded that ApEn is able to discern subtle system changes and to provide insights separate from those given by a number of widely employed pulse detection algorithms (discussed further below), thus providing a complementary statistic to such algorithms. In particular, it was shown that ApEn can potentially distinguish systems given upwards of 180 data points and an intraassay coefficient of variation of 8%, suggesting ApEn is applicable to clinical hormone secretion data within the near future. We next discuss results of ApEn analyses to actual hormone secretion time-series data.

Sample Application—Growth Hormone Data

Time series of serum GH in samples were obtained at 5-min intervals for 24 hr (N = 288 points) in three groups of subjects (2): (i) (N = 16) normal young adults, both in fed and fasted (physiologically enhanced GH secretion) states; (ii) (N = 19) acromegalics with active disease; (iii) (N = 9) acromegalics in biochemical remission following therapy (transsphenoidal surgery, radiation, bromocriptine). For each subject, we calculated normalized ApEn with both (i) m = 1, r = 0.2SD and (ii) m = 2, r = 0.02SD (SD here is individual subject time-series standard deviation). As indicated above, because normalized regularity is scale invariant, this allows sensible regularity comparisons given the dissimilar mean and individual subject SD levels. We discuss the ApEn results with m = 2 here, displayed in Fig. 4; the m = 1 results are similar (2).

Acromegalics with active disease had ApEn values (0.97 ± 0.18) which

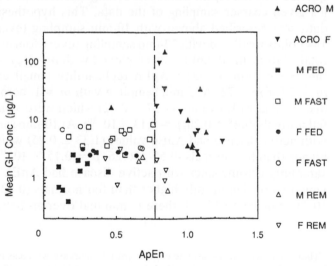

FIG. 4   Individual subject normalized ApEn (2, 0.2 SD) values versus mean GH concentrations, logarithmic scale. Symbols are indicated on the right-hand side: M and F stand for male and female, FED and FAST refer to controls, REM to remissions, and ACRO to active acromegalics. The dashed line separates all but one acromegalic with active disease from normal (both fed and fasted) subjects.

were significantly higher than normals in either the fed (0.33 ± 0.18, $p = 0.18 \times 10^{-12}$) or fasted state (0.48 ± 0.20, $p = 0.12 \times 10^{-7}$). All normal subjects, fed and fasted, had ApEn values < 0.77; all but one acromegalic patient had ApEn values above 0.77. Thus ApEn separated acromegalic from normal GH secretion with both high sensitivity (95%) and specificity (100%). Furthermore, the one acromegalic patient with a lower ApEn value was somewhat unusual in that he had McCune–Albright syndrome and was unable to undergo transsphenoidal surgery due to severe sphenoid bone hyperostosis. Acromegalic patients in biochemical remission had ApEn values (0.68 ± 0.16) that were intermediate between those of acromegalics with active disease and normal fed subjects ($p < 0.001$). Also, note that there is significant overlap between (normal) fasted state and acromegalic GH mean level individual subject distributions, which reinforces the utility of analysis beyond a first-order mean level in determining clinical state.

Full Pulse Reconstruction May Be Unnecessary to Establish Distinction

As many laboratories and/or protocols do not afford 5-min sampling, we queried whether ApEn would still distinguish normals from active acromegalics given coarser sampling of the data. This hypothesis was tested on the data sets described above, with 10-min sampling (every other point, $N = 144$ points total) and with 20-min sampling (every fourth point, $N = 72$ points total). Normalized ApEn was employed with $m = 1$, $r = 20\%$ SD here, with the expectation of better ApEn replicability (small error bars) than with $m = 2$ for $N = 72$. At 5-min sampling with $m = 1$, acromegalics with active disease had ApEn values (1.37 ± 0.32) which were significantly higher than fed normals (0.45 ± 0.20, $p = 0.13 \times 10^{-10}$). At 10-min sampling, acromegalics with active disease had ApEn values (1.33 ± 0.25) which were significantly higher than fed normals (0.55 ± 0.26, $p = 0.15 \times 10^{-10}$). Finally, at 20-min sampling, acromegalics with active disease had ApEn values (1.27 ± 0.21) which were significantly higher than fed normals (0.62 ± 0.31, $p = 0.53 \times 10^{-8}$).[2] Similar to Fig. 4, there is minimal overlap between the normal and

---

[2] There is a theoretical reason to expect ApEn to increase with less frequent sampling, as seen here for the normal subjects; analytics can be established for specified test processes to prove this. Heuristically, one expects to lose information, or predictability, with less frequent sampling, manifested in greater ApEn. The most transparent analytics, thought not at all biological, come from analyzing the function $f_a(x) = ax$ (mod 1) from the interval [0,1] to itself, for fixed $a > 1$. For instance, if $a = 2$, $f_a(2^{1/2}/2) = 2$ $(2^{1/2}/2)$ mod $1 = 2^{1/2} - 1$. These maps are called Bernoulli transformations in ergodic theory, and for $a > 1$ they have (Kolmogorov–Sinai) entropy rate log a. Halving the sampling rate would be equivalent to composing two such maps f_a back to

active acromegalic subject ApEn values, both at 10- and 20-min sampling rates. ApEn thus provides scale-independent regularity distinctions of normals from acromegalics at these coarser sampling regimens with high sensitivity and specificity.

Also, in comparing Figs. 1 and 5, which plot GH time series for three subjects at 5-min (Fig. 1) and 20-min (Fig. 5) intervals, note that Fig. 5 (particularly Fig. 5C) might confound some pulse-identification algorithms. The message is that one need not necessarily fully reconstruct pulses to quantify (relative) regularity, in order to establish high-probability distinctions on the basis of same.

Relationship to Other Approaches

Moment Statistics

ApEn is a regularity, not a magnitude statistic. Although it affords a new approach to data analysis, it does not replace moment statistics, such as the mean and *SD*. Epistemologically, ApEn addresses the change from orderly to random, not changes in average (mean) level, or the degree of spread about a central value (*SD*). As such, we recommend use of ApEn in conjunction with other statistics, not as a sole indicator of system characteristics.

Feature Recognition Algorithms

The orientation of ApEn is to quantify the amount of regularity in time-series data as a (single) number. This approach involves a different and complementary philosophy than do algorithms that search for particular pattern features in data. Representative of these latter algorithms are the pulse detection algorithms central to endocrine hormone analysis which identify the number of peaks in pulsatile data and their locations (1). When applied to clearly pulsatile data, such pulse-detection algorithms have provided significant capability in the detection of abnormal hormone secretory

---

back into a single map f_{2a}, with entropy rate log $2a$, which equals log 2 + log a. Quartering the sampling rate would transform the original map to a single map f_{4a}, with entropy rate log $4a$, which equals 2 log 2 + log a. In other words, for these pure mathematical processes, halving the sampling rate should increase the entropy by log 2, and quartering the sampling rate should increase the entropy by 2 log 2. For small ApEn values, ApEn approximately equals this entropy, so this gives the aforementioned theoretical justification for normal subject ApEn increase with less frequent sampling.

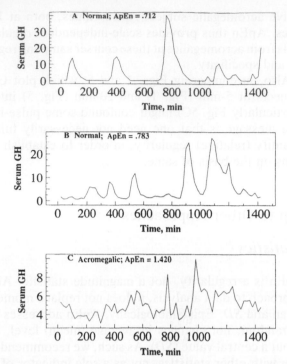

FIG. 5   Growth hormone (GH) serum concentrations for the same subjects as in Fig. 1, measured at 20-min intervals for 24 hr. (A) Normal subject, mean concentration = 4.743, ApEn(1, 0.20SD, 72) = 0.712. (B) Normal subject, mean concentration = 5.617, ApEn(1, 0.20SD, 72) = 0.783. (C) Acromegalic subject, mean concentration = 4.981, ApEn(1, 0.20SD, 72) = 1.420.

patterns. However, such algorithms often ignore secondary features whose evolution may provide substantial information. For instance, ApEn will identify changes in underlying episodic behavior that do not reflect in changes in peak occurrences or amplitudes, whereas the aforementioned pulse-identification algorithms generally ignore such information. Also, ApEn can be applied to those signals in which the notion of a particular feature, such as a pulse is not at all clear (e.g., an EEG time series). We recommend applying feature recognition algorithms in conjunction with ApEn when there is some physical basis to anticipate repetitive presence of the feature.

Statistics Related to Chaos

The historical development of mathematics to quantify regularity has centered around various types of entropy measures. Entropy is a concept addressing system randomness and predictability, with greater entropy often associated with more randomness and less system order. Unfortunately, there are numerous entropy formulations, and many entropy definitions cannot be related to one another (4). K-S entropy, developed by Kolmogorov and expanded on by Sinai, allows one to classify deterministic dynamical systems by rates of information generation (10). It is this form of entropy that algorithms such as those given by Grassberger and Procaccia (11) and by Eckmann and Ruelle (19) estimate. There has been keen interest in the development of these and related algorithms (12), since entropy has been shown to be a parameter that characterizes chaotic behavior (22).

However, the K-S entropy was not developed for statistical applications, and has major debits in this regard. The original and primary motivation for the K-S entropy was to handle a highly theoretical mathematics problem, determining when two Bernoulli shifts are isomorphic. In its proper context, this form of entropy is primarily applied by ergodic theorists to well-defined theoretical transformations, for which no noise and an infinite amount of data are standard mathematical assumptions. Attempts to utilize K-S entropy for practical data analysis represent an out-of-context application, which often generates serious difficulties, as it does here. K-S entropy is badly compromised by steady, even very small amounts of noise, generally requires a vast amount of input data to achieve convergence (17, 23), and is usually infinite for stochastic (random) processes. Hence a blind application of the K-S entropy to practical time series will only evaluate system noise, not underlying system properties. All these debits are key in the present context, as most biological time series likely are comprised of both stochastic and deterministic components.

ApEn was constructed along thematically similar lines to the K-S entropy, though with a different focus, namely, to provide a widely applicable, statistically valid formula that will distinguish data sets by a measure of regularity (4, 8). The technical observation motivating ApEn is that if joint probability measures for reconstructed dynamics which describe each of two systems are different, then their marginal probability distributions on a fixed partition, given by conditional probabilities as in Eq. (4), are likely different. We typically need orders of magnitude fewer points to accurately estimate these marginal probabilities than to accurately reconstruct the attractor measure defining the process. ApEn has several technical advantages in comparison to K-S entropy for statistical usage. ApEn is nearly unaffected by noise of magnitude below r (the filter level), gives meaningful information with a reasonable number of data points, and is finite for both stochastic and deterministic processes. This last point allows ApEn the capability to distinguish versions of composite and stochastic processes from each other, while K-S entropy would be unable to do so.

There exists an extensive literature about understanding (chaotic) deterministic dynamical systems through reconstructed dynamics. Parameters such as correlation dimension (18), K-S entropy, and the Lyapunov spectrum have been much studied, as have techniques to utilize related algorithms in the presence of noise and limited data (24–26). More recently, prediction (forecasting) techniques have been developed for chaotic systems (27–29). Most of these methods successfully employ embedding dimensions larger than $m = 2$, as is typically employed with ApEn. Thus in the purely deterministic dynamical system setting, for which these methods were developed, they are more powerful than ApEn, in that they reconstruct the probability structure of the space with greater detail. However, in the general (stochastic, especially correlated stochastic process) setting, the statistical accuracy of the aforementioned parameters and methods appears to be poor, and the prediction techniques are no longer sensibly defined. Complex, correlated stochastic and composite processes are typically not evaluated, as they are not truly chaotic systems. The relevant point here is that as dynamical mechanisms of most biological signals remain undefined, a suitable statistic of regularity for these signals must be more cautious, to accommodate general classes of processes and their much more diffuse reconstructed dynamics.

Generally, changes in ApEn agree with changes in dimension and entropy algorithms for low-dimensional, deterministic systems. The essential points here, assuring broad utility, are that (i) ApEn can potentially distinguish a wide variety of systems, such as low-dimensional deterministic systems, periodic and multiply periodic systems, high-dimensional chaotic systems, and stochastic and mixed (stochastic and deterministic) systems (4, 16); and (ii) ApEn is applicable to noisy, medium-sized data sets, such as those typically encountered in biological time-series analysis. Thus ApEn can be applied to settings for which the K-S entropy and correlation dimension are either undefined or infinite, with good replicability properties as discussed below.

Power Spectra, Phase Space Plots

Generally, smaller ApEn and greater regularity corresponds in the spectral domain to more total power concentrated in a narrow frequency range, in contrast to greater irregularity, which typically produces broader banded spectra with more power spread over a greater frequency range. The two opposing extremes are (i) periodic and linear deterministic models, which produce very peaked narrow banded spectra with low ApEn values, and (ii) sequences of independent random variables, for which time series yield intuitively highly erratic behavior, and for which spectra are very broad-

banded with high ApEn values. Intermediate to these extremes are autocorrelated processes, which can exhibit complicated spectral behavior. These
autocorrelated aperiodic processes can be either stochastic or deterministic
chaotic. In some instances, evaluation of the spectral domain will be in-
sightful, when pronounced differences occur in a particular frequency band.
In other instances, there is oftentimes more of an ensemble difference between the time series, viewed both in the time domain and in the frequency
domain, and the need remains to encapsulate the ensemble information into
a single value to distinguish the data sets.

Also, greater regularity (lower ApEn) generally corresponds to greater
ensemble correlation in phase space diagrams. Such diagrams typically display plots of some system variable $x(t)$ versus $x(t - T)$, for a fixed time-lag
T. These plots are quite voguish, in that they are often associated with claims
that correlation, in conjunction with aperiodicity, implies chaos. A cautionary
note is strongly indicated here. The labeling of bounded, aperiodic, yet
correlated output as deterministic chaos has become a false cognate. This
is incorrect; application of Theorem 6 in Ref. 5 proves that any n-dimensional
steady-state measure arising from a deterministic dynamical system model
can be approximated to arbitrary accuracy by that from a stochastic Markov
chain. This then implies that any given phase space plot could have been
generated by a (possibly correlated) stochastic model. The correlation seen
in such diagrams is typically real, as is geometric change that reflects a shift
in ensemble process autocorrelation in some comparisons. However, these
observations are entirely distinct from any claims regarding underlying model
form (chaos versus stochastic process). Similarly, in power spectra, decreasing power with increasing frequency (oftentimes labeled $1/f$ decay) is also
a property of process correlation, rather than underlying determinism or
chaos (21).

MIX(p)—A Family of Stochastic Processes
with Increasing Irregularity

Above, we indicated that statistics developed for truly chaotic settings are
often inappropriate for general time-series application. Analysis of the
MIX(p) processes (4) vividly indicates some of the difficulties realized in
applying such statistics out of context, and emphasizes the need to calibrate
statistical analysis to intuitive sensibility. MIX, a family of stochastic processes that samples a sine wave for $p = 0$, consists of independent (IID)
samples selected uniformly (completely randomly) from an interval for $p =$
1, intuitively becoming more random with increasing p. The four time series

shown in Fig. 3 represent sample realizations of MIX(p) for the indicated values of p. Formally, to define MIX(p), first fix $0 \leq p \leq 1$. For all j, define

$$X_j = 2^{1/2} \sin(2\pi j/12), \tag{5}$$
$$Y_j = \text{IID uniform random variables on } [-(3^{1/2}), 3^{1/2}], \tag{6}$$
$$Z_j = \text{IID random variables, } Z_j = 1 \text{ with probability } p,$$
$$Z_j = 0 \text{ with probability } 1 - p. \tag{7}$$

Then define

$$\text{MIX}(p)_j = (1 - Z_j)X_j + Z_jY_j. \tag{8}$$

The MIX(p) family is motivated by considering an autonomous unit that produces sinusoidal output, surrounded by a network of interacting processes that in ensemble produces output that resembles noise relative to the timing of the unit. The extent to which the surrounding world interacts with the unit could be controlled by a gateway between the two with a larger gateway admitting greater apparent noise to compete with the base signal. Note that the MIX process has mean 0 and SD 1 for all p, so these moments do not discriminate members of MIX from one another. In conjunction with intuition, ApEn monotonically increases with increasing p, as indicated in the legend of Fig. 3.

In contrast, the correlation dimension of MIX(p) = 0 for $p < 1$, and the correlation dimension of MIX(1) = ∞ (4). As well, the K-S entropy of MIX(p) = ∞ for $p > 0$, and = 0 for MIX(0). Thus both of these chaos statistics perform inadequately for this template process, even given an infinite amount of theoretical data: they do not distinguish members of MIX(p) from one another, and do not capture the evolving complexity change. We anticipate that many of the difficulties of the correlation dimension and the K-S entropy applied to MIX(p) are mirrored for other correlated stochastic processes. Stated differently, even if an infinite number of points were available, with no added noise atop MIX(p), we still could not use the correlation dimension or K-S entropy to distinguish members of this family. The difficulty here is that the parameters are identical, not that we have insufficient data or too much noise.

Mechanistic Hypothesis for Altered Regularity

It seems important to determine a unifying theme suggesting greater signal regularity in a diverse range of complicated neuroendocrine systems. We would hardly expect a single mathematical model, or even a single family of models, to govern a wide range of systems; furthermore, we would expect that *in vivo*, each physiologic signal would usually represent the output of a complex, multinodal network with both stochastic and deterministic components. Our mechanistic hypothesis is that in a variety of systems, greater regularity (lower ApEn) corresponds to greater component and subsystem autonomy. This hypothesis has been mathematically established via analysis of several very different, representational (stochastic and deterministic) mathematical model forms, conferring a robustness to model form of the hypothesis (16, 21). Restated contrapuntally, ApEn typically increases with greater system coupling and feedback, and greater external influences, thus providing an explicit barometer of autonomy in many coupled, complicated systems.

The growth hormone results above, consistent with many other endocrine hormone findings, suggest that hormone secretion pathology usually corresponds to greater signal irregularity. Accordingly, a possible mechanistic understanding of such pathology, given this hypothesis, is that healthy, normal endocrine systems function best as relatively closed, autonomous systems (marked by regularity and low ApEn values), and that accelerated feedback and too many external influences (marked by irregularity and high ApEn values) corrupt proper endocrine system function. It would be very interesting to attempt to experimentally verify this hypothesis in settings where some of the crucial network nodes and connections are known, via appropriate interventions to normal neuroendocrine flow, coupled with signal analysis at one or more output sites.

Representative Model Confirming the Hypothesis

To give an explicit sense of how this hypothesis relates network and signal, we discuss, in part, one of the models considered in Ref. 21. This model is an archetypal form of a multinodal, interconnected network, defined by a coupled linear stochastic differential equation [linear ODE driven by random process input $\xi(t)$], from which one realizes bounded, autocorrelated, yet aperiodic output. Stochastic differential equations are fundamental to a variety of settings (e.g., Einstein's modeling of molecular bombardment), digital signal processing, and system identification in neurophysiology. A typical assumption, made below, is to let $\xi(t) = W'(\sigma^2)(t)$, white noise with parameter σ^2 (Brownian motion). In particular, we want to study the effects on component output of system coupling, and analyze the following reduced version of a two-node network:

$$X''(t) + aX'(t) + bX(t) + KY(t) = W'(\sigma^2)(t),$$
$$Y''(t) + aY'(t) + bY(t) + KX(t) = W_1'(\sigma^2)(t). \tag{9}$$

Equation (9) is a coupled system of linear differential equations, with a and b constants, and $W'(\sigma^2)(t)$ and $W_1'(\sigma^2)(t)$ white noise. We distinguish W and W_1 simply to indicate that, although they have the same distribution, they are distinct processes. The K indicates the extent of the coupling by which X and Y are linked; when $K = 0$, the X and Y systems are free running, independent of one another. Increased coupling K imposes greater X contribution to the Y system, and conversely. Because we are interested in studying component behavior, we analyze, for example, $X(t)$ solutions, viewing the system from the $X(t)$ perspective. We reduce Eq. (9) to a single differential equation by exploiting symmetry: for any a, b, K, and σ, the (stationary) solutions $X(t)$ and $Y(t)$ must have identical distributions, hence the stationary X process must satisfy

$$X''(t) + aX'(t) + (b + K)X(t) = W'(\sigma^2)(t). \qquad (10)$$

Denoting $X''(t) + cX'(t) + dX(t) = W'(1)(t)$ by StoDE(c, d), the mechanistic hypothesis will be confirmed if we show that ApEn of StoDE(a, $b + K$) increases with increasing K, for fixed a and b: greater coupling will correspond to increased ApEn, decoupling and greater component autonomy will correspond to decreased ApEn. We illustrate this in Fig. 6 for $a = 1$; the procedure indicating how we numerically produced sample output time-series solutions (realizations) for StoDE(c,d) is indicated in Ref. 21. Time series of 1000 points for ApEn analysis were generated from realizations by sampling every 0.5 t units.

Figure 6a–c displays sample StoDE(1,c) realizations for three values of c, with greater apparent regularity corresponding to smaller values of c. This is quantified in Fig. 6d, with ApEn monotonically increasing as c increases, confirming the general hypothesis. The assumption of $a = 1$ for studying StoDE(a,c) evolution is purely representational; similar monotonicity of ApEn with increasing c follows for other choices of $a > 0$.

We understand the monotonicity of ApEn in Fig. 6d by analyzing the autocorrelation functions $acf(t)$ of solutions to Eq. (10) analytically (21, 30). Figure 6e shows the $acf(t)$ for StoDE(1,c) for the indicated values of c. The point is that for StoDE(a,c) with fixed $a > 0$, for any fixed, small value of t, $acf(t)$ monotonically decreases for a large range of increasing c—there is smaller autocorrelation of StoDE(a,c) at any specified lag with increasing c. This observation is again consistent with the interpretation of ApEn as an ensemble parameter of autocorrelation.

Caveat—Distinct Nodes Can Produce Widely Disparate Signals

It is important to note, however, that although greater coupling will (generally) increase signal irregularity at each system node, the signal outputs from

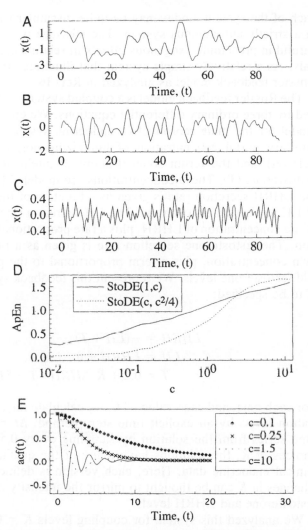

FIG. 6 Sample path time series for the stochastic differential equation $X''(t) + aX'(t) + cX(t) = W'(1)(t)$, denoted StoDE($a,c$), with $a = 1$ and (a) $c = 0.25$, (b) $c = 1.5$, or (c) $c = 10$. (d) ApEn(2, 0.2SD, 1000) versus control parameters c in StoDE(1,c) and c in StoDE($c,c^2/4$), the latter a linear-derivative term coupling analytically studied in Ref. 21. (e) Autocorrelation functions $acf(t)$ of stationary solution to StoDE(1,c) for values $c = 0.1$, $c = 0.25$, $c = 1.5$, and $c = 10$.

each of the system modes may be quite dissimilar from one another, even in a strictly autonomous system. The message is that one should be very careful in presuming exogenous effects in relating apparently disparate signals. As a relatively simple system that illustrates this point, consider the Rossler feedback model as analyzed in Ref. 16.

The Rossler feedback model is a coupled system of three variables, quantified by three ordinary differential equations. We consider this as a putative model for the male reproductive endocrine system, with variables of the pituitary portal concentration of luteinizing hormone-releasing hormone (LHRH), and the serum concentrations of luteinizing hormone (LH) and testosterone (T). These concentrations are modeled by a feedback system: the LHRH secretion rate is given as a function of the local concentrations of LH and serum testosterone. The LH secretion rate is given as a function of the concentration of LHRH, plus a rate proportional to its own concentration. The testosterone secretion rate is given as a rate proportional to its own concentration, plus a term proportional to the product of the LHRH and testosterone levels. We represent this feedback system as follows, with K to be specified:

$$
\begin{aligned}
L\dot{H}RH &= -(LH + T), \\
\dot{L}H &= LHRH + 0.2\,LH, \\
\dot{T} &= 0.2 + K(LHRH * T - 5T).
\end{aligned}
\tag{11}
$$

For each time, and each value of K, we calculate the corresponding concentration levels by an explicit time step method, $\Delta t = 0.005$. We extract a time series from the solution by sampling every $0.5\ t$ units. For suitable choices of K, the solutions have many of the qualitative features seen in clinical endocrine data. Here, each version is defined by a choice for K. Changes in K can be thought to mirror the intensity of interaction between testosterone and LHRH levels.

We analyzed this system for coupling levels $K = 0.4$, 0.7, 0.8, 0.9, and 1.0 (16). Note that we postprocessed each solution time series as follows, to ensure positive values: we converted $LHRH$ to $0.1\,LHRH + 3.0$, converted LH to $0.1\,LH + 3.0$, and converted T to $0.1\,T + 3.0$. ApEn was seen to monotonically increase with increasing K (16), confirming the mechanistic hypothesis.

Figure 7 compares the LHRH, LH, and T time series from the $K = 1.0$ version of this model, and illustrates our point. The LHRH and LH time series are visually similar; both have 16 pulses, similar amplitudes, and general pulse characteristics. One could easily believe that these hormones belonged to a common autonomous system. The behavior of T, however, is

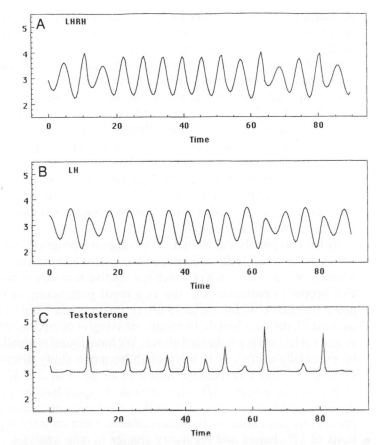

FIG. 7   Rossler coupled differential equation model, showing comparison of time series for $K = 1.0$ for (a) LHRH data, (b) LH data, and (c) testosterone data.

visually discordant with the behavior of LHRH and LH; there are 12 pulses, long stretches of flat tracings, spiked pulses, and three pulses that are much greater in amplitude than the others. We thus conclude that dissimilar pulsatile characteristics of hormonal plasma concentrations do not eliminate the possibility that the hormones may be derived from a single system, with no external influences.

Summary and Conclusion

The principal focus of this chapter has been the description of a recently introduced regularity statistic, ApEn, that quantifies the continuum from perfectly orderly to completely random in time-series data. Several properties of ApEn facilitate its utility for biological time-series analysis: (i) ApEn is nearly unaffected by noise of magnitude below a de facto specified filter level; (ii) ApEn is robust to outliers; (iii) ApEn can be applied to time series of 100 or more points, with good confidence (established by standard deviation calculations); (iv) ApEn is finite for stochastic, noisy deterministic, and composite (mixed) processes, the last of which are likely models for complicated biological systems; and (v) increasing ApEn corresponds to intuitively increasing process complexity in the settings of (iv). This applicability to medium-sized data sets and general stochastic processes is in marked contrast to capabilities of chaos algorithms such as the correlation dimension, which are properly applied to low-dimensional iterated deterministic dynamical systems. The potential uses of ApEn to provide new insights in biological settings are thus myriad, from a complementary perspective to that given by classic statistical methods.

ApEn is typically calculated by a computer program, with a FORTRAN listing for a basic code referenced above. It is imperative to view ApEn as a family of statistics, each of which is a relative measure of process regularity. For proper implementation, the two input parameters m (window length) and r (tolerance width, de facto filter) must remain fixed in all calculations, as must N, the data length, to ensure meaningful comparisons. Guidelines for m and r selection are indicated above. We have found normalized regularity to be especially useful, as in the growth hormone study discussed above; r is chosen as a fixed percentage (often 15 or 20%) of the SD of the subject, rather than of a group SD. This version of ApEn has the property that it is decorrelated from process SD, in that it remains unchanged under uniform process magnification or reduction; thus we can entirely separate the questions of SD change and regularity change in data analysis.

Regarding immediate applicability to (neuro)endocrine time series, ApEn has been shown to clearly discern abnormal pulsatility in clinical hormone secretion data, given data set lengths as small as $N = 72$ points. ApEn has been shown to provide insights separate from those given by pulse-detection algorithms, thus providing a complementary statistic to such algorithms. Finally, in part to explain the endocrine findings, we proposed a mechanistic hypothesis: in a variety of systems, greater regularity (lower ApEn) corresponds to greater component and subsystem autonomy. This hypothesis has been mathematically established via analysis of several very different, representational mathematical model forms, conferring a robustness to model form of the hypothesis. The growth hormone results above, consistent with many other endocrine hormone findings, suggest that hormone secretion pathology usually corresponds to greater signal irregularity. Accordingly, a possible mechanistic understanding of such pathology is that healthy, normal endocrine systems function best as relatively closed, autonomous systems, and that accelerated feedback and too many external influences (marked by irregularity and high ApEn values) corrupt proper endocrine system function.

References

1. R. J. Urban, W. S. Evans, A. D. Rogol, D. L. Kaiser, M. L. Johnson, and J. D. Veldhuis, *Endocr. Rev.* **9,** 3 (1988).
2. M. L. Hartman, S. M. Pincus, M. L. Johnson, D. H. Matthews, L. M. Faunt, M. L. Vance, M. O. Thorner, and J. D. Veldhuis, *J. Clin. Invest.* **94,** 1277 (1994).
3. S. M. Pincus, T. R. Cummins, and G. G. Haddad, *Am. J. Physiol.* **264** (*Regul. Integrative* **33**), R638 (1993).
4. S. M. Pincus, *Proc. Natl. Acad. Sci. U.S.A.* **88,** 2297 (1991).
5. S. M. Pincus, *Proc. Natl. Acad. Sci. U.S.A.* **89,** 4432 (1992).
6. L. A. Fleisher, S. M. Pincus, and S. H. Rosenbaum, *Anesthesiology* **78,** 683 (1993).
7. D. T. Kaplan, M. I. Furman, S. M. Pincus, S. M. Ryan, L. A. Lipsitz, and A. L. Goldberger, *Biophys. J.* **59,** 945 (1991).
8. S. M. Pincus, I. M. Gladstone, and R. A. Ehrenkranz, *J. Clin. Monit.* **7,** 335 (1991).
9. S. M. Pincus and R. R. Viscarello, *Obstet. Gynecol.* **79,** 249 (1992).
10. A. N. Kolmogorov, *Dokl. Akad. Nauk SSSR* **119,** 861 (1958).
11. P. Grassberger and I. Procaccia, *Phys. Rev. A* **28,** 2591 (1983).
12. F. Takens, *in* "Atas do 13.Col. Brasiliero de Matematicas." Rio de Janerio, 1983.
13. W. J. Parer, J. T. Parer, R. H. Holbrook, and B. S. B. Block, *Am. J. Obstet. Gynecol.* **153,** 402 (1985).
14. S. M. Pincus and A. L. Goldberger, *Am. J. Physiol.* (*Heart and Circ.*) **266(35),** H1643 (1994).
15. S. M. Pincus and W. M. Huang, *Commun. Stat. Theory Methods* **21,** 3061 (1992).
16. S. M. Pincus and D. L. Keefe, *Am. J. Physiol.* **262** (*Endocrinol. Metab.* **25**), E741 (1992).
17. A. Wolf, J. B. Swift, H. L. Swinney, and J. A. Vastano, *Physica D* **16,** 285 (1985).
18. P. Grassberger and I. Procaccia, *Physica D* **9,** 189 (1983).
19. J. P. Eckmann and D. Ruelle, *Rev. Mod. Phys.* **57,** 617 (1985).
20. V. Barnett and T. Lewis, "Outliers in Statistical Data." Wiley, New York, 1978.
21. S. M. Pincus, *Math. Biosci.* **122,** 161 (1994).
22. R. Shaw, *Z. Naturforsch. A.* **36,** 80 (1981).
23. D. S. Ornstein and B. Weiss, *Ann. Prob.* **18,** 905 (1990).
24. D. S. Broomhead and G. P. King, *Physica D* **20,** 217 (1986).
25. A. M. Fraser and H. L. Swinney, *Phys. Rev. A* **33,** 1134 (1986).
26. G. Mayer-Kress, F. E. Yates, L. Benton, M. Keidel, W. Tirsch, S. J. Poppl, and K. Geist, *Math. Biosci.* **90,** 155 (1988).
27. M. Casdagli, *Physica D* **35,** 335 (1989).
28. J. D. Farmer and J. J. Sidorowich, *Phys. Rev. Lett.* **59,** 845 (1987).
29. G. Sugihara and R. M. May, *Nature* (*London*) **344,** 734 (1990).
30. P. G. Hoel, S. C. Port, and C. J. Stone, "Introduction to Stochastic Processes." Houghton Mifflin, Boston, 1972.

[15] Methods for the Evaluation of Saltatory Growth in Infants

Michael L. Johnson and Michelle Lampl

Introduction

We have reported evidence for saltatory growth as a characteristic feature of normal human growth and development from the analysis of time-intensive data during the first 2 years of life (Lampl *et al.*, 1992). These results derive from a novel mathematical descriptor (Lampl *et al.*, 1992; Johnson, 1993; Lampl, 1993) applied to anthropometric data collected at weekly, semi-weekly, and daily intervals for durations ranging from 4 to 18 consecutive months, significantly more intensive protocols than had previously been published. Discrete growth pulses that occur episodically and aperiodically were identified to punctuate variable durations of stasis, times during which no statistically significant growth was documented. These times of stasis were not positively correlated with episodes of illness, and the total growth of the infants during the study was equal to the sum of growth events, within measurement error. A description of this model is incomplete without a concomitant explication and analysis of how the saltatory model compares mathematically with more traditional continuous models of human growth applied to the same data, and a validation of the saltatory method itself.

The analysis of serial growth data is challenging due to the nature of the data: Longitudinal growth data comprise an assessment of a biological process characterized by an increasing trend of known inter- and intraindividual variability in velocity with age-related temporal characteristics. The identification of the serial temporal changes is complicated by the serial autocorrelation in experimental observations. Each data point contains errors of measurement deriving from the equipment, the observers, the protocol for measurement involving the positioning of the subjects, and the cooperation of the subject. Attempts to separate growth process from noise must consider these features of the data.

Traditional approaches to the identification of the biology of growth as described by serial growth data involved the curvilinear smoothing of serial data points by fitting equations based on the assumption that the best fit of a continuously increasing mathematical function is a good descriptor (Johnson, 1993). A number of such equations have been fit to different data series, and it has been found that different functions best fit serial data encompassing

Methods in Neurosciences, Volume 28

different developmental ages. Karlberg (1987) proposed a three part equation to describe the changing rates of growth followed during infancy, childhood, and adolescence. Mathematically, an exponential (infancy), a polynomial (childhood), and a sigmoid (adolescent) are the bases of this model.

These previous preferred mathematical descriptors for human growth have been based on infrequently collected data (observations made at annual and semiannual intervals) in studies that were conducted over many years. When assessed at such infrequent intervals, growth appears to be a continuous function with a few rapid components. In contrast, data on total body length and height, as well as lower leg length, collected at more frequent intervals (monthly and weekly) suggested to a number of observers that simple continuous functions (both linear and nonlinear) may not best capture the processes contained in the serial data, and alternative procedures have been employed. Emphasizing that individual serial data are, in fact, a time series, formal time series analyses (Fourier methods) have been employed in the attempt to identify what appeared to be the presence of periodic processes in human growth (Togo and Togo, 1982; Hermanussen *et al.*, 1988; Hermanussen and Burmeister, 1993). One problem with the approach is that the physiological mechanisms of growth are not readily describable in terms of the sums of periodic sine and/or cosine waves (an assumption of Fourier methods). Thus, once the Fourier coefficients are evaluated they are difficult if not impossible to interpret in terms of molecular mechanisms of growth (Johnson and Lampl, 1994). Another problem with the classic time series methods is that they commonly utilize moving average boxcar filters. The results from such methods are difficult, if not impossible, to interpret because running average boxcar filters introduce extraneous harmonics into the time series. These harmonics give the impression that there are cyclic, or pulsatile, characteristics in data series that may or may not be real (Harris, 1978; Johnson, 1993; Johnson and Lampl, 1994).

Understanding the biology (molecular mechanisms) of growth requires a comprehension of changes in growth velocity and acceleration patterns for individuals. Methods aiming to describe dynamical growth patterns in sample data have been published (Gasser *et al.*, 1990). A traditional interest in auxology has been to generate growth curves that have general validity and characterize average growth to facilitate sample comparisons between groups of children. Thus, the substantial interindividual variability in growth velocity and individual peculiarities in growth dynamics have been considered to detract from the construction of average growth curves and a substantial amount of literature discusses the problems of individual growth curve characteristics in determining average growth profiles.

A considerable problem in previous attempts to analyze growth velocity changes is that it is frequently impossible to determine from published methods if the reported values are statistically significant. The analysis of serial growth increments, which is in essence growth velocity change with time, is fraught with methodological problems. The process of calculating serial growth increments (i.e., subtracting serial measurements) introduces a negative autocorrelation in the growth increments themselves (e.g., errors in a positive direction on one day will be reflected in a negative growth increment subsequently). Thus, no simple analysis of growth increments can validly represent growth.

In response to the inherent problems involved in the analysis of serial growth data when seeking to identify the biological process of growth itself, it is preferable to analyze the original series of growth data of an individual (rather than increments) by a mathematical descriptor that makes no assumptions regarding the underlying temporal structure of the process. To this end, we have focused our attention on developing and testing methods of analysis for serial growth data that aim to identify the validity of apparent discontinuities in time-intensive data series, and thus to provide further hypotheses regarding the underlying process of growth itself.

The objective of this chapter is to present a validation of the analysis methods utilized for the determination of saltatory growth in infants and to identify statistical criteria for how these methods perform. First, we describe the criteria by which the methods identify the growth of a particular infant as saltatory. This requires a comparison of the saltatory descriptors with more traditional, continuous mathematical models of growth applied to the same data series. Second, we identify the predictive validity of the models, that is, how well the proposed saltatory analytical procedures identify the saltations. In order to address these issues, Monte Carlo methods are employed to identify false positive rates of the saltatory model when applied to random number series, generated with and without saltations.

Methods

It is critical to demonstrate how the saltatory model compares as a description of the experimental data to that provided by a smooth continuous function, such as an exponential or polynomial, that are traditional descriptors of the biological process previously assumed to characterize infant growth. While it is clearly impossible to test every possible smooth continuous function for consistency with the experimental data, the most generic smooth continuous function will be employed. Polynomials are a sufficiently general form that can provide a good approximation to most other continuous functions. If the saltatory model is truly a better description of the experimental data than a continuous model, then the saltatory model should provide a significantly lower variance-of-fit. Furthermore, the ability of the saltatory analysis procedure to identify individual saltations is tested by the Monte Carlo procedure and predictive false positive rates identified.

The Saltation and Stasis Model

An outline of the saltatory analysis procedure has been previously presented as it was applied to a single individual (Johnson, 1993), but the details of the methodology are presented here for the first time. The major assumption of the saltatory model is that growth occurs as a series of distinct positive growth events separated by extended periods of stasis, during which no growth occurs. According to this model an experimental measure of growth, such as height, can be mathematically expressed as

$$Y_i = \sum_{k=1}^{i} G_k + e_k,$$ (1)

where Y_i is the observed cumulative growth on the ith day, or, height in the current example, the G_k is the amount of growth that occurs on the kth day. The G_k is positive for a measurement interval during which growth occurs (saltations), and G_k is 0 for intervals (stasis periods) during which growth does not occur. The e_k term represents the experimental uncertainty in the measurement.

The saltatory model iteratively assigns discrete growth events by three sequential steps: the evaluation of gradient, addition of a presumptive peak, and a least-squares parameter estimation to evaluate saltation amplitudes. These steps are repeated until the assignment of a growth event does not provide a significant decrease in the variance-of-fit. Subsequent to the assignment process each discrete growth event is statistically tested for significance ($p < 0.05$).

In practice, the process is as follows: Initially it is assumed that no growth events occur. The gradient of the variance with respect to a unit amount of growth during the kth measurement period is calculated as

$$\text{gradient}_k = \frac{\partial X^2}{\partial G_k}.$$ (2)

X^2 is the weighted variance-of-fit of the experimental data as determined by a least-squares parameter estimation (Johnson and Frasier, 1985) that determines the heights of each of the assigned saltations. A presumptive new saltatory event is assigned to the measurement period, k, that corresponds to the largest negative gradient as in Eq. (2). This corresponds to the point where the largest decrease in the variance-of-fit can be achieved with the smallest positive amount of growth. This procedure is analogous to a hormonal pulse detection method proposed by Munson and Rodbard (1989). The least-squares parameter estimation procedure is repeated to simultaneously evaluate the current presumptive saltation size as well as those of all previous saltations. These steps (evaluation of gradient, addition of a presumptive saltation and the new least-squares parameter estimation) are repeated until the addition of a presumptive peak does not result in a significant ($p <$ 20%) decrease in the variance-of-fit. As additional saltations are identified the apparent significance levels of the previously identified saltations change. Consequently, the use of the test for an additional term, Eq. (3), is performed at a particularly low level of significance ($p <$ 20%) so that all reasonably significant saltations will be included in the list of presumptive saltations.

The significance for including the $j + 1$ presumptive peak is determined by a test for an additional term:

$$\frac{X_j^2}{X_{j+1}^2} = 1 + \frac{2}{N - 2J - 2} F(2, N - 2J - 2, 1 - PROB). \tag{3}$$

In this equation, the J is the current number of saltatory events, N is the number of observations, and F is the F-statistic for 2 and $N - 2J - 2$ degrees of freedom. The 2s in Eq. (3) are a consequence of two degrees of freedom being lost for each of the presumptive saltations, one for the saltation height and one for the saltation position. An argument might be made that only one degree of freedom is lost for each saltation as the least-squares procedure estimates only one parameter (the saltation height) for each presumptive saltation. However, the loss of two degrees of freedom per saltation is a more conservative and correct choice and is thus less likely to find saltations that do not actually exist. We chose this conservative approach to reduce false positive identifications.

Once the list of presumptive saltations is completed each of these presumptive saltations is tested for significance a second time. That is, if there are J presumptive saltations then J additional least-squares parameter estimations are performed. Each of the presumptive saltations is removed one at a time in sequence and the amplitudes of $J - 1$ remaining saltations and the variance-of-fit are determined. The significance of each of the presumptive saltations is then evaluated in the presence of the remaining $J - 1$ saltations according to Eq. (3) with two degrees of freedom lost for each presumptive saltation. Only saltations with a significance of 5% or less are finally accepted as significant saltations. It should be noted that this procedure will generally find an apparent saltation at the very first measurement period. This corresponds to the growth that occurred before the first measurement and is not reported as a saltation.

The procedure outlined in this section was designed to answer the specific question, If intervals of saltations and stasis occur in a data series, where are they? This procedure is one of a number of mathematical descriptors of the data series, and the goodness-of-fit of this model can be judged only in a relative sense by a statistical comparison of the fit of other models of growth. Thus, we present the results of a comparative analytic approach employing previously published mathematical descriptors of growth during infancy.

Analysis of Distributions of Daily Growth Increments

An alternative method for the analysis of saltatory growth based on distribution functions of daily growth increments has recently been reported (Oerter Klein *et al.*, 1994). The basic concept of this method is that a smooth continuous growth model would predict a Gaussian-shaped distribution of daily growth increments with a mean and median greater than 0 and a skewness and kurtosis of 0. The saltatory model would predict a skewed, possibly bimodal, distribution with a median of 0, a mean greater than 0 and nonzero values for the skewness and kurtosis. The method is to evaluate the median, skewness, and kurtosis of the distribution of daily growth increments and then compare them with the values expected for smooth continuous growth models and the saltatory model. We evaluated the skewness and kurtosis with the basis univariate statistics routine (UVSTA) in the International Mathematics and Statistics Library (IMSL) (IMSL, 1989).

Other Models of Growth

Mathematical models of human growth consist predominantly of smoothly varying functions that can accommodate only very small daily changes in growth velocity. For example, the Karlberg (1987) Infancy Childhood Puberty (ICP) model as applied to infants is essentially a single exponential function that spans the first few years of life. Such a smooth continuous function is inconsistent with the episodes of saltation punctuating intervals of stasis proposed by the saltatory model. The Preece–Baines (1978) equations also model growth as a smooth continuous curve, incompatible with intervals of stasis and episodic saltations. It is impossible to test all possible smooth continuous function for consistency with the experimental data. Furthermore, no consensus exists about the best choice of mathematical function to describe the first few years of growth. Polynomials are a sufficiently general form in that they provide a good approximation to most other continuous functions. If the saltatory model is truly a better description of the experimental data than a continuous model, then the saltatory model should provide a significantly lower variance-of-fit than a sixth-degree polynomial. We arbitrarily used a sixth-degree polynomial because it can model most continuous functions with good precision and yet cannot describe saltations of growth.

The use of a sixth-degree polynomials for this test is an extremely stringent requirement. It is expected that some of the saltatory nature of the data can be accommodated by the flexibility of a sixth-degree polynomial. This would imply that both the false positive and the true positive rates for the detection of saltatory behavior by comparison to a sixth-degree polynomial would be smaller than would be found for smooth continuous curves with less flexibility. Such less flexible growth curves are often utilized. For example Karlberg (1987) modeled the growth curve during childhood as a quadratic polynomial. However, we felt that the use of the sixth-degree polynomial was important because of its lower false-positive rate.

Possibility of Saltations and Stasis Intervals within Serial Growth Data

One approach to the issue of growth is to ask, "Are the data better described by a saltatory process, or by a smooth continuous curve?" This question is addressed by analyzing the data with mathematical functions that represent both possibilities and comparing them statistically. We employed three statistical tests to ascertain which of these mathematical alternatives better represented the underlying biological process as represented in our data sets. Least-squares procedures statistically compare the fit of each model to the experimental data. The relative quality of the two fits is compared by F-test:

$$X_A^2 / X_B^2 = F(NDF_A, NDF_B, 1 - PROB). \tag{4}$$

The A and B subscripts refer to the two different models, where the model with the lowest variance-of-fit corresponds to B. In addition, we utilized other statistical procedures for the comparison of competing models by evaluating and comparing various goodness-of-fit criteria (Straume and Johnson, 1992a) concerning the residuals, namely, comparisons of the autocorrelation of the residuals and results of runs tests of the residuals. The residuals are simply the differences between the experimental observations and the function that has been fit through the data points. Tests such as autocorrelation and runs test answer the question, "Are the residuals random?" When least-squares parameter estimation procedures are correctly applied (Johnson and Frasier, 1985) and experimental uncertainties follow the correct Gaussian distributions (Johnson and Frasier, 1985) the residuals should consist of a normally distributed (Gaussian) random variable. If the residuals are not random then one or more of the assumptions of a fitted mathematical model is not valid. Thus, if the residuals are random for one model and not random for a second model, then most likely the second model does not correctly describe the central tendencies of the data and can be ruled out by comparison with the first model.

Thus, to compare the saltatory model with other models of human growth we should fit both the equations of Karlberg (1987) and Preece–Baines (1978). These models were designed to describe human growth over a period from birth to adulthood, that is, about 18 years. Our daily, semiweekly, and weekly observations of growth during the first 2 years of life span a substantially smaller range of time, typically 1 year or less. It is not possible to estimate uniquely all of the parameters of these models for our data because of the relatively limited total time range. As a consequence, we chose to compare the fit of the saltatory model of our data to a simpler smooth function, a fourth- and/or sixth-degree polynomial. Over the limited time span of the data, any relatively smoothly varying model can be adequately described by a sixth-degree polynomial. It should be noted that polynomials have been previously utilized to describe continuous growth prenatally (Jeanty *et al.*, 1984) as well as during childhood (Karlberg, 1987).

The actual polynomial that we used has a scaled time axis, ξ_i, where

$$\xi_i = 2 \frac{time_i - time_{first}}{time_{last} - time_{first}} - 1. \tag{5}$$

The *first* and *last* subscripts refer to the first and last time values. The net effect of the transformation is to force all of the scaled time values to be between $+1$ and -1. The objective in the scaling process is to minimize truncation and roundoff errors within the computer algorithms. The sixth-degree polynomial that we utilized was

$$Y_i = \sum_{j=0}^{6} A_j(\xi_i)^j. \tag{6}$$

The As are the polynomial coefficients to be estimated. The fourth-degree polynomial is analogous with the upper limit of the summation being four.

Monte Carlo Simulations

A general class of test procedures known as Monte Carlo simulations were used to test for the validity of the saltatory method (Straume and Johnson, 1992b). The Monte Carlo methods employ random numbers to simulate a large (≥ 100) series of data sets. It is critical for any Monte Carlo procedure that the simulated data be representative of the actual experimental observations, including a realistic amount of pseudorandom experimental uncertainty. These simulations include data sets with and without saltations. The analytic procedures are applied to these simulations to evaluate the performance characteristics of the procedures.

It is important to note that the results of such a validation procedure apply only to the specific set of experimental observations being simulated. There are too many variables to model all possible growth patterns. The validation procedure should be repeated for every change in the experimental protocol, such as timing of measurement assessment (in our data, weekly, semiweekly, and daily data collection protocols), age of the subject, duration of the data series, magnitude of experimental uncertainty, and genetic background of the subject. We employ several individual sets of actual experimental observations as models for the simulations.

After each set of data was simulated, pseudorandom experimental uncertainties were added. The magnitude of these uncertainties corresponded to the variance-of-fit based on the saltatory model analysis of each data set. This variance-of-fit includes positive contributions from the actual measurement error, the variations due the variability of the infant, and the possibility of variation due to a less than perfect fit of the experimental observations.

In our Monte Carlo analysis, we modeled the experimental data series for two infants measured daily during 4 consecutive months and one infant measured semiweekly for 18 consecutive months. The protocols for data acquisition have been previously published (Lampl *et al.*, 1992; Lampl, 1993).

One series of simulations was designed to answer the question, "Can saltatory growth be distinguished from smooth continuous growth?" To address this question an actual series of experimental observations is analyzed according to the saltatory model and also analyzed as a sixth-degree polynomial. The results of these analyses provide two simulated growth curves that closely match the experimental observations but without experimental uncertainties, one with saltations and one smooth continuous curve. These analyses also provide two sets of residuals that correspond to the distribution of experimental uncertainties that exist in the actual experimental observations and assuming the particular growth model. The Monte Carlo approach can be applied to generate a series (≥ 100) of realistic smooth continuous curves by repeatedly shuffling the appropriate series of residuals and adding them to the calculated smooth continuous curve evaluated above. A series of realistic saltatory data sets is also generated by shuffling the appropriate residuals and adding them to the above calculated saltatory curve. These two series of data sets are then analyzed by various methods. Ideally the analysis method should always indicate that the simulated saltatory data are saltatory and the smooth simulated data are not saltatory. The true positive rate from this analysis is the percentage of the simulated saltatory data sets found to be saltatory. The other performance statistics are evaluated analogously.

The derived true-positive, etc., rates derived from a Monte Carlo simulation of this type are critically dependent on the assumed form of the smooth continuous function. What is actually derived by this simulation is the true-positive, etc., rate for deciding if data is saltatory as compared to a sixth-degree polynomial. As noted earlier, polynomials can be utilized as a reasonable approximation of a wide variety of the other smooth continues curves.

A second series of simulations is designed to answer the question, "If the data are saltatory how well can the locations of the saltations be identified?" To address this question, 100 sets of simulated data are generated and analyzed. These generated data are a close approximation to the actual experimental observations. The number of simulated daily measurements is the same as the experimental observations, however the locations of the simulated saltations were randomly assigned. The performance of the analysis procedures will not be markedly altered by a nonrandom placement of the saltations (the actual data series). The amplitude of the simulated saltations are fixed at the total observed growth during the interval divided by the number of simulated saltations. Although assigning the saltations a constant amplitude is not consistent with the actual experimental observations it will not alter the conclusions about the overall performance of the algorithms. The true-positive rate is the fraction of the simulated saltations that are correctly located. The false-positive rate is the fraction of identified saltations within the simulated data that do not correspond to a simulated saltation.

Random Number Generator

The validity of the Monte Carlo approach is critically dependent on the quality of the series of Gaussian pseudorandom numbers. Thus it should be noted that we generated Gaussian pseudorandom numbers as the sum of 12 evenly distributed pseudorandom numbers in the range ±0.5. This procedure provides a Gaussian distribution with a mean of 0 and a standard deviation of 1. These numbers can then be scaled to any desired standard deviation and mean. The evenly distributed pseudorandom numbers were evaluated by the procedure RAN3 (Press *et al.*, 1986).

Results

Saltatory versus Continuous Models Compared with Experimental Observations

The results of analysis of the experimental data by the saltatory model and the continuous model are presented first. The data shown in the top graphs of Figs. 1 and 2 are daily measurements of the length of a normal infant female approximately 1 year of age. An analysis according to the saltatory model indicated that 15 significant ($p < 0.05$) saltatory events are evident. These saltations range in size from 0.21 cm to 0.80 cm (mean = 0.543, SD = 0.193, median = 0.589 cm). The solid line in the top graph of Fig. 1 is the calculated saltatory pattern and the solid line in the top graph of Fig. 2 is the calculated sixth-degree polynomial least-squares fit to the same data.

The middle graph of Fig. 1 presents the residuals of the fit of the saltatory model to these data. The observed number of runs in these residuals is 62 and the expected number of runs is 59.0 ± 5.3. Clearly the runs test does not demonstrate that the residuals do not follow a Gaussian distribution. Therefore, the runs test indicates Gaussian distributed residuals and thus that the saltatory model is consistent with the experimental observations.

The corresponding residuals for the same data fit to a sixth-degree polynomial are shown in the middle graph of Fig. 2. The magnitude of these residuals is significantly ($p < 1\%$) greater than the residuals from the saltatory fit (middle graph of Fig. 1). Thus, the saltatory model is a better description of the experimental observations than is a sixth-degree polynomial. The observed number of runs is 67 with an expected number of runs of 60.3 ± 5.4. While the runs test for the residuals of the saltatory model is better than the runs test for the corresponding residuals from the sixth-degree polynomial, the difference is not statistically significant.

The bottom graphs of Figs. 1 and 2 present the corresponding autocorrelations of the residuals: The autocorrelation analysis does not show any significantly nonzero autocorrelations for the saltatory model, Fig. 1, but does show significant deviations for the sixth-degree polynomial analysis, Fig. 2. The autocorrelations from the saltatory analysis, Fig. 1, fail to demonstrate that the residuals do not follow a Gaussian distribution. Thus, the saltatory

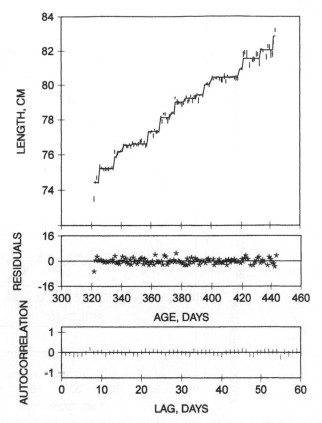

FIG. 1   Saltatory analysis of daily observations of stature for an infant female between the ages of 321 and 443 days. The data were collected as described elsewhere (Lampl *et al.*, 1992; Lampl, 1993). (Top) Experimental observations with measurement uncertainties. The solid line results from an analysis of the data according to the saltatory model. (Middle) Weighted residuals of the fit. (Bottom) Autocorrelation function of the weighted residuals.

model is consistent with the experimental observations. Conversely, the autocorrelations from the smooth continuous polynomial analysis, Fig. 2, demonstrates that the residuals do not follow a Gaussian distribution. Thus, the smooth polynomial model is not consistent with the experimental observations.

If the saltatory model is truly a better description of the experimental data than a continuous model, then the saltatory model will provide a significantly lower variance-of-fit. For the data presented in Figs. 1 and 2, the overall variance-of-fit provided by the saltatory analysis corresponded to an average

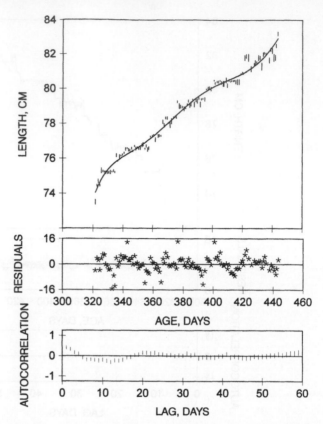

FIG. 2   The analysis of the same data as in Fig. 1 with a sixth-degree polynomial growth function. (Top) Experimental observations with measurement uncertainties. The solid line results from an analysis of the data according to a sixth-degree polynomial. (Middle) Weighted residuals of the fit. (Bottom) Autocorrelation function of the weighted residuals.

measurement error of 0.156 cm. It should be noted that this uncertainty level is approximately twice the magnitude of the actual measurement error of the experimental observations (Lampl, 1993). The observed ratio for the variance-of-fit of the two models, the saltatory model versus a continuous mathematical function, is 5.77 with 112 degrees of freedom in the numerator and 85 degrees of freedom in the denominator (variance-of-fit for the sixth-degree polynomial divided by the variance-of-fit for the saltatory model). This observed variance ratio corresponds to a very small probability ($p < 1\%$) that the sixth-degree polynomial provides a description of experimental observations that is as good as, or better than, the saltatory model.

For the experimental observations shown in the upper graphs of Figs. 1 and 2 the distribution of daily growth increments statistics are: median 0.025, mean 0.0873, skewness 1.265, and kurtosis 1.653. It appears that the skewness and kurtosis indicate a saltatory model, but without a knowledge of the expected ranges of these values as predicted for the saltatory and smooth continuous models no conclusions can be made.

Performance Characteristics of Saltatory Analysis Methods

First we address the question, Can saltatory growth be distinguished from smooth continuous growth? As previously outlined this question is addressed by a Monte Carlo simulation. For this, two series of simulated sets of data are generated. One series of simulated data sets are based on the saltatory solid curve in the top graph of Fig. 1 and the other based on the smooth continuous solid curve in the top graph of Fig. 2. For the present example each series contained 200 simulated data sets. All 200 simulated saltatory data sets were described significantly ($p < 0.01$) better by the saltatory model as compared to the sixth-degree polynomial when our saltatory analysis method was used. Our analysis method indicated that all 200 simulated polynomial data sets were consistent with either the saltatory model or the polynomial model. The results were ambiguous and the models could not be distinguished with the polynomial simulated growth curves. This indicates that the true-positive rate for identifying the existence of saltatory growth for our saltatory method is approximately 100% and that the false positive rate for identifying the existence of saltatory growth for our saltatory method is approximately 0%. This is consistent with our conservative statistical tests as outlined in the methods section. However, please note that these true-positive and false-positive rates apply only to the specific data set being simulated. To reach a general conclusion by this method an infinite number of growth curves would need to be simulated.

An alternative approach that has been proposed to validate the analysis of saltatory growth is to examine the distribution of growth increments. It might be expected that the concept of saltations and stasis would predict that a distribution of growth increments would be bimodal with one large peak having a mean of 0 corresponding to the periods of stasis and a second smaller peak having a mean greater than 0 that corresponds to the saltations (Oerter Klein *et al.*, 1994).

To evaluate the feasibility of such an analysis protocol we evaluated the expected distributions of daily growth increments for the data presented in Fig. 1 and in Lampl *et al.* (1992) in the presence of observational uncertainties. For these calculations the observed saltatory patterns (e.g., solid line in Fig.

1 and the relevant figure of Lampl *et al.*, 1992) were used as known saltatory patterns. A realistic amount of Gaussian distributed pseudorandom noise was added to the known saltatory patterns and the distribution of daily growth increments was evaluated. The noise levels are given in Tables II and III. This process was repeated until 10^5 growth increments were accumulated to generate a relatively smooth distribution function. The top graph of Fig. 3 corresponds to the expected distribution for the data presented in Fig. 1. The bottom graph is the analogous distribution for the observations presented in Lampl *et al.* (1992). It is clear from these simulations that the distribution is expected to be skewed and non-Gaussian. However, in neither

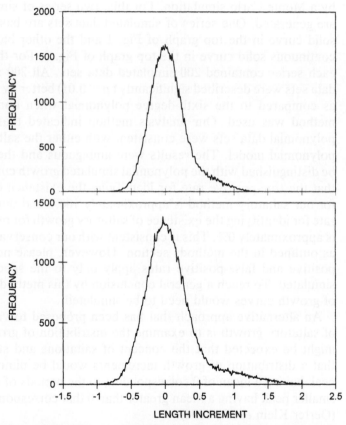

FIG. 3   Expected distributions of daily growth increments based on the saltatory analysis results. (Top) Distribution for the data shown in Fig. 1. (Bottom) Distribution for the data shown in Fig. 1 of Lampl *et al.* (1992).

case is the observed distribution of daily growth increments bimodal as was postulated by Oerter Klein *et al.* (1994).

There are three reasons that the distributions in Fig. 3 are not bimodal. First, because all of the observed saltations are not the same size, the smaller peak of the distribution is expected to be spread. Second, the presence of observational uncertainties will, of course, significantly spread both peaks. Third, the peak corresponding to the saltations will be significantly smaller than the peak corresponding to the stasis periods. The combination of these three generates a skewed unimodal distribution. Thus, the presence of a unimodal distribution that is skewed toward positive daily growth increments is consistent with the saltatory model. Furthermore, a bimodal distribution of daily growth increments is clearly not a requirement of saltatory growth.

Conversely, the concept of smooth continuous growth predicts that the distribution function should be unimodal, Gaussian in shape, and a maximum of the distribution that is slightly positive approximately reflecting the daily growth rate. The mean of the two distributions (saltatory and continuous) is not a viable method for distinguishing the two models as both models predict a distribution mean that is slightly positive.

The same simulations used to test our saltatory method (outlined above) can also be used to test the significance of the median, mean, skewness, and kurtosis which are the basis of the distribution of growth increments method for the detection of saltatory growth. For this the median, mean, skewness, and kurtosis are evaluated for each of the simulated data sets. For the 200 simulated saltatory data sets the average skewness was 0.702 ± 0.314 *(SD)* and for the 200 simulated polynomial data sets the average skewness was -0.003 ± 0.265 *(SD)*. Thus the skewness of 1.265 for the actual experimental observations indicates that the growth curve is saltatory ($p < 0.01$).

A detailed analysis of performance of the distribution of growth increments method was also performed, using the skewness values as an example. The true-positive rate for the detection of saltatory behavior is calculated as the percentage of skewness values from the simulated saltatory data sets that fall within ($p < 0.05$) the distribution of the simulated saltatory data sets but not within ($p < 0.05$) the distribution of the simulated polynomial data sets: in the present case 70.5%. The false-positive rate for the detection of saltatory behavior is calculated as the percentage of skewness values from the simulated polynomial data sets that fall within ($p < 0.05$) the distribution of the simulated saltatory data sets but not within ($p < 0.05$) the distribution of the simulated polynomial data sets: in the present case 1.5%. The false-negative rate for the detection of saltatory behavior is calculated as the percentage of skewness values from the simulated saltatory data sets that fall within ($p < 0.05$) the distribution of the simulated polynomial data sets but not within ($p < 0.05$) the distribution of the simulated saltatory data sets: in the present case 2.5%.

The true-negative rate for the detection of saltatory behavior is calculated as the percentage of skewness values from the simulated polynomial data sets that fall within ($p < 0.05$) the distribution of the simulated polynomial data sets but not within ($p < 0.05$) the distribution of the simulated saltatory data sets: in the present case 61%. The ambiguous rate for the saltatory data is the percent of the skewness values from the simulated saltatory data that fall in both or neither distribution ($p < 0.05$): in the present case 27%. The ambiguous rate for the polynomial data is the percent of the skewness values from the simulated polynomial data that fall in both or neither distribution ($p < 0.05$): in the present case 37.5%. The results for the median, mean, skewness, and kurtosis are presented in Table I. Clearly, distribution functions are not a sensitive diagnostic of saltatory growth, for this example.

A standard recommendation for any analysis of experimental data is that the analysis should be performed on the original untransformed data (Johnson and Frasier, 1985). The origin of the difficulty in the use of distribution functions may arise from the analysis of the derived distribution function instead of the original length data. The saltatory model has two basic concepts, the saltations and the periods of stasis between the saltations. The distribution function clearly preserves the information about the magnitudes of the saltations, but it removes the temporal information pertaining to the sequence of the growth increments. Thus, it contains no information about the stasis periods. Consider the 119 data points shown in Fig. 1. This data set contains 118 growth increments and the corresponding distribution function can easily be generated. However, given the distribution function it is impossible to recreate the same unique growth pattern. The reason being that the order of the growth increments is not included in the distribution function. As the growth increments could occur in any order there are 118 factorial, or around 5×10^{194} different growth patterns that will generate precisely the same distribution function. Clearly, the distribution function

TABLE I   Performance of Distribution Method

	Median	Mean	Skewness	Kurtosis
True positive	20.5	0.0	70.5	56.0
False negative	3.0	4.5	2.5	0.0
False positive	1.5	0.0	1.5	3.5
True negative	78.0	28.0	61.0	20.0
Ambiguous saltatory	76.5	95.5	27.0	44.0
Ambiguous polynomial	20.5	72.0	37.5	76.5

contains less information than the original data and thus the distribution should not be used for analysis. For comparison, Fig. 4 presents two addition growth functions that correspond to the same distribution function as the data in Fig. 1. However, these two growth functions are not saltatory by our analysis procedure or by visual inspection, yet they both have a skewness of 1.265.

Having demonstrated that our saltatory method can distinguish saltatory growth from smooth continuous growth we can now address the question, If the data are saltatory, how well can the locations of the saltations be identified? The ability of the saltatory analysis procedure to identify saltations was tested by another Monte Carlo procedure. In this procedure data sets are simulated with saltations at known positions and our saltatory method is used to identify the locations of the saltations. Table II presents the results from a simulation of 15 saltations per data set and a noise level of 0.156 cm. This uncertainty level was a conservative choice because it exceeds the

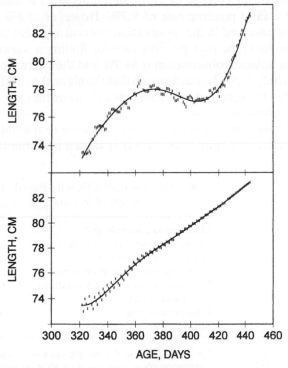

FIG. 4   Two possible growth patterns that will generate the same distribution of daily growth increments as the experimental data presented in Fig. 1.

TABLE II   Simulated Results Based on Data
Shown in Figure 1

SD of added uncertainties[a]	0.156 cm
Simulated number of saltations	15
Number of observations	119
Growth during observation period	9.7 cm
Mean number of observed saltations	12.78[b]
True-positive rate	0.769
False-positive rate	0.097
Adjacent true-positive rate	0.078

[a] It should be noted that the simulated uncertainty level is approximately twice the magnitude of the measurement error of the experimental observations (Lampl, 1993).
[b] $SD = 1.34$, $N = 100$.

actual technical error of the measurements (Lampl, 1993). This simulation mimics the data shown in Fig. 1. For these parameters, the analysis procedure reported an average of 12.78 saltations with a true positive rate of 76.9% and a false positive rate of 9.7%. However, 7.8% of these false positives were observed in the observation interval adjacent to the simulated position. Therefore, the true positive rate for finding a saltation correctly placed or in an adjacent observation is 84.7% and the corresponding false positive rate was only 1.9%. It thus appears that the algorithm itself finds saltations present within the data and rarely reports saltations that are not present, for the data set shown in Fig. 1.

An analogous simulation corresponding to the data which were presented in Fig. 1 of Lampl *et al.* (1992) is shown in Table III. These data are daily

TABLE III   Simulated Results Based on Data
Presented in Lampl *et al.* (1992)

SD of added uncertainties[a]	0.191 cm
Simulated number of saltations	13
Number of observations	118
Growth during observation period	12.6 cm
Mean number of observed saltations	12.28[b]
True-positive rate	0.866
False-positive rate	0.083
Adjacent true-positive rate	0.070

[a] It should be noted that the simulated uncertainty level is approximately twice the magnitude of the measurement error of the experimental observations (Lampl, 1993).
[b] $SD = 10.96$ $N = 100$.

assessments of total body length in an infant over 4 months old, between 5 and 9 months of age. By comparison with the first infant described above, 13 instead of 15 saltations are simulated with 12.6 cm of growth instead of 9.7 cm, and a slightly higher *SD* of added uncertainties. These apparently small changes in terms of amount of growth and frequency of growth events resulted in profound changes in the validation statistics of the algorithm. The true-positive rate increased to either 86.6%, or, if adjacent saltations are accepted, 93.6%. The false-positive rate decreased to 8.3%, or, if adjacent saltations are accepted, 1.3%. Furthermore, a larger percentage of the simulated saltations were located in the second simulation. Clearly, the 50% increase in the size of the simulated saltation markedly improves the ability of the algorithm to locate saltations.

It is also of interest to test the algorithm with simulated data that do not contain saltations, that is, simulated data for a smooth continuous function. Clearly, if saltations are not present, the algorithm should not indicate that saltations are present. To this end, 100 sets of simulated data are generated where the simulated function is a straight line (i.e., no saltations) and with an experimental uncertainty level equal to that shown in Table III. It should be noted that this simulated uncertainty level is approximately twice the actual technical measurement error of the experimental observations (Lampl, 1993). Such data sets are exactly equivalent to data sets where there is a small saltation at every observation point and no stasis intervals. The computer algorithms are robust enough that the saltatory analysis procedure will model the straight line as a series of saltations and stasis intervals, some of which may appear to be significant. However, the most important result of this simulation is the comparison of the variance-of-fit for the polynomial compared to the saltatory model. For this simulation, the average observed variance ratio was 0.95 for both the fourth- and sixth-degree polynomials as compared to the saltatory model. The fourth- and sixth-degree polynomials both fit the data better than the saltatory model. Thus, it cannot be concluded that the saltatory model is better for this simulated data series; and, it cannot be concluded that saltations exist for these simulated data. It also cannot be concluded that the polynomials provide a significantly better description of the data than the saltatory model based on the better variance-of-fit. They provide nondistinguishable equivalent descriptions of the data.

All of the simulations to this point have been for data sets with daily observations of the length of the subject. Table IV presents an analogous analysis for semiweekly measurements of a female infant between ages 3 and 530 days, one of the longest data sets in our sample. For this data set the variance ratio of the fourth-degree polynomial compared to the saltatory model is 2.714 and of the sixth-degree polynomial is 2.73. Both variance ratios are substantially larger than the 1.63 variance ratio corresponding to

TABLE IV   Simulated Results Based on a
Typical Data Set with Weekly
Observations

SD of added uncertainties[a]	0.257 cm
Simulated number of saltations	21
Number of observations	134
Growth during observation period	30.7 cm
Mean number of observed saltations	19.91[b]
True-positive rate	0.900
False-positive rate	0.051
Adjacent true-positive rate	0.041

[a] It should be noted that the simulated uncertainty level is
approximately twice the magnitude of the measurement error
of the experimental observations (Lampl, 1993).
[b] $SD = 1.58$, $N = 100$.

a 1% confidence indicating that these data are distinctly saltatory ($p < 1\%$). This simulation clearly demonstrates that the saltatory model and algorithm work well under these observation conditions: The true-positive rate for the saltatory model was 0.900 and the false-positive rate was 0.051. If adjacent saltations are accepted, the true-positive rate was 0.941 and the false-positive rate was 0.010.

Two conclusions are abundantly clear from the above simulations. First, our saltatory analysis protocol performs very well and is extremely unlikely to identify saltatory growth that does not exist. Second, the use of distribution functions of growth increments is not recommended as it is shown to be a nonspecific and insensitive method.

Discussion

We have outlined the methodology by which we are analyzing time-intensive growth data collected on a sample of normal human infants. The frequency of temporal assessment of these infants provides data that are challenging in terms of traditional approaches to the interpretation and modeling of human growth, as they have been primarily aimed at describing the average growth of groups of children through curvilinear modeling.

We are presently engaged in developing and testing the validity of novel mathematical techniques by which to further characterize the biological processes involved in human growth as it occurs in individuals. The data base for such investigations is necessarily presently constrained to noninvasive technology that can be repeatably applied to individual subjects over time. The importance of valid data acquisition techniques that are independently validated prior to their employment in longitudinal studies must be emphasized. Attention to observer training, assessment of measurement error, equipment and positioning protocol, diurnal variability in measurements due to physiology, and reliability/validity assessments must be carefully designed in order to interpret serial dynamical processes.

From a historical perspective, we are particularly interested in the peculiarities of individual growth dynamics for what these may tell us about more general mechanisms of growth itself in a time when molecular biology and cell cycle research has provided us with substantially more detail on potential control mechanisms. We are investigating the utility of pulse identification methods originally designed for the description of hormonal physiological systems for what they may contribute to the characterization of individual growth which by now has been noted to be distinctly pulsatile by a number of observers (Veldhuis and Johnson, 1992; Johnson and Veldhuis, 1995). Our own data go further and suggest that there may be not only merely nonlinear dynamics involved in the growth process, but also distinctly discontinuous processes: saltatory growth events aperiodically punctuating stasis intervals. It can be speculated that such a process reflects a putative, but as yet unknown interface between hormonal pulses and the cell cycle itself, which has been clearly documented to be discontinuous. Thus, advances in cell cycle controls may not only be essential to our understanding of aberrant growth, but be at the very basis of normal organismic growth during development. It is not surprising that the hormonal processes contributing to these processes are as yet unknown, considering the relatively new appreciation of cell cycle control mechanisms themselves.

We believe that further collaboration between diverse disciplines can be most productive in raising questions concerning basic processes in human biology and in generating new paradigms for research in developmental processes.

Acknowledgments

The authors acknowledge the support of the National Science Foundation Science and Technology Center for Biological Timing at the University of Virginia, the Diabetes Endocrinology Research Center at the University of Virginia NIH DK-38942, the University of Maryland at Baltimore Center for Fluorescence Spectroscopy NIH RR-08119, and National Institutes of Health Grant GM-35154.

References

T. Gasser, A. Kneip, P. Ziegler, R. Largo, and A. Prader (1990). A method for determining the dynamics and intensity of average growth. *Ann. Hum. Biol.* **13,** 129–141.

F. J. Harris (1978). On the use of windows for harmonic analysis with the discrete Fourier transform. *Proc. IEEE* **66,** 51–83.

M. Hermanussen and J. Burmeister (1993). Children do not grow continuously but in spurts. *Am. J. Hum. Biol.* **5,** 615–622.

M. Hermanussen, K. Geiger-Benoit, J. Burmeister, and W. G. Sippell, (1988). Periodical changes of short term growth velocity ("mini growth spurts") in human growth. *Ann. Hum. Biol.* **15,** 103–109.

International Mathematics and Statistics Library (1989). "User's Manual IMSL STAT/LIBRARY Version 1.1." IMSL, Houston, Texas.

P. Jeanty, E. Cousaert, F. Cantraine, J. C. Hobbins, B. Tack, and J. Struyven (1984). A longitudinal study of fetal limb growth. *Am. J. Perinatol.* **1,** 136–144.

M. L. Johnson (1993). Analysis of serial growth data. *Am. J. Hum. Biol.* **5,** 633–640.

M. L. Johnson and S. G. Frasier (1985). Nonlinear least-squares analysis. *In* "Methods in Enzymology" (C. H. W. Hirs and S. N. Timasheff, eds.), Vol. 117, pp. 301–342. Academic Press, San Diego.

M. L. Johnson and M. Lampl (1994). Artifacts of fourier series analysis. *In* "Methods in Enzymology" (M. L. Johnson and L. Brand, eds.), Vol. 240, pp. 51–68. Academic Press, San Diego.

M. L. Johnson and J. D. Veldhuis (1995). Evolution of deconvolution analysis as a hormone pulse detection method. this volume, Chapter 1.

J. Karlberg (1987). On the modelling of human growth. *Stat. Med.* **6,** 185–192.

M. Lampl (1993). Evidence for saltatory growth in infancy. *Am. J. Hum. Biol.* **5,** 641–652.

M. Lampl, J. D. Veldhuis, and M. L. Johnson (1992). Saltation and stasis: A Model of human growth. *Science* **258,** 801–803.

P. J. Munson and D. Rodbard (1989). Pulse detection in hormone data: Simplified, efficient algorithm. *Proceedings of the Statistical Computing Section of the American Statistical Association*, pp. 295–300. American Statistical Association, Alexandria, Virginia.

K. Oerter Klein, P. J. Munson, J. D. Bacher, G. B. Cutler, Jr., and J. Baron (1994). *Endocrinology (Baltimore)* **134,** 1317–1320.

M. A. Preece and M. J. Baines (1978). A new family of mathematical models describing the human growth curve. *Ann. Hum. Biol.* **5,** 1–24.

W. H. Press, B. P. Flannery, S. A. Teukolsky, and W. T. Vetterling (1986). "Numerical Recipes: The Art of Scientific Computing." Cambridge Univ. Press, Cambridge.

M. Straume and M. L. Johnson (1992a). Analysis of residuals: criteria for determining goodness-of-fit. *In* "Methods in Enzymology" (L. Brand and M. L. Johnson, eds.), Vol. 210, pp. 87–105. Academic Press, San Diego.

M. Straume and M. L. Johnson (1992b). Monte Carlo method for determining complete confidence probability distributions of estimated parameters. *In* "Methods in Enzymology" (L. Brand and M. L. Johnson, eds.), Vol. 210, pp. 117–129. Academic Press, San Diego.

M. Togo and T. Togo (1982). Time-series analysis of stature and body weight in five siblings. *Ann. Hum. Biol.* **9,** 425–440.

J. D. Veldhuis and M. L. Johnson (1992). Deconvolution analysis of hormone data. *In* "Methods in Enzymology" (L. Brand and M. L. Johnson, eds.), Vol. 210, pp. 539–575. Academic Press, San Diego.

[16] Analysis of Calcium Fertilization Transients in Mouse Oocytes

W. Otto Friesen, Timothy R. Cheek, Orla M. McGuinness, Roger B. Moreton, and Michael J. Berridge

Introduction

Alterations in the concentration of free calcium ions inside cells is one means of regulating many cellular activities, from fertilization in oocytes to the secretion of neurotransmitters. Because of the central role of intracellular calcium ions in controlling cell function, it is not surprising that calcium concentrations are under the control of multiple regulatory mechanisms, including membrane calcium channels and calcium pumps. These mechanisms depend on multiple negative feedback loops to maintain homeostasis in cytosolic calcium concentrations. Because systems with delayed, negative feedback tend to give rise to oscillations (1, 2), one might expect cytosolic calcium concentrations to be unstable. In fact, oscillations in calcium concentrations now have been observed in many cell types, including gonadotropes, hepatocytes, and both invertebrate and vertebrate oocytes. The waveforms range from sinusoidal oscillations to widely spaced, spikelike, calcium concentration transients (3–5).

The specific mechanisms by which fusion of the sperm and oocyte lead to the generation of calcium spikes are unknown; however, it is generally recognized that the spikes are caused by the rapid release of calcium ions from internal stores. Release from these stores occurs not only as a consequence of fertilization, but can also be induced by injection of inositol 1,4,5-trisphosphate (IP_3) (6–9), application of thimerosal, a sulfhydryl reagent (10, 11), and by injection of calcium itself (6, 8). Injection of a sperm factor from hamster and boar also induces calcium oscillations in mature mouse oocytes (8). Moreover, Miyazaki et al. (12) showed that local injection of IP_3 (13) generates a calcium wave in hamster oocytes that spreads across the egg in about 2 sec, equivalent to the transit times for calcium waves observed in mammalian oocytes following fertilization. They also found that injection of calcium ions was as effective as IP_3 in eliciting calcium spikes in mouse eggs.

A reasonable explanation for these results is that IP_3 is generated during the fertilization process. Release of calcium ions is mediated by an IP_3 receptor that is sensitive to both Ca^{2+} and IP_3 (14, 15). This model is supported

Methods in Neurosciences, Volume 28

by Miyazaki *et al.* (16), who suggest that calcium-induced calcium release (CICR; Refs. 17 and 18) may be indistinguishable from calcium-sensitized, IP_3-induced calcium release (IICR). They found that both CICR and IICR are enhanced by thimerosal. Moreover, injection of a monoclonal antibody (MAb) to the IP_3 receptor blocked calcium spikes otherwise obtained by either IP_3 or Ca^{2+} injection (12). Efflux from the stores is likely to occur via channels formed by the IP_3 receptor. Because of the dual control of this receptor, either an increase in the concentration of calcium ions or an increase in the intracellular IP_3 concentration could trigger the release of calcium from internal stores.

At least four distinct quantitative models for mechanisms generating intracellular calcium oscillations are found in the scientific literature (4, 17, 19–21). All assume that both IP_3 and Ca^{2+} concentrations play important roles in generating the oscillations in cytosolic Ca^{2+} concentration. The models differ in that some assume a positive feedback role for the action of Ca^{2+}, whereas others assume a negative one. They also differ in the specific mechanisms involved in Ca^{2+} release from intracellular stores.

Berridge and collaborators (5, 22, 23) have described and evaluated a conceptual model based on the assumption that two intracellular pools of Ca^{2+} exist: an IP_3-sensitive pool, which provides a steady influx of Ca^{2+} on fertilization, and an IP_3-insensitive pool. This two-pool model does not assume oscillations in the IP_3 concentration and includes a CICR mechanism for the release of Ca^{2+} from the IP_3-insensitive pool. It is therefore a positive feedback model in that an increase in cytosolic calcium engenders additional release from vesicular stores. Negative feedback is provided by calcium pumps in the stores membrane, which fill and refill the stores, and in the plasma membrane, to return the cytosolic calcium concentration to baseline following calcium spikes.

The simulations described here extend this conceptual model, which was previously explored through computer simulations by Goldbeter and colleagues (22, 23). Our model, like the Goldbeter simulations, is based on the filling and emptying of calcium stores, but differs from it in several respects. First, our model is scaled specifically for the size of the mouse oocyte. Therefore, all fluxes predicted by model output are directly comparable to fluxes measurable in mouse oocytes. Second, the equation for efflux of calcium ions via CICR channels has been modified so that the conductance of these channels is independent of the calcium concentration of the deep stores. The rate of calcium efflux in our model is a function of the permeability of CICR channels, which depends solely on the concentration of calcium in the cytosol, and on the difference in the calcium concentrations in the deep stores and the cytosol. Third, the CICR channel is assumed to be a calcium/ IP_3 receptor whose sensitivity is increased by the sulfhydryl agent thimerosal.

Finally, our model explicitly includes the buffering capacities of both the cytosol and the deep stores.

Both the model and the results of our experiments are presented in detail below. This model is plausible in that it generates traveling calcium transients that resemble closely those observed in mouse eggs. Further exploration of model properties leads to several predictions described below. Our most interesting result is that changes in the relative buffering capacities of the cytosol and deep stores alter the waveforms of calcium transients from smooth sinusoidal oscillations to sharp concentration spikes.

Physiological Results

Calcium Spikes Induced by Fertilization in Mouse Oocytes

Methods for measuring of the calcium concentrations in mouse eggs have been described previously (11). Briefly, freshly ovulated oocytes (about 60 μm in diameter) were loaded with fura-2, exposed to spermatozoa, and then monitored photometrically while alternately illuminated with 340- and 380-nm light. Fura-2 fluorescence was recorded on videotape and subsequently analyzed to obtain average calcium concentrations throughout the oocyte or in five optical slices made at right angles to a line extending from the site of fertilization to the opposite pole of the egg. These photometric techniques generate 4-μm optical sections of the spherical mouse egg. The calcium transients described below were obtained in one of a series of experiments described previously (11).

Examination of false-color representations of fertilization in mouse oocytes reveals stereotypical alterations in calcium concentrations (see Fig. 2 in Ref. 11). First, a calcium wave is initiated at the point of fertilization. Within a space of 1 to 2 sec the wave sweeps across the oocyte, leading to an elevation in cytosolic calcium concentration ([Ca]$_i$) that persists for tens of seconds. A graph showing the temporal sequence of [Ca]$_i$ values averaged over an entire oocyte can be seen in Fig. 1a. Figure 1a shows the first three spikes in [Ca]$_i$ induced by fertilization; these arise from a nearly flat baseline of less than 30 nM, increase very rapidly to a prolonged peak of about 300 to 400 nM, and then decay rapidly to baseline.

Data from the oocyte were also evaluated spatiotemporally by dividing the image into five longitudinal strips (compartments) aligned at right angles to the direction of wave motion. The averaged [Ca]$_i$ within each of these compartments was used to measure the progression of the calcium wave across the oocyte. Data from the first calcium spike are illustrated in Fig. 1b, which shows [Ca]$_i$ in the first, third, and fifth compartments on an expanded time scale. This illustration demonstrates that oscillations are sometimes superimposed on the spike plateau [also observed by Cuthbertson and Cobbold (24) and by Swann (8)] and that the half-maximum value in $[Ca]_i$ occurs sequentially in compartments 1, 3, and 5. In this example (the initial spike), the calcium wave swept from compartment 1 to 5 in about 1.5 sec; the propagation velocity was therefore about 35 μm/sec. The return to half-maximal values is nearly synchronous in the five compartments.

The initial spike was followed by a series of similar spikes with even more rapid rise and decay kinetics but with durations of the plateaus reduced to about one-half that of the initial spike. The expanded view of the second calcium transient (Fig. 1c) shows the same pattern as for the first—a rapid rise to peak, a slowly decaying plateau, and a rapid return to baseline. Three noteworthy differences between this calcium transient and the first one are: (1) no large oscillations during the plateau phase, (2) much shorter duration of the spike (42 sec compared to 91 sec), and (3) nearly synchronously $[Ca]_i$ rise throughout the oocyte. The third spike in the train has a similar appearance to the second one.

From a dynamics point of view, the most striking features of the calcium transients in the mouse oocyte are the nearly constant levels of $[Ca]_i$ during the interspike intervals, nearly constant levels of $[Ca]_i$ during the spike plateau, and remarkably rapid alterations in calcium concentrations associated with the rise (about 100 nM/sec) and decay (about 40 nM/sec) of the spikes. The data demonstrate that covert dynamic processes occur during both the interspike interval and the spike plateau. These covert processes cannot now be observed through physiological experiments, therefore modeling studies are required to elucidate their nature.

Overviews of the Model

One-Compartment Model

Our model for the interactions that generate calcium transients in mouse oocytes comprises four spaces: the extracellular space, the cytosol of the oocyte, and two calcium-containing stores: a superficial, submembrane calcium store, and deep calcium stores (5, 15). In this scheme (Fig. 2a), the efflux of calcium from the submembrane store is controlled by the IP_3 concentration near the plasma membrane at the point of sperm entry. This concentration is in turn regulated by a receptor sensitive to sperm entry (or to external hormone concentrations). For simplicity, cellular mechanisms for refilling of this calcium pool are not considered; we simply assume that the calcium concentration in this pool is maintained at some constant level, so there can

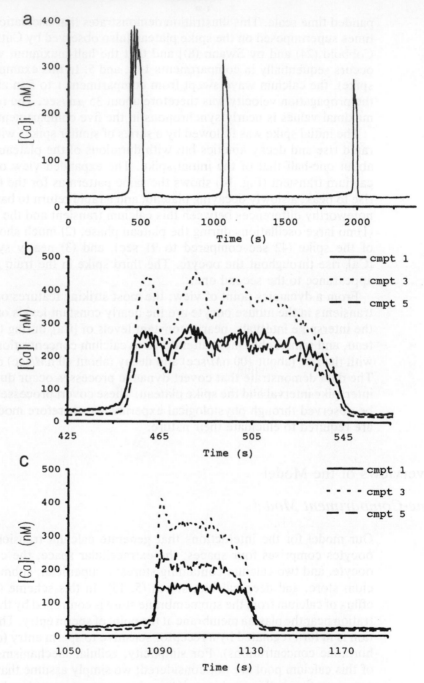

be continuous efflux from this pool. The concentration of calcium in the extracellular space also is assumed to be constant.

Only two of the four calcium spaces undergo dynamic changes in calcium concentration: the cytosol and the deep stores. Concentrations in these two spaces are determined by six calcium fluxes. Each of these fluxes contributes to the cytosolic calcium concentration ($[Ca]_i$), and three of them set the deep stores calcium concentration ($[Ca]_s$). Two calcium fluxes pass through the membrane: a leakage flux (J_{LkM}) that contributes to rises in $[Ca]_i$ and a second flux (J_{PmpM}) generated by a calcium pump that extrudes calcium and lowers it. (Descriptions of all the model parameters are summarized in Table I.) The extrusion rate of the pump is a nonlinear function of $[Ca]_i$. Calcium efflux from the superficial, IP_3-sensitive store (J_{IP3}) acts in parallel with J_{LkM} to elevate $[Ca]_i$. The value of J_{IP3} is regulated by local IP_3 concentrations and does not oscillate.

Three additional calcium fluxes, which set both $[Ca]_i$ and $[Ca]_s$, are associated with the deep stores membrane. These fluxes have opposing effects on the calcium concentration in the deep stores ($[Ca]_s$); those that raise $[Ca]_i$ concurrently lower $[Ca]_s$, and the reverse. The leakage flux (J_{LkS}) from the deep stores elevates $[Ca]_i$ and lowers $[Ca]_s$. The amplitude of this flux is determined by the permeability of the deep stores membrane and by the difference between $[Ca]_i$ and $[Ca]_s$. A second calcium pump, similar to the one in the cell membrane, pumps calcium into the deep stores from the cytosol (J_{PmpS}). The activity of this pump also is regulated by $[Ca]_i$ alone; it acts independently of $[Ca]_s$. The last flux (J_{CICR}) is generated via the opening of channels in the deep stores membrane by the CICR mechanism. In our model, the permeability of the CICR channels is determined in a nonlinear manner by $[Ca]_i$ alone. This Ca^{2+} flux does depend on the level of $[Ca]_s$, which provides for the movement of calcium ions down a concentration gradient.

FIG. 1   Calcium oscillations in a mouse oocyte induced by fertilization. (a) Calcium concentration averaged over the entire oocyte. The trace shows the calcium concentration as determined from fura-2 fluorescence ratios at 340 versus 380 nm excitation wavelengths, sampled at intervals of 0.8 sec. Sperm were added to the medium bathing the oocyte at about 100 sec from time zero. (b, c) The image of the oocyte was divided into five equispaced compartments, perpendicular to the direction of wave progression. The traces are from compartments 1 (solid), 3 (dotted), and 5 (dashed). Part (b) shows the timing and the rates of rise during the first transient and (c) the second transient.

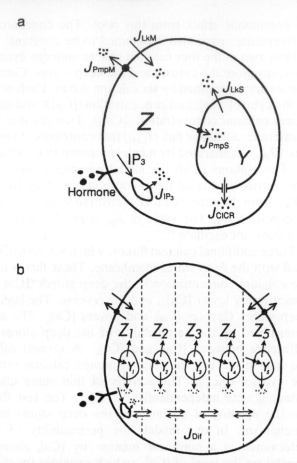

FIG. 2   Schematic diagram of the conceptual model. (a) One-compartment model. (b) Five-compartment model. Because the mechanism by which fertilization may elevate IP_3 concentrations near the fertilization site is not well characterized, the value of J_{IP3} is set directly during simulation experiments. The cytosolic calcium concentrations are denoted by Z; the deep stores concentration are denoted by Y.

Five-Compartment Model

As shown in Fig. 1, the rapid rise in calcium concentration inside mouse oocytes during the first calcium transient is not spatially uniform. To analyze and to simulate spatial inhomogeneity requires a model that explicitly includes space as a variable. We have formulated an extension of the one-

TABLE I Glossary of Model Parameters

Name	Description
J_{IP3M} (0 amol/sec)[a]	Ca^{2+} flux from submembrane IP_3-sensitive stores
J_{LkM} (68 amol/sec)	Inward Ca^{2+} flux through plasma membrane channels
$RPmpM$ (0.5 fmol/sec)	Maximum flux generated by the plasma membrane pump
$KPmpM$ (200 nM)	Concentration at which pump flux is half-maximum
$PermS$ (1 × 10^{-9}/sec)	Permeability of the stores leakage channels for Ca^{2+}
$RPmpS$ (50 fmol/sec)	Maximum flux generated by the deep stores plasma membrane pump
$KPmpS$ (200 nM)	Concentration at which pump flux is half-maximum stores pump
$PermCICR$ (6 × 10^{-9}/sec)	Maximum permeability of the CICR channels
$KCICR$ (100 nM)	$[Ca]_i$ at which CICR channels are half-activated
$CBufCy$ (1 mM)	Concentration of calcium buffer in the cytosol
$KBufCy$ (500 nM)	Concentration at which cytosolic calcium buffer is half-saturated
$CBufS$ (2 mM)	Concentration of calcium buffer in deep stores
$KBufS$ (2000 nM)	Concentration at which calcium buffers in deep stores are half-saturated
$DifConst$ (10^{-8}/sec)	Rate of diffusion between compartments
$CellVol$ (0.1 nl)	Total volume of cell

[a] Nominal value and unit of measure.

compartment model that comprises five compartments (15, 23). These compartments, which form a linear one-dimensional array, are of three distinct types (Fig. 2b). Compartment 1 simulates the cell membrane and cytosol near the site of fertilization. It includes all of the fluxes incorporated into the one-compartment model, including calcium flux from a superficial, IP_3-sensitive store. In addition, model fluxes account for calcium ion diffusion within the cytosol by simulating diffusion from the near-membrane region toward the center of the oocyte and from the center of the oocyte toward the cell membrane. The central region of the oocyte is simulated by three identical compartments. The cytosolic calcium concentrations in each of these compartments is determined by fluxes between the cytosol and deep stores and by diffusion between compartments. The third type of compartment, compartment 5, includes both cell membrane fluxes and deep-stores fluxes. Because compartment 5 simulates cell membrane remote from the fertilization site, however, there is no superficial, IP_3-sensitive store. The differences between compartments 1 and 5 are crucial for simulating the asymmetric spread of the calcium wave across the oocyte from the point of fertilization.

Model Equations

A detailed description of the model is provided in the appendix; we summarize here the two differential equations that describe the dynamics of cytosolic calcium concentration ($[Ca]_i \equiv Z$) and the calcium concentration in the deep stores ($[Ca]_s \equiv Y$). The formulation of our model in terms of calcium fluxes follows the approach of Friel and Tsien (25). As demonstrated by these researchers, calcium fluxes can be measured or approximated in living cells, so this approach allows for direct comparisons between experimental measurements and model quantities. Two differential equations for the one-compartment model are

$$dZ/dt = -(BufFactZ/CellVol)$$
$$* (J_{LkM} + J_{IP3} + J_{PmpM} + J_{LkS} + J_{PmpS} + J_{CICR}) \quad (1)$$

and

$$dY/dt = (BufFactY/CellVol) * (J_{LkS} + J_{PmpS} + J_{CICR}), \quad (2)$$

where the flux terms (J's) are those described earlier. The sign convention for the fluxes is that calcium fluxes into the cytosol are assigned a negative value (25). Therefore J_{LkM}, J_{IP3}, J_{LkS}, and J_{CICR} are negative, whereas J_{PmpM} and J_{PmpS} are positive. Calcium buffering in the cytosol and the deep stores is described by the terms $BufFactZ$ and $BufFactY$, respectively. Because the $CellVol$ term (volume of the oocyte) is included, this model simulates actual flux magnitudes in mouse oocytes.

Modeling Results

Origin of Calcium Spikes in Conceptual Model

The primary features of cytosolic calcium concentrations in the mouse oocytes are (A) the nearly flat baseline, (B) the very rapid rise and subsequent decay of the transients, and (C) the sustained plateau. The model accounts for these features as follows. (1) Prior to fertilization, calcium leakage into the oocyte is balanced via extrusion by the cell membrane pump, and calcium leakage from the deep stores is balanced by the deep stores pump. At this time, the value of $[Ca]_i$ is low, and that of $[Ca]_s$ is at a high level. (2) Fertilization generates an efflux of calcium (J_{IP3}) from near-membrane stores (conceptually, via an elevation of the IP_3 near the membrane). (3) The calcium ions provided by this additional influx are sequestered into the deep stores by the stores calcium pump, so that $[Ca]_s$ gradually rises to even higher values. As described later, the action of calcium buffers initially slows the rate of this rise in $[Ca]_s$, but as these become saturated $[Ca]_s$ begins to rise more rapidly. (4) Both J_{LkS} and J_{CICR} increase because the concentration difference between the deep stores and the cytosol rises. (5) The deep-stores fluxes initiate a gradual rise in $[Ca]_i$. (6) When sufficiently large (50–60 nM), $[Ca]_i$ induces explosive CICR from the deep stores, closely followed by the strong activation of both the plasma membrane and stores calcium pumps. (7) During the plateau phase of the transient, CICR and calcium pumping nearly balance, leading to a nearly constant value in $[Ca]_i$. (8) The continued action of the calcium pumps and a reduction in CICR return both $[Ca]_i$ and $[Ca]_s$ to their initial levels. (9) This state is not stable because continued release of calcium from the submembrane stores once again leads to the sequestration of calcium in the deep stores, setting the stage for additional calcium transients.

Quantitative Simulation of Calcium Spikes for One Compartment

The challenge to the modeler is to replicate these features with a simple model. We approximated the physiological profile of $[Ca]_i$ (Fig. 3a) by incorporating two large fluxes into the model, J_{PmpS} and J_{CICR}. Between transients, when $[Ca]_i$ is very low (near 40 nM), J_{CIRC} is small and nearly balances the other deep-stores fluxes, J_{PmpS} and J_{LkS} (Fig. 3b). As the transient begins, the CICR flux becomes explosively large because of positive feedback from the rapidly increasing cytosolic calcium concentration. This increase in $[Ca]_i$ activates the deep stores calcium pump, so that with very little delay J_{PmpS} also becomes large—enough to balance the sum of J_{LkS} and J_{CICR} (Fig. 3b). As long as these fluxes remain in balance, the spike plateau is maintained. Note that the dynamic changes in the amplitudes of J_{CICR} and J_{PmpS} occur against a background of efflux of calcium from the deep stores via J_{LkS}. Leakage from deep stores does not vary greatly despite large alterations in $[Ca]_i$ because $[Ca]_s$, which is much larger than $[Ca]_i$, does not vary greatly. (See Tables I and II for glossaries of variable and parameter names.)

In the model, calcium transients occur only for some intermediate rates of calcium influx. At the low, resting rates, the cytosolic calcium concentration remains at its basal level near 40 nM. When calcium influx is stepped to a high and constant level, the cytosolic calcium concentration rapidly rises to a maintained and high level (not illustrated). It should be noted that because the superficial, IP$_3$-sensitive stores are continually refilled by external calcium

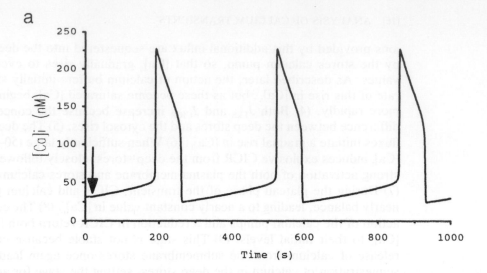

a

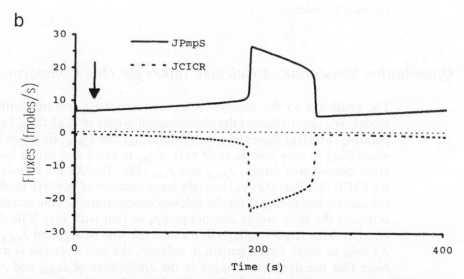

b

FIG. 3 Simulated fertilization transient in the one-compartment model. (a) Time course of [Ca]$_i$ following simulated fertilization. At the arrow J_{IP_3} was set from 0 to 41 amol/sec. (b) Deep-stores calcium fluxes during a calcium spike on an expanded time scale. Two fluxes are illustrated, J_{PmpS} and J_{CICR}. J_{PmpS} is positive because it removes calcium ions from the cytosol, whereas J_{CICR} is a negative flux. The cytosolic calcium spike begins with the rapid onset of J_{CICR} and terminates when this flux declines to near zero values. Note that the background value of J_{PmpS} is large to balance the leakage of calcium (J_{LkS}, not shown) from the deep stores. Unless explicit values are given in the legend, model parameters in this and succeeding illustrations are set to the nominal values listed in Table I.

TABLE II   Glossary of Variable Names

Name	Description
State variables (units)	
Z (nM)	Concentration of intracellular (cytosolic) free Ca^{2+}
Y (nM)	Concentration of free Ca^{2+} in deep stores
Z- and Y-dependent variables (units)	
J_{PmpM} (fmol/sec)	Outward Ca^{2+} flux via plasma membrane pump
J_{LkS} (fmol/sec)	Ca^{2+} flux through nongated channels in the deep stores membrane
J_{PmpS} (fmol/sec)	Ca^{2+} flux into the deep stores via a membrane pump
J_{CICR} (fmol/sec)	Ca^{2+} flux from deep stores via CICR channels
J_{Dif} (fmol/sec)	Ca^{2+} flux between cell compartments

(by a capacitative calcium entry mechanism), elevation of IP_3 also effectively increases calcium influx.

Calcium Transients in the Five-Compartment Model

Extension of the one-compartment model to five compartments requires only a few modifications in Eqs. (1) and (2). First, diffusion fluxes between compartments (Fig. 2b) require the addition of appropriate diffusion terms to Eqs. (1) and (2) (see Appendix). Second, with the oocyte divided conceptually into five compartments, the volume of individual compartments is set to one-fifth of the volume of the one-compartment model. The values of deep-stores pump rates and leakage permeabilities are reduced by a corresponding factor of 5. The IP_3-sensitive efflux of calcium from superficial stores is unaltered but is confined to compartment 1. Both the cell membrane leakage flux and the rate of calcium extrusions are reduced by a factor of two and limited to the end compartments, numbers 1 and 5. Other model parameters are unchanged.

Simulated fertilization, with efflux of calcium from superficial calcium stores in compartment 1, generates calcium transients in the five-compartment model that resemble very closely those generated by the one-compartment model. Again, a gradual increase in [Ca]$_i$ of compartment 1 leads to a filling of the deep stores, an explosive release of calcium ions from the deep stores causing a transient increase in [Ca]$_i$, followed by a rapid return to near baseline conditions. In fact, each of the five compartments generates a series of calcium transients that are nearly identical to those of the one-compartment model (Fig. 4a). These transients spread from compartment 1 to compartment 5 in the form of a very broad calcium wave (Fig. 4b). The model replicates the time required for the leading edge of the wave to traverse the diameter of the oocyte (about 1.5 sec, compare Fig. 1b and Fig. 4b; see

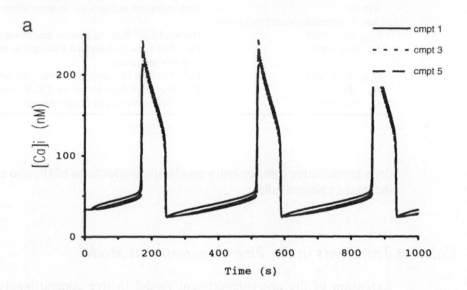

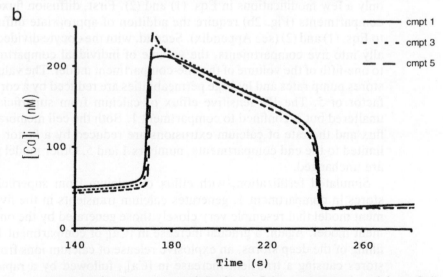

FIG. 4 Simulated fertilization transients in the five-compartment model. (a) Spike train at low temporal resolution. Simulated fertilization (increase in J_{IP3} to 41 amol/sec) was initiated at 20 sec. (b) Details of first transient in five-compartment model.

below). (This was accomplished by selecting an appropriate value for the diffusion parameter *DifConst*.) Although the rates of rise and fall of the calcium transients are realistically simulated, the plateau region of the mouse transient is more nearly flat than is simulated by the model. Moreover, the very flat baseline calcium concentrations observed in mouse eggs are not faithfully replicated by the model.

Simulation of Variations in Calcium Influx

Our model generates calcium transients only if the rate of calcium influx to the cytosol, either from submembrane stores or via the leakage conductance, has an intermediate value. Influx is subthreshold if it can be counterbalanced by the cell membrane pump to prevent a suprathreshold rise in $[Ca]_i$. As noted above, a large influx generates a rapid increase in $[Ca]_i$, but this high level is maintained indefinitely. At intermediate rates of calcium influx, the characteristics of calcium transients are determined by the influx rate. Three examples of transient behavior for model J_{IP3} fluxes of 36, 72, and 119 amol/sec are shown in Fig. 5. For the lowest influx level, the latency to the first transient is relatively long, its duration is brief, and the interval between successive transients is relatively long (Fig. 5a). Similar to results of Dupont and colleagues (23), our simulations show that when the influx is set to larger values, latencies to the first transients are smaller, their durations are larger, and the intervals between transients is reduced (Fig. 5b,c).

Although not apparent in Fig. 5 because of the low time resolution, the velocity of the calcium wave also is sensitive to the rate of calcium influx. Surprisingly, the propagation velocity of the waves is inversely related to the rate of influx (Fig. 6). As influx is increased (weak to moderate to strong stimulation), the travel time of the calcium front from compartment 1 to compartment 5 increases from 1 to 1.9 to 2.5 sec, respectively. These values correspond to conduction velocities of 50, 26, and 20 μm/sec (Fig. 6b–c). In this simulation, moderate stimulation yields wave propagation that mostly closely resembles that observed in the mouse oocyte (Fig. 6a).

Simulation of Thimerosal Treatment

Application of thimerosal to mouse oocytes induces repeated calcium transients that resemble those observed following fertilization (Fig. 7a), except that the changes in calcium concentrations occur synchronously throughout the oocyte (11). We simulated the effect of thimerosal application by manipulating the parameter (*KCICR*) which sets the calcium sensitivity of the CICR conductance. In the control state (as in the simulations described above),

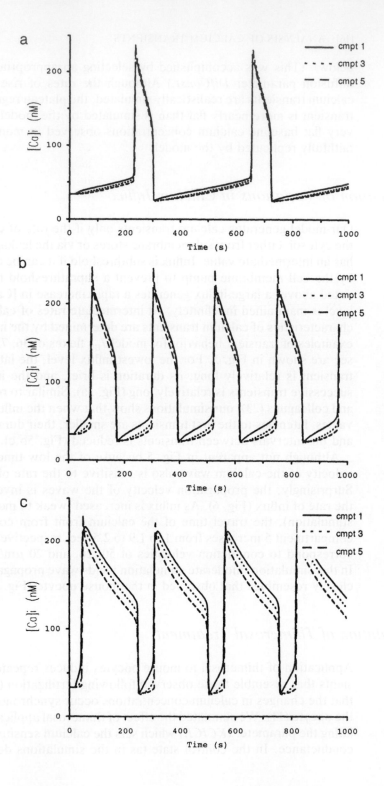

KCICR was set to 100 n*M*. With calcium leakage into the oocyte simulated by a low level of J_{LkM} and with J_{IP3} set to 0, no calcium transients occur (not illustrated). Without increasing calcium influx, we then mimicked the application of a low level of thimerosal by increasing the calcium sensitivity of the CICR channels from 100 to 80 n*M*. This sensitization resulted in the generation of a single calcium spike, followed by a prolonged interval of lowered cytosolic calcium (Fig. 7b). The calcium undershoot in some of these experiments was as large as 20 n*M*. Similar undershoot patterns in calcium concentration were observed by Nohmi and co-workers (26) in bullfrog sympathetic neurons after applying caffeine. Because both caffeine and thimerosal applications increase the sensitivity of CICR to cytosolic calcium concentrations, our model can account for the undershoot observed in both preparations.

A further experiment in which *KCICR* was lowered from the control value to 60 revealed that such increased CICR sensitization leads to repeated calcium transients. As in the physiological data (Fig. 7a), the first of these transients was more prolonged and its amplitude greater than those of subsequent ones (Fig. 7c). The rate of rises (about 150 n*M*/sec) and decay (about 50 n*M*/sec) of the calcium transients induced by this (simulated) thimerosal application were similar to those seen following both fertilization and application of thimerosal to mouse oocytes (11). A further increase in CICR sensitivity led to very long calcium spikes and a greater disparity in the amplitudes of the first and subsequent spikes (Fig. 7d).

Effects of Calcium Buffering

Most of the calcium in the cytosol and within membranous stores is bound to specific calcium binding proteins such as calsequestrin, calreticulin, calcineurin, and calmodulin. Therefore, in order to generate model fluxes that can be compared directly to fluxes in animal cells we not only incorporated appropriate size scaling for model parameters but also simulated calcium buffering. As described in more detail in the Appendix at the end of this chapter, we assumed that calcium buffering rates are very rapid when com-

FIG. 5   Effect of alterations in calcium influx. (a) Weak stimulation (J_{IP3} = 36 amol/sec). (b) Moderate stimulation (J_{IP3} = 72 amol/sec). (c) Strong stimulation (J_{IP3} = 119 amol/sec). Data obtained with five-compartment model. The value of calcium release from the superficial, IP$_3$-sensitive stores was increased from 0 to the indicated values at about 20 sec for each trace.

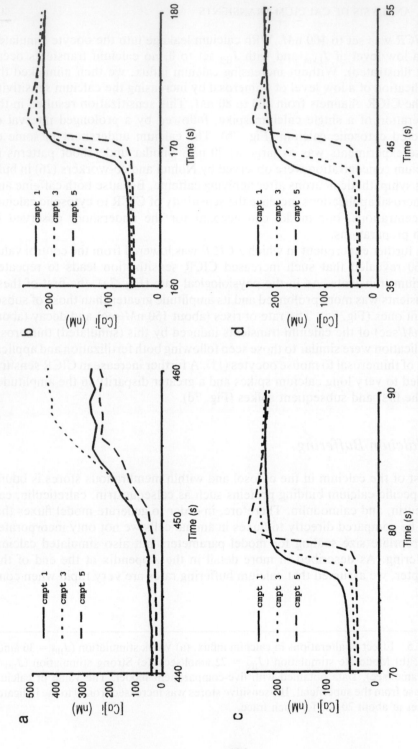

FIG. 6 Conduction velocity. (a) Experimentally observed calcium spikes for Fig. 5a–c on expanded time scales. The delay between the half-rise of the spike in successive compartments increases as the stimulation level is increased, therefore conduction velocity is inversely related to stimulus intensively.

Model output showing the rise of first calcium spikes. (b–d)

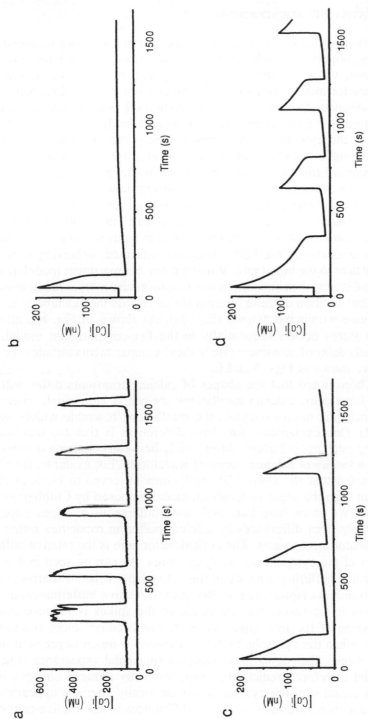

FIG. 7 Comparison of experimental and model results following application of thimerosal. (a) Calcium transients induced in mouse oocytes by application of 10 μM thimerosal. (b) Simulation results of a step change in *KCICR* from 100 to 80 n*M*. (c) Simulation results of a step change in *KCICR* from 100 to 60 n*M*. (d) Simulation results of a step change in *KCICR* from 100 to 40 n*M*. In these modeling experiments reductions in *KCICR* immediately elicited one or more calcium spikes.

405

pared with the dynamics of calcium fluxes. Furthermore, we assumed first-order, noncooperative binding between calcium and the buffering proteins. Approximate values for the calcium affinity and concentrations that appear appropriate for mouse oocytes were derived from published reports (27).

In the absence of active extrusion by membrane pumps, constant calcium influx (through leakage channels, for example) leads to a constant rate of increase in the cytosolic concentration. The presence of calcium buffers alters this profile because then the rate of increase is no longer constant. We investigated this nonlinear rise with our model by setting the cytosolic concentration to 0 and then initiating a constant leak into the cell (Fig. 8a). For the simplest one-compartment cell (without any membrane-delimited intracellular stores), the concentration of free calcium initially rises at a very low rate because most of the inflowing calcium binds to the buffer. The rate increases gradually as the buffer becomes saturated, achieving very large values relative to the initial rate. When the one-compartment model includes the internal stores, the concentration in the stores also increases in a nonlinear manner, but addition of these internal stores retards the calcium concentration increase within the cytosol (Fig. 8a). As shown in Fig. 8b, filling of the deep stores occurs sequentially in the five-compartment model. The sequentially delayed, nonlinear rise in these compartments initiates the traveling wave shown in Figs. 5 and 6.

It has been noted that the shapes of calcium transients differ with cell type. In some cells, calcium oscillations are nearly sinusoidal, whereas in others, including mouse oocytes, the oscillations resemble widely spaced spikes (3). One explanation for these differences is that the mechanisms generating calcium oscillations differ widely between cell types and thus will generate a variety of calcium transient waveforms. For example, Bird *et al.* (28) proposed that the sinusoidal oscillations observed in lacrimal glands may result from the negative feedback model proposed by Cuthbertson and Chay (20). We show here that such wide differences in appearance may result simply from differences in calcium buffering capacities rather than from oscillator mechanisms. The critical factor here is the relative buffering capacities of the cytosol and the deep stores. As can be seen in Fig. 9, a relatively low buffering capacity in the cytosol gives rise to sharp cytosolic calcium transients (spikes in Fig. 9a). As the relative buffering capacity of the cytosol is increased, the amplitude of the spikes is reduced and the upward slope of the interspike concentration becomes more pronounced (Fig. 9b). When the cytosolic buffering capacity is much larger than that of the deep stores, the oscillations assume a sinusoidal appearance (Fig. 9c). The model therefore predicts that experimentally induced changes in the buffering capacity of the cytosol will dramatically alter the appearance of intracellular calcium transients. Thus, differences in the relative concentra-

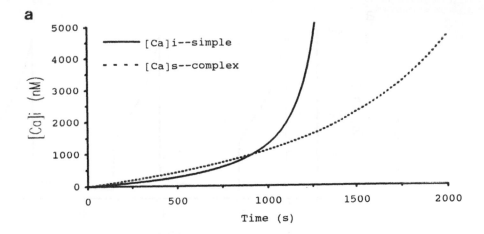

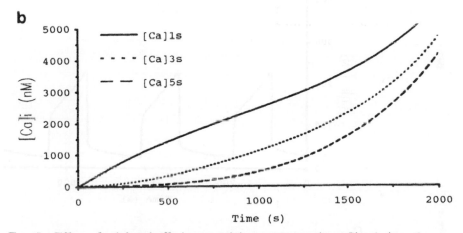

FIG. 8   Effect of calcium buffering on calcium concentrations. Simulation of a constant rate of calcium influx which raises the total amount of intracellular calcium at a constant rate. (a) One-compartment model. The trace labeled [Ca]$_i$—simple illustrates the rise in [Ca]$_i$ when deep stores are eliminated. Trace [Ca]s—complex shows rise in concentration of deep-stores calcium for a one-compartment model that includes buffered, deep calcium stores. For this more complex model the calcium concentration in the cytosol does not rise visibly above the x axis (with ordinate scale used here). (b) Five-compartment model. The simulation is as in (a) except that the model now includes five compartments. Traces [Ca]1s, [Ca]3s, and [Ca]5s illustrate the calcium concentrations of the deep stores in compartments 1, 3, and 5, respectively. The calcium concentrations of all compartments and the rate for the membrane calcium pump were first set to 0. Then calcium influx through the leakage conductance was set to 36 amol/sec, beginning at time 0. Other parameters are nominal.

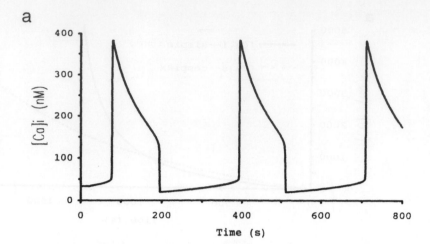

a

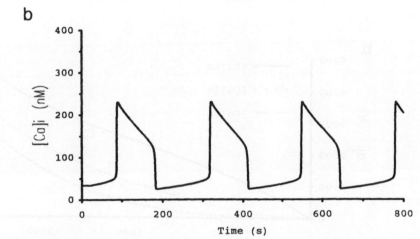

b

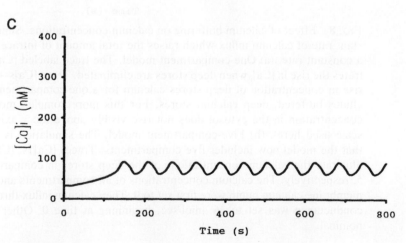

c

tions of calcium buffers in cell compartments can adequately account for some of the disparity in the shapes of calcium transients without needing to invoke alternative mechanisms for the underlying rhythmicity.

Prediction That Rapid Reduction of Calcium Extrusion from Cells Generates Spikes

In an early experiment on intracellular calcium concentration oscillations, Kuba (29) showed that sympathetic cells from bullfrogs could be induced to generate a single calcium transient (detected as a hyperpolarization in the cell membrane due to gK_{Ca}) by dropping the temperature by several degrees. Furthermore, Kuba found that repeated stepwise drops in temperature each elicited further individual transients in calcium concentration. A plausible explanation for generating calcium spikes by reducing temperatures is that the activity of the deep-stores pump is temperature-dependent. Reduced calcium sequestration into the deep pool caused by a reduced pumping rate could then give rise to a suprathreshold increase in $[Ca]_i$. We tested the plausibility of this mechanism by reducing the maximum rate ($RPmpS$) of the deep stores pump stepwise and did indeed trigger calcium transients repeatedly (Fig. 10). These transients are obtained, however, only if the pump rate is reduced abruptly; gradual changes in pump rate result in overall redistributions of calcium ions between the cytosol and the deep stores but do not generate calcium spikes. Therefore, experiments to replicate the results of Kuba in mouse oocytes require that the temperature be lowered quickly—within the span of several seconds.

Discussion

We have extended the Berridge, two-pool model (15, 22, 23) for describing calcium transients in mouse eggs by (1) restricting the calcium sensitivity of CICR channels solely to $[Ca]_i$, (2) using appropriate scaling for cell size, and

---

FIG. 9   Effects of the relative calcium buffering capacities of cytosol and deep stores on calcium oscillation waveforms. (a) Relatively low buffering capacity in the cytosol (relative buffering of 0.2). (b) Moderate buffering capacity in the cytosol (relative buffering of 0.5). (c) Relatively high buffering capacity in the cytosol (relative buffering of 1.7). Relative buffering is defined as $CBufCy/CBufS$. The flux J_{IP3} was set from 0 to 72 amol/sec at 20 sec from the beginning of each trace.

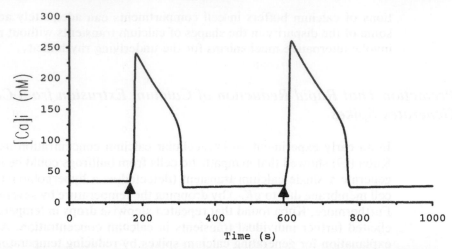

FIG. 10   Calcium spikes induced by step reductions in temperature. From time 0 until the first arrow the simulated cell was in a steady-state condition with all parameters nominal except that J_{LkM} was set to 72 amol/sec. At the first arrow, the maximum rate of the deep stores pump ($RPmpS$) was reduced from 50 to 40 fmol/sec (to simulate a step reduction in temperature), eliciting a calcium spike. A further reduction in $RPmpS$ from 40 to 25 fmol/sec (at the second arrow) elicited a second spike.

(3) including cytosolic and stores buffering. Our simulation studies demonstrate that this extended model can account for many physiological results. In particular, they show that this model does indeed generate a series of calcium transients (spikes) that closely resemble those observed in mouse oocytes; moreover our model also can account for calcium spikes generated in mouse oocytes in response to thimerosal application. Because thimerosal and intracellular IP_3 concentrations act identically to alter the calcium sensitivity of CICR, our results also account for calcium oscillations observed in other cells following the application of IP_3.

Our results demonstrate that generation of calcium transients by the two-pool model does not require CICR channels to be sensitive to the calcium concentration in the deep stores. In the light of evidence indicating that IP_3 sensitivity is enhanced by luminal calcium (30), however, it is possible that the onset of CICR might be influenced by store loading. We theorize that the explosive release of calcium from the deep stores by CICR is triggered by a rise in cytosolic calcium concentrations because of the limited sequestration of calcium in the deep stores. In the model, this limitation is due to calcium efflux through CICR channels and through leakage channels in deep stores membranes. These channels prevent unlimited increases in $[Ca]_s$ by the activity of membrane calcium pumps.

The two-pool model, combined with simple one-dimensional diffusion of calcium ions, accounts in some detail for the traveling wavefront of the first fertilization-induced calcium transient. Indeed, not only does the model correctly predict dynamics of the traveling wave, but it also predicts the nearly simultaneous reduction of the cytosolic calcium concentration throughout the oocyte at the termination of calcium spikes. Whereas the traveling wave is generated by the asymmetric rise of calcium concentrations at the point of sperm entry, the concurrent decline in $[Ca]_i$ is largely due to the symmetric distribution of calcium pumps at the two poles of the oocyte (compartments 1 and 5 in the model). Surprisingly, the model predicts that the conduction velocity of calcium spikes is related inversely to the strength of the trigger stimulus. This prediction may be readily tested in physiological preparations. The model also explains why traveling waves are not generated after exposure to thimerosal, because this agent acts symmetrically on deep stores CICR channels. It must be noted, however, that the model fails to account for the lack of a traveling-wave component in later spikes of the fertilization response. One explanation is that the effect of the sperm may become delocalized; as its influence spreads around the egg surface, all submembrane stores might be activated and the response may become nearly synchronous.

Perhaps the most interesting result of these modeling experiments is the new finding that the relative buffering capacities of the cytosol and deep stores determine the shape of the cytosolic calcium concentration waveforms. Specifically, this implementation of the two-pool model predicts that if the buffering capacity of the deep stores is relatively low, the calcium transients will resemble the spikes observed in mouse oocytes. On the other hand, relatively high buffering capacity in cells should give rise to more sinusoidal oscillations, such as those observed in exocrine cells (5). These predictions can be tested through studies of the buffering capacities of cells exhibiting calcium transients with markedly differing shapes. More directly, the predictions could be tested by manipulating the cytosolic buffering capacity of a single cell type.

Model Limitations

Our model approximates the physiological processes described for oocyte fertilization; yet it certainly does not encompass all that is known about the fertilization transients. Thus it comes as no surprise that the model output differs in several ways from calcium transients recorded in mouse oocytes. One shortcoming of the model is that it fails to generate the oscillations that are sometimes superimposed on the calcium plateau of the first calcium spike. Another is that it does not accurately reproduce the very flat baseline values and spike plateaus induced by thimerosal application and fertilization. During fertilization, flat plateaus may result from adaptation processes that gradually diminish the rate of calcium release from superficial stores (23). Our preliminary studies demonstrate that such flat plateaus can be achieved when receptor adaptation is incorporated into the model. In addition, adaptation generates a decelerating series of calcium spikes much as observed in oocytes (not illustrated).

Another limitation of our model is that it does not incorporate time-dependent calcium fluxes through the cell membrane. In both mouse (11) and pancreatic acinar cells (31), the rate of calcium influx as observed in manganese quenching experiments with fura-2 varies with time and is altered by chemical agents. Therefore, the simulation of calcium fluxes into the cytosol from the extracellular space as simple, constant leakage is an oversimplification. It would be interesting to explore how calcium influx dynamics could alter the shape of calcium transients.

Conclusions

Several different types of models have been devised to account for calcium transients. All of these generate calcium concentration dynamics resembling those of specific cells and therefore cannot be rejected on that basis as inappropriate. Clearly, more subtle tests are required to determine which model mostly closely approximates the underlying physiology.

Our unpublished exploration of the output of the Goldbeter et al. (22) model using simulated pulses of injected calcium yielded qualitatively similar results to those obtained with simulated calcium pulses applied to the Somogyi and Stucki model (14). However, we found that the experiment of dropping temperature to evoke calcium spikes (29) does distinguish the one-pool Somogyi and Stucki (14) and the Meyer and Stryer models (19, 32) from our two-pool model. Both of these earlier models fail to generate spikes when the maximum rate for the endoplasmic reticulum (ER) pump is reduced (to simulate the effect of temperature). In fact, increasing pump rate in both of these models generates calcium spikes! On that basis alone, our model is more appropriate for describing calcium concentrations in frog sympathetic neurons. Whether mouse eggs also generate calcium spikes in response to sudden temperature reductions remains to be determined.

Further knowledge of the physiological details underlying calcium transients will enable us to construct increasingly detailed, physiology-based models. As more detailed perturbation studies on calcium transients become available, more rigorous tests will enable us to reject inappropriate models.

Appendix: Details of the Model

Biological Foundations

Our model aims to simulate the concentration of free calcium within cells, particularly in mouse oocytes, as measured through the use of the photometric dye fura-2. Modeling considerations are based on experiments in which the oocyte is surrounded by a medium containing a specified concentration of free calcium. In this resting state the cell maintains a low cytosolic calcium concentration whose value (less than 50 nM) is determined by calcium influx through membrane channels and calcium extrusion by pumps in the plasma membrane. When exposed to mouse sperm, hormones, injected IP$_3$, high levels of external calcium, or the sulfhydryl agent thimerosal, the calcium concentration becomes unstable, often undergoing repeated spiking. Although the specific characteristics of these spikes depend on the stimulus, the response includes a latent interval, a very rapid rise to a peak value (300–500 nM), then a plateau (nearly flat or with small downward slope). Tens of seconds later the calcium concentration rapidly returns to near-resting levels. Often repeated spikes occur, with a cycle period of hundreds of seconds, as a result of the explosive release of calcium from deep calcium stores. In mouse oocytes, as in other preparations, the first fertilization spike travels as a wave of increased concentration originating at the point of stimulation, the point of sperm binding. Subsequent spikes arise nearly synchronously across the cell.

Conceptual Model

Our model for the mechanisms which generate calcium transients in mouse oocytes is that the oocyte includes, in addition to the cytosol, two intracellular calcium pools (15). First, we posit an IP$_3$-sensitive pool that is closely associated with the cell membrane at one pole of the oocyte. Release of Ca^{2+} ions from this pool is regulated by the interaction of IP$_3$ near the membrane with IP$_3$ receptors in the pool membrane. These stores are assumed to be continually replenished from the extracellular milieu and thus maintain a constant level of Ca^{2+} ions. Calcium efflux from this pool therefore is functionally equivalent to leakage through the plasma membrane.

Second, the model includes a distributed calcium pool deep in the cell. This pool is filled by the action of a calcium pump whose activity is regulated by [Ca]$_i$. Calcium is released from this deep pool via a calcium channel that is sensitive both to IP$_3$ deep within the cell and to the concentration of calcium ions within the cytosol. This model differs from that of Cuthbertson and Chay (20), in which intracellular oscillations of IP$_3$ rather than calcium-induced calcium release (CICR) is the critical factor controlling release of calcium ions from the intracellular stores.

Justification for the Structure of the Model

The fundamental structure of our model is that of the successful Goldbeter *et al.* (22) and Dupont *et al.* (23) models. We have added a term to control the sensitivity of CICR to agents such as thimerosal and IP$_3$. (This was considered to be an IP$_3$-insensitive store in the Goldbeter and Dupont models.) We also assume an IP$_3$-sensitive store localized at one point near the oocyte membrane. These two stores might have different receptors (33), or they might just behave differently because they are subjected to different IP$_3$ concentrations. Our five-dimensional model extends the one-compartment model as described in Dupont *et al.* (23). In our model, however, both the cell membrane and deep-stores pumps are first-order and saturatable, whereas in Dupont *et al.* (23) the cell membrane pump is linear and not saturatable, and the deep-stores pump is first-order and saturatable.

Our model includes CICR channels that are sensitive to intracellular IP$_3$ because, as reviewed by Meyer and Stryer (32), most mammalian cells have an IP$_3$-gated calcium channel. Moreover, Miyazaki *et al.* (16) suggest that CICR may be indistinguishable from Ca^{2+}-sensitized, IP$_3$-induced calcium release (IICR). They found that both CICR and IICR are enhanced by thimerosal. In addition, Miyazaki *et al.* (12) showed that local injection of IP$_3$ generates a calcium wave in hamster oocytes. They also found that injection of calcium ions was as effective as IP$_3$ in eliciting calcium spikes in mouse eggs. Therefore, release of calcium ions from the deep stores in our model is controlled by a complex interaction between IP$_3$, cytosolic Ca^{2+}, and the IP$_3$ receptor responsible for CICR (14). Even this complex formulation represents a simplification because calcium ions are likely to act via calmodulin.

Limit cycle oscillations require nonlinear differential equations. One source of this nonlinearity in our model is a fourth-order dependence of CICR on the cytosolic calcium concentration. In choosing a fourth-order relationship, we followed the precedents set by Goldbeter *et al.* (22), Dupont *et al.* (23), and Meyer and Stryer (32). In the absence of detailed descriptions of calcium pumps in mouse oocytes, we incorporated linear pumps—as suggested by the physiological data on sympathetic neurons (25)—rather than second-order pumps (22, 32). Our estimate of the calcium sensitivity of calcium pumps is based on values obtained for the calcium ATPase in plasma membrane (34). We employed a value of 200 nM for the K_M (27).

We simulated the action of the sulfhydryl agent thimerosal [or of acetylcholine (ACh)] by increasing the sensitivity of CICR as suggested by physiological studies on oocytes by Swann (10) and by Cheek and co-workers (11). Although some modelers have included calcium inactivation of the IP_3-sensitive calcium ER channel (32, 35, 36), we did not incorporate calcium inactivation of CICR because this appears to occur at concentrations greater than those observed in mouse oocytes.

Our simulation of the fertilization-induced traveling calcium wave follows the examples of Dupont *et al.* (23) and of Girard *et al.* (37). In early versions of our model, we used up to 21 cellular compartments to simulate spatial features of the calcium wave. We found that five-model compartments, which match the five compartments employed in fura-2 measurements on mouse oocytes, gave results that were nearly identical to those obtained with 21 compartments. Our simulation of the mouse data with five equal-sized compartments may be a rather crude representation because the two end calcium compartments in mouse egg optical sections are not actually identical to the central compartments.

We included calcium buffers in both the cytosol and the deep-stores pool. Because detailed data are not available for the properties or concentrations of buffers in mouse oocytes, we based our parameter values on the buffering capacity in pancreatic acinar cells (27). In these cells the buffering capacity for calcium is about 1 mM, with about 20% of the binding sites on calcium buffers occupied at resting concentrations. The dissociation constant for cytosolic buffer is estimated to be about 0.5 μM, based on the value for K_m of calmodulin (38). We set deep-stores calcium concentrations relatively high, as suggested by data obtained with targeted aequorin (39). Kendall *et al.* (39) found that the calcium concentration in the ER is 5–20 times that of the cytosol, or about 1–5 μM. The K_m for calcium buffering in the deep stores (2 μM) is based on values of K_m estimated for calsequestrin (38).

Model Overview

The model is designed to simulate the cellular calcium spikes either as a one-compartment model or as a series of five compartments. In the multicompartment version, the two end compartments represent calcium concentrations near the membrane (compartment 1 simulates calcium concentrations at the point of stimulus via sperm or hormone, whereas the other end compartment simulates near-membrane calcium concentrations far from the stimulus site). The remaining compartments represent calcium concentrations in cytoplasm and stores that are far removed from the membrane. The cytosolic calcium concentration is controlled by seven fluxes: (1) leakage into the cell through the plasma membrane (J_{LkM}), (2) release of calcium from submembrane calcium stores (J_{IP3}), (3) efflux from the cell via plasma membrane pumps (J_{PmpM}), (4) calcium leakage from the deep stores (J_{LkS}), (5) sequestration of calcium into the deep stores by pumps (J_{PmpS}), (6) calcium-induced calcium-release from the deep stores (J_{CICR}) and, for multicompartment realizations of the model, (7) diffusional fluxes between compartments (J_{Dif}).

Equations for the One-Compartment Model

We simulate the calcium concentrations in the cytosol and the deep stores by numerical integration of two equations that link these calcium fluxes with the rates of change in calcium concentrations. In abbreviated notation, the differential equation that describes cytosolic calcium concentration (Z) is

$$dZ/dt = -(BufFactZ/CellVol)$$
$$* (J_{LkM} + J_{IP3} + J_{PmpM} + J_{LkS} + J_{PmpS} + J_{CICR}), \quad (A.1)$$

where dZ/dt is the rate at which cytosolic calcium concentration changes and where the terms on the right are the fluxes described above. The term *BufFactZ* incorporates the calcium buffering capacity of the cytosol scaled appropriately for the number of compartments and the cytosolic volume of the cell (equal to 1/2 of the cell volume). Similarly, the deep-stores calcium concentration (Y) is given by

$$dY/dt = (BufFactY/CellVol) * (J_{LkS} + J_{PmpS} + J_{CICR}), \quad (A.2)$$

where dY/dt is the rate at which the calcium concentration in the deep stores changes, and where *BufFactY* incorporates the calcium buffering capacity of the deep stores, scaled appropriately for the number of compartments and the deep stores volume (one-half of the cell volume).

Calculation of Cell Membrane Fluxes

The total calcium within the oocyte is determined by two plasma membrane fluxes—leakage of calcium into the cell through nongated channels and extrusion of calcium by a first-order saturating pump. The leakage flux J_{LkM}, which depends on the number of leakage channels and the external calcium concentration, is set as a parameter. The rate of calcium extrusion depends on Z as follows

$$J_{PmpM} = RPmpM * Z/(KPmpM + Z), \quad\quad (A.3)$$

where *RPmpM* is the maximum rate of calcium extrusion by the membrane pump and *KPmpM* is the calcium concentration at which the pump rate is half maximal. Note that pump rate is linear at low values of Z and saturates when Z is large. Prior to fertilization (in the absence of calcium transients) J_{LkM} and J_{PmpM} are equal.

Calcium Release from the Submembrane, IP₃-Sensitive Store

We assume that calcium within cells is found within two membrane-bounded stores as well as in the cytoplasm. Localized near the membrane in a restricted region of the cell is a submembrane store from which calcium is released by the action of IP_3, but the details of how this release occurs are not specified in the model. We simply use a parameter J_{IP3} to specify the magnitude of this flux. For the one-compartment implementation of this model, the release of calcium from the submembrane store simply acts to increase the cytosolic concentration; for the five-compartment implementation, submembrane IP_3 release is restricted to compartment 1.

Regulation of Calcium-Induced Calcium Release Sensitivity to Cytosolic Calcium

Release of calcium from the deep stores occurs primarily via a CICR mechanism. Our model, like that of Somogyi and Stucki (14) and Goldbeter *et al.* (22), describes CICR permeability with a fourth-order dependence on the cytosolic calcium concentration. Thus, the calcium flux from the deep stores is given by

$$J_{CICR} = -PermCICR * \{Z^4/(KCICR^4 + Z^4)\} * (Y - Z), \quad (A.4)$$

where *PermCICR* is the maximum permeability of the deep stores membrane for calcium and where *KCICR* is the concentration at which the permeability is at one-half its maximum value. The concentration difference term ($Y - Z$) drives this calcium flux. Although release from the deep stores does not depend directly on IP_3 concentrations, the sensitivity of CIRC to the cytosolic calcium concentration is determined by IP_3 (or by agents such as thimerosal). We simulate such alterations in CICR sensitivity by setting *KCICR* to values higher or lower than the nominal value of 100 nM. Therefore in our model IP$_3$ and thimerosal reduce the calcium concentration required to activate CICR.

Calculation of Two Additional Fluxes through the Stores Membrane

Two additional fluxes, similar to those of the plasma membrane, control the concentration of calcium ions in the stores: the leakage of calcium through nongated channels and extrusion of calcium by a pump. They are described, respectively, by

$$J_{LkS} = -PermS * (Y - Z) \text{ and} \tag{A.5}$$
$$J_{PmpS} = RPmpS * Y/(KPmpS + Y), \tag{A.6}$$

where $PermS$ is the membrane calcium permeability, $RPmpS$ is the maximum rate for the pump, and $KPmpS$ is the calcium concentration at which the pump rate is half-maximal.

Equations for the Effects of Calcium Buffers

In the absence of calcium buffering, calcium concentrations in the deep stores and in the cytosol would rise linearly for a constant calcium influx. The effect of calcium buffers is to introduce a nonlinear relationship between calcium influx and the concentration of free calcium. Although no detailed description of calcium buffering in mouse oocytes presently exists, we can assume, in analogy to other cells, that calcium is buffered in the cytosol by a high-affinity calcium binding protein and by lower affinity calcium binding proteins in the deep stores.

The effect of these buffers on the rate of change in calcium concentration can be deduced from the following considerations. Let X and X_b be the concentrations of free and bound calcium ions, respectively. Then the total concentration of calcium, X_T, is equal to $X + X_b$. We assume simple buffering, so

$$X_b = CBuf * X/(K_m + X), \tag{A.7}$$

where $CBuf$ is the concentration of the buffer (or of the total number of calcium binding sites) and K_m is the concentration at which the buffer is half-saturated. The value of X_T is given by the initial concentration (X_0) plus the integral of all fluxes until the present time; that is,

$$X_T = X_0 + (1/V) * \int J(t) * dt, \tag{A.8}$$

where V is the volume and $\int J(t) * dt$ is the integral from time 0 to the present of all fluxes that control calcium concentrations. Combining Eqs. (A.7) and (A.8) with the knowledge that $X = X_T - X_b$, we have

$$X = X_0 + \int J(t) * dt - CBuf * X/(K_m + X). \tag{A.9}$$

We can now differentiate both sides of Eq. (A.9) to get

$$dX/dt = \{(K_m + X)^2/[(K_m + X)^2 + CBuf * K_m)]\} * (1/V) * J(t). \tag{A.10}$$

The term $(K_m + X)^2/[(K_m + X)^2 + CBuf * K_m)]$ in Eq. (A.10) is given the variable names *BufFactZ* (to designate the instantaneous effects of calcium buffering in the cytosol) and *BufFactY* (for similar buffering in the deep stores) in Eqs. (A.1) and (A.2), respectively, with an appropriate selection of parameter values.

Differential Equations Describing the Five-Compartment Model

Compartment 1

The two differential equations that describe the calcium concentrations in this compartment are very similar to those for the one-compartment model. The cytosolic term simply includes a diffusion flux such that

$$dZ/dt = -(BufFactZ/CellVol)$$
$$* (J_{LkM} + J_{IP3} + J_{PmpM} + J_{LkS} + J_{PmpS} + J_{CICR} + J_{Dif}). \tag{A.11}$$

In general, the term for calcium diffusion J_{Dif} is described by

$$J_{Dif} = DifConst * \{(Z_n - Z_{n-1}) + (Z_n - Z_{n+1})\}, \tag{A.12}$$

where the subscripts designate compartments to the left $(n - 1)$ or right $(n + 1)$ of a specific compartment (n). The deep-stores equation is independent of the number of compartments except that the variable *BufFactY* is scaled by the number of compartments. We obtain Eq. (A.2):

$$dY/dt = (BufFactY/CellVol) * (J_{LkS} + J_{PmpS} + J_{CICR}).$$

Mid-Cell Compartments

The compartments simulating the middle of the cell have no plasma membrane fluxes but do communicate with compartments on either side via diffusion. For these compartments

$$dZ/dt = -(BufFactZ/CellVol) * (J_{LkS} + J_{PmpS} + J_{CICR} + J_{Dif}), \quad (A.13)$$

where diffusion in both directions is included in the J_{Dif} term [Eq. (A.12)].

Compartment 5

This final compartment is identical to compartment 1 except that it lacks the superficial, IP_3-sensitive store. Thus the equation describing the cytosolic calcium concentration reduces to

$$dZ/dt = -(BufFactZ/CellVol)$$
$$* (J_{LkM} + J_{PmpM} + J_{LkS} + J_{PmpS} + J_{CICR} + J_{Dif}). \quad (A.14)$$

Units of Measure

The equations are written with the units for volume in nanoliters (nl); the cytosolic and stores calcium concentrations in nanomolar (nM); the rate of change in the calcium concentrations in nM/sec; the calcium fluxes in attomoles/sec (amol/sec); and time in seconds. For convenience of plotting model results, the fluxes are expressed in femtomoles/sec (fmol/sec). With these units, the volume of a mouse oocyte (diameter of 60 μm) is about 0.1 nl, the resting calcium concentration of the cytosol is about 40 nM, and that of the deep stores is several thousand nanomolar. Membrane fluxes range from about 0.04 fmol/sec for J_{LkM} to about 30 fmol/sec for J_{PmpS} and J_{CICR}. As a consequence of fertilization, the calcium concentration rises to about 300 nM within a time span of several seconds. The maximum rate of change in the calcium concentration at the onset of a spike can exceed 100 nM/sec. See Table I for a complete list of units for parameters and their nominal values.

NeuroDynamix

The computer simulations described here were carried out with NeuroDynamix, a general-purpose, graphically oriented, and computer-based modeling program. Designed to aid scientific intuition, this program is particularly useful for modeling biological oscillations because it includes graphical windows both for plotting time series and phase plane data. Model parameters can be altered online so that manipulations of model parameters are immediately expressed in model output, either in the time series graphs or on the phase plane. This modeling system has proved itself as a highly effective tool for simulating oscillations in the leech nervous system (40), for studying cellular oscillators described in the scientific literature, and for formally modeling circadian clocks (41). NeuroDynamix is written in C++.

Summary

Fusion of sperm with the cell membrane of mouse oocytes during fertilization elicits a series of intracellular calcium spikes that initiate the development of the egg. These calcium transients arise abruptly from a constant baseline, persist as a plateau for tens of seconds, and then return very rapidly to baseline levels. Transients repeat with a period of about 600 sec. We examined the mechanisms that give rise to such calcium spikes on fertilization or after the addition of thimerosal by comparing intracellular calcium concentrations in the oocytes with values predicted by a mathematical model that includes both IP_3-sensitive calcium stores near the cell surface and deeper stores that release calcium via CICR.

A novel feature of this model is that it can mimic changes in CICR sensitivity to cytosolic calcium concentrations caused by IP_3 or the sulfhydryl reagent thimerosal. In addition, it incorporates calcium buffering in the cytosol and deep stores and realistic volumes for cell compartments to generate realistic values for calcium fluxes. The model successfully simulates trains of calcium spikes and calcium waves as previously observed in mouse oocytes. Without additional parameter adjustments, it also generates oscillations via simulated sensitization of CICR as might be obtained from thimerosal or increased IP_3 levels.

In an extension of previous studies, we demonstrated that a wide range of calcium concentration waveforms can result from changes in the ratios of cytosolic to deep stores buffering capacities. These studies show that a single mechanism can generate both sinusoidal and spikelike calcium transients in animal cells.

Acknowledgments

Research was supported by the National Science Foundation Center for Biological Timing (NSF Grant DIR8920162), a National Institutes of Health Research Grant (RO1-NS21778), and by a Fogarty Senior International Fellowship (FO6-TWO1924); the latter supported the Sesquicentennial leave, sponsored by the Center for Advanced Studies at University of Virginia, for one of us (W.O.F.). We thank Peter Mathias, Master, and the Fellows of Downing College for selecting one of us (W.O.F.) as the Thomas Jefferson Visiting Fellow during the spring of 1993. We also thank Dennis Bray for stimulating discussions.

References

1. W. O. Friesen and G. D. Block, *Am. J. Physiol.* **246,** R847 (1984).
2. W. O. Friesen, G. D. Block, and C. G. Hocker, *Annu. Rev. Physiol.* **55,** 661 (1993).
3. M. J. Berridge and A. Galione, *FASEB J.* **2,** 3074 (1988).
4. R. W. Tsien and R. Y. Tsien, *Annu. Rev. Cell Biol.* **6,** 715 (1990).
5. M. J. Berridge, *Cell Calcium* **12,** 63 (1991).
6. A. Peres, *FEBS Lett.* **275,** 213 (1990).
7. I. Parker and I. Ivorra, *J. Physiol. (London)* **433,** 229 (1991).
8. K. Swann, *Biochem. J.* **287,** 79 (1992).
9. J. Carroll and K. Swann, *J. Biol. Chem.* **267,** 11196 (1992).
10. K. Swann, *FEBS Lett.* **278,** 175 (1991).
11. T. R. Cheek, O. R. McGuinness, C. Vincent, R. B. Moreton, and M. J. Berridge, *Development (Cambridge, UK)* **119,** 179 (1993).
12. S.-I. Miyazaki, Y. Michisuke, K. Makada, H. Shirakawa, S. Nakanishi, S. Nakade, and K. Mikoshiba, *Science* **257,** 251 (1992).
13. M. J. Berridge and R. F. Irvine, *Nature (London)* **312,** 315 (1984).
14. R. Somogyi and J. W. Stucki, *J. Biol. Chem.* **266,** 11068 (1991).
15. M. J. Berridge, *Nature (London)* **361,** 315 (1993).
16. S.-I. Miyazaki, H. Shirakawa, K. Nakada, Y. Honda, M. Yuzaki, S. Nakade, and K. Mikoshiba, *FEBS Lett.* **309,** 180 (1992).
17. T. J. Rink and J. E. Merritt, *Curr. Opin. Cell Biol.* **2,** 198 (1990).
18. A. Galione, H. C. Lee, and W. B. Busa, *Science* **253,** 1143 (1991).
19. T. Meyer and L. Stryer, *Proc. Natl. Acad. Sci. U.S.A.* **85,** 5051 (1988).
20. K. S. R. Cuthbertson and T. R. Chay, *Cell Calcium* **12,** 97 (1991).
21. C. Fewtrell, *Annu. Rev. Physiol.* **55,** 427 (1993).
22. A. Goldbeter, G. Dupont, and M. J. Berridge, *Proc. Natl. Acad, Sci. U.S.A.* **87,** 1461 (1990).
23. G. Dupont, M. J. Berridge, and A. Goldbeter, *Cell Calcium* **12,** 73 (1991).
24. K. S. R. Cuthbertson and P. H. Cobbold, *Nature (London)* **316,** 541 (1985).
25. D. D. Friel and R. W. Tsien, *Neuron* **8,** 1109 (1992).
26. M. Nohmi, S.-Y. Hua, and K. Kuba, *J. Physiol. (London)* **450,** 513 (1992).
27. S. Muallem, *Annu. Rev. Physiol.* **51,** 83 (1989).
28. G. St. J. Bird, M. F. Rossier, J. F. Obie, and J. W. Putney, Jr., *J. Biol. Chem.* **268,** 8425 (1993).
29. K. Kuba, *J. Physiol. (London)* **298,** 251 (1980).
30. L. Missiaen, H. Desmedt, J. B. Parys, and R. Casteels, *J. Biol. Chem.* **269,** 7238 (1994).
31. P. A. Loessberg, H. Zhoa, and S. Muallem, *J. Biol. Chem.* **266,** 1363 (1991).
32. T. Meyer and L. Stryer, *Annu. Rev. Biophys. Biophys. Chem.* **20,** 153 (1991).
33. M. H. Nathanson, M. B. Fallon, P. J. Padfield, and A. R. Maranto, *J. Biol. Chem.* **269,** 4693 (1994).
34. E. Carafoli, *Annu. Rev. Physiol.* **53,** 531 (1991).
35. G. W. De Young and J. Keizer, *Proc. Natl. Acad. Sci. U.S.A.* **89,** 9895 (1992).

36. A. Atri, J. Amundson, D. Clapham, and J. Sneyd, *Biophys. J.* **65,** 1727 (1993).
37. S. Girard, A. Luckhoff, J. Lechleiter, J. Sneyd, and D. Clapham, *Biophys. J.* **61,** 509 (1992).
38. U. B. Koupp and K.-W. Koch, *Annu. Rev. Physiol.* **54,** 153 (1992).
39. J. M. Kendall, R. L. Dormer, and A. K. Campbell, *Biochem. Biophys. Res. Commun.* **189,** 1008 (1992).
40. J. D. Angstadt and W. O. Friesen, *J. Neurophysiol.* **66,** 1858 (1991).
41. W. O. Friesen and J. A. Friesen, "NeuroDynamix, A Computer-Based System for Simulating Neuronal Properties." Oxford Univ. Press, New York, 1994.

Index

Printed and bound by CPI Group (UK) Ltd, Croydon, CR0 4YY

03/10/2024

01040317-0011